Essays in Decision Making

Springer
Berlin
Heidelberg
New York
Barcelona
Budapest
Hong Kong
London
Milan
Paris
Santa Clara
Singapore
Tokyo

Professor Stanley Zionts

Mark H. Karwan · Jaap Spronk
Jyrki Wallenius (Eds.)

Essays in Decision Making

A Volume in Honour of Stanley Zionts

With 50 Figures

Springer

Professor Mark H. Karwan
State University of New York at Buffalo
School of Engineering and Applied Sciences
412 Bonner Hall
Buffalo, NY 14260
USA

Professor Jaap Spronk
Erasmus University Rotterdam
Department of Finance
P. O. Box 1738
3000 DR Rotterdam
The Netherlands

Professor Jyrki Wallenius
Helsinki School of Economics
and Business Administration
Runeberginkatu 14–16
00100 Helsinki
Finland

Cataloging-in-Publication Data applied for
Die Deutsche Bibliothek - CIP-Einheitsaufnahme

Essays in decision making : a volume in honour of Stanley Zionts / Mark H. Karwan ... (ed.) With contributions by numerous experts. - Berlin ; Heidelberg ; New York ; Barcelona ; Hong Kong ; London ; Milan ; Paris ; Santa Clara ; Singapore ; Tokyo : Springer, 1997
ISBN-13: 978-3-642-64499-3
NE: Karwan, Mark H. [hrsg.]; Zionts, Stanley: Festschrift

ISBN-13: 978-3-642-64499-3 e-ISBN-13: 978-3-642-60663-2
DOI: 10.1007/978-3-642-60663-2

SPIN 10560971 42/2202-5 4 3 2 1 0 – Printed on acid-free paper

Table of Contents

Part 2 Preferences and Learning

Part 3 Applications

Quot homines, tot sententiae

This volume is a Festschrift in honour of Stanley Zionts. The occasion is his upcoming 60th birthday in January, 1997. Dr. Zionts, Alumni Professor of Decision Support Systems, State University of New York at Buffalo, is one of the leaders of the field of Multiple Criteria Decision Making. He is the founder of the International Society of Multiple Criteria Decision Making and its first President. In the seventeen years since its formation, under his leadership and under the leadership of Professor Ralph Steuer, the Society has grown and become very international, reflecting the nature of the field. Today it encompasses some 1200 members representing 82 different countries.

The personal contributions to research of Stanley Zionts are far reaching. They range from mathematical programming to financial modelling to multiple criteria decision making to negotiation modelling. He has written important contributions to the theory and practice in all of these fields, but particularly to the interrelated fields of multiple criteria decision making and negotiation modelling – and undoubtedly will continue producing such contributions. In his research, Stanley Zionts has always maintained a healthy balance between theory and the needs of practice.

Stanley Zionts has over the years served in many capacities. He has acted as the adviser and mentor to numerous graduate students. He has been a stimulating colleague and co-author to many of us. His quest for the highest standards of excellence in research has served – and will serve – as an example for all of us.

This volume consists of more than twenty original contributions written by some of Stanley Zionts' former students, colleagues, co-authors and friends, of whom more than a few are in leading positions in the academic world in different countries all over the world. In fact we cannot recall a research monograph in our field written by such a distinguished group of scholars. The contributions discuss recent research in the theory and practice of multiple criteria decision making and related fields.

Our idea of writing a Festschrift in honour of Stanley Zionts was met with enthusiasm. Our friends and colleagues expressed their willingness to contribute to such a volume, even though at that point we had no idea of the publisher.
The editors wish to thank all contributors to this volume and the Springer Verlag for their efficient and professional publishing of this volume. We hope that this volume will not only be a fitting birthday present for Stanley Zionts, but that

many graduate students and scholars working in the field would find it a valuable and stimulating reference. We also want to express our gratitude to our secretaries. Their kind assistance was indispensable in managing the globetrotting streams of information and papers, both in electronic and hardcopy format, going between the authors and the editors. A special word of thanks goes to Hélène Molenaar who took the responsibility for preparing the manuscript of the total volume as it was sent to the publishers.

Apart from this introduction and a final part in which the life and work of Professor Zionts is highlighted, this volume consists of three parts. The first part has been devoted to multiple objective linear programming and interactive methods. Stewart discusses the convergence and validation of interactive methods in MCDM. In his paper he reviews simulation studies, in which interactive methods of both the value function and goal programming types have been implemented in hypothetical computer-generated decision contexts. The results of these simulation studies provide a substantial level of validation for both types of interactive methods, but do also provide warnings of how careless implementation of these methods can lead to very poor results. Also Wierzbicki has studied the issues of convergence of interactive methods in multiobjective optimization and decision support. Known procedures with guaranteed convergence under classic assumptions are reviewed. An alternative approach to convergence based on an indifference threshold for increases of value functions or on outranking relations is proposed and illustrated by a new procedure called Outranking Trials. Kaliszewski, Michalowski and Kersten propose a new hybrid interactive technique, which allows the decision maker to use different search principles depending on his/her perception of the achieved values of the objectives and trade-offs. Angur and Lotfi compare the aspiration level interactive method (AIM) and conjoint analysis. They present the results of an empirical analysis based on subjects' preferences for a multiattribute product (buying a house) and a service (selecting an MBA program for study). Korhonen provides a historical overview and state-of-art review of the development of the reference direction approach to multiple objective linear programming. Steuer implements the Tchebycheff method in a spreadsheet available on almost everyone's PC, thus greatly increasing the transportability of interactive multiple objective programming software.

The second part of this volume has been devoted to the analysis of preferences and learning. Zeleny develops the notion of optimum conceived as a balance among multiple criteria. He proposes a classificational scheme of eight different, separate and mutually irreducible optimality concepts, with the traditional single-objective 'optimality' representing a one special case. Bouyssou, Pirlot and Vincke present a general model which encompasses many procedures used for aggregating preferences in multicriteria decision making (or decision aid) methods. They cover MAUT, ELECTRE and several other outranking methods. Atherton and French highlight some of the anomalies between how people 'should' make and how they 'do' make intertemporal decisions: i.e. between findings in the normative and descriptive literatures. The intention is to identify some of the bridges which need to be built between descriptive and normative ideas if decision makers are to be supported effectively in making intertemporal decisions. Fishburn discusses cancellation conditions for multiattribute preferences on finite sets in the broad context of the theory of additive conjoint measurement. Axiomatic theories for additive utilities are well developed but are not without gaps. In this contribution, a number of interesting new insights are delivered. Larichev investigates how to measure differences in the preferences of different decision makers or the same decision maker (DM) over time when using the Weighted Sum of Criteria Estimates method. A new measure is suggested: the number of alternative pairs in complete order given by WSCE for which the superiority of one alternative upon the other depends on DM's preferences. Yu and Liu describe basic concepts of the habitual domain theory. They present three of the most commonly used principles of expanding habitual domains. In addition, they present a theoretical interpretation of the principles using the notion of activation probabilities and attention spectra. Nakayama writes that one of the main themes in artificial intelligence is to simulate brains of human beings. Techniques in artificial intelligence are usually discussed for different brain functions separately. For example, decision making is a major brain function. This paper shows several kinds of multi-objective problems which appear in artificial intelligence along with some trials for solving them.

The third part of this volume consists of a number of real life problems and cases approached by different types of decision-oriented methods. Kim, Moskowitz and Shin propose a decomposition algorithm that can also be applied to forming machine cells and part families in flexible manufacturing systems when the design incidence matrix is nonbinary with no diagonal structure and the column entities are correlated. Brockett, Cooper, Kwon and Ruefli compare ex ante and ex post evaluations of mutual fund investment strategies. Ex ante evaluations of risk and return are found to be positively correlated – as posited in the finance, decision theory and economics literatures – but their ex post pairings are negatively related. Somewhat surprisingly, ex ante to ex post evaluations of risk are positively correlated while ex ante to ex post evaluations of return are negatively correlated. Hallerbach and Spronk present a multi-dimensional framework that can serve as a decision aid in the portfolio selection and management process. The framework yields room for different settings of the portfolio management problem and offers an alternative to both unstructured ad hoc approaches and complex approaches that severely restrict the decision process. Cohon starts from the position that sustainable development has emerged as the centerpiece of natural resources management and environmental protection. The purpose of his paper is to demonstrate the potential role of MCDM in planning for sustainable development, using the example of water resources in India. Teich, Korhonen, Phillips and Wallenius contribute to the discussion on BATNA, the Best Alternative to a Negotiated Agreement by offering some extensions. In addition they improve the methodology of conducting negotiation experiments. They illustrate this by means of an experiment testing several hypotheses concerning the role of BATNA. Köksalan and Kondakci briefly review various categories of approaches to multiple criteria scheduling on a single machine. They then present a general approach to find the most preferred schedule in a bicriteria environment for any given nondecreasing composite function of the criteria. Boffey and Narula review research on point location covering problems and their development from the Maximal Covering Location Problem. With this as a basis, a particular path problem (the Maximal Covering Shortest Path Problem) is chosen to serve as a prototype combined covering – location – routing problem and possible developments are suggested. Keeney and Von Winterfeldt evaluate the U.S. policy to manage nuclear waste from power plants. With many different assumptions about uncertainties and objectives, this strategy is shown to be the equivalent of $10,000 million to $50,000 million inferior to other available strategies. The implications of the analysis strongly suggest that the national policy to manage nuclear waste should be changed.

The editors join the authors in wishing Stan a happy birthday! We are looking forward to new joint adventures with Stan. We trust that you will always share an enthusiasm and interest in research – and travelling! As the Chinese proverb says, don't count your birthdays, enjoy them! Long live MCDM – and we don't mean the Roman numeral!

In Buffalo, Rotterdam and Helsinki, October 1996
Mark Karwan, Jaap Spronk and Jyrki Wallenius

PART 1:

MULTIPLE OBJECTIVE LINEAR PROGRAMMING AND INTERACTIVE METHODS

Convergence and Validation of Interactive Methods in MCDM: Simulation Studies

Theodor J. Stewart

Department of Statistical Sciences, University of Cape Town, Rondebosch 7700, South Africa

Abstract. A wide array of interactive methods, or (more correctly perhaps) methods of progressive articulation of preference, are available to the MCDM practitioner. In some cases (such as the Zionts-Wallenius method) it can be proved that the procedures converge in the sense of terminating after a finite number of iterations. In other instances, no such convergence can be demonstrated, and such methodologies must be classed as heuristics. It is, however, by no means evident that mathematical convergence is either necessary or sufficient to demonstrate the practical validity of a procedure. Practical validity requires that within a very small number of iterations, the decision maker is enabled to gain sufficient understanding of the decision space and the necessary trade-offs, to have confidence in and be satisfied with the solution selected at that stage with the help of the method. In this paper we review simulation studies, in which interactive methods of both the value function and goal programming types have been implemented in hypothetical computer-generated decision contexts. The results of these simulation studies provide a substantial level of validation for both types of interactive methods, but do also provide warnings of how careless implementation of these methods can lead to very poor results.

1. Introduction

As I began to develop an interest in the field of multiple criteria decision making during the later 1970's, one of the first papers that attracted my attention was that of Stan Zionts and Jyrki Wallenius~[1976]. What attracted me particularly was the combination of mathematical rigour and pragmatism, which I believe is the hallmark of much of the work emanating from Stan Zionts and his co-workers. On the one hand, there was careful attention paid to the practicalities of formulating relatively small numbers of simple unambiguous choices for the decision maker ("DM"), and the recognition of inconsistencies which may arise in the learning process of MCDM. On the other hand, the method was based on a well-specified preference model, and convergence was guaranteed provided that the DM was not too wildly inconsistent with the assumptions of the model.

This exposure to the Zionts-Wallenius procedure led me to a wider interest in the use of "interactive" methods (or more correctly "progressive articulation of preferences") in MCDM. Key questions seemed to be: Do such procedures possess any form of provable convergence properties? And, if so, can we assess the quality of the solution obtained, or in other words, does the procedure converge to a "best" solution in some sense? or does the procedure itself favour certain types of solution over others? Of course, the concept of "best" in the MCDM context is arguable. If we simply rely on how good the DM feels about the solution and/or the process, then this may well be confounded by many issues not related to the long run consequences of the decision. On the other hand, even in the long run, we may never be able to assess the DM's true satisfaction with the outcome. Nevertheless, in work related to fisheries management (Stewart [1988]), we were able to provide some evidence that the decision reached using a method which converged quickly, and which was perceived to be very user-friendly, was poorer in terms of the apparent aspirations of the DM than a much more slowly converging procedure.

To some, perhaps, concern with convergence in MCDM may be considered a futile exercise, as the prime purpose could be seen to be to provide the DM with a tool that facilitates learning rather than an "optimal answer". To a large extent I concur with this sentiment, but nevertheless, MCDM tools are used, especially in a group setting, to provide a methodology for selecting a course of action by a procedure which is perceived to be fair and just to all interests. In this sense, there is reason to be concerned that the procedure itself does converge to a recommendation that is not, in the long run, demonstrably a poorer compromise between the conflicting criteria (according to the DM's long term preference structure) than another; and that the procedure does not of itself introduce biasses into the selections made. The work summarized in the remainder of this paper was undertaken to address such concerns.

2 Assessment of Convergence Properties by Simulation

True DM preferences are unobservable, and some would say do not even exist (as preferences evolve, and are in fact "constructed" during the decision support process — *cf.* Roy [1993]). We can nevertheless construct a hypothetical model of an idealized preference structure, representing some unkown, or perhaps even unknowable, ultimate towards which the DM is striving, and use this to test the behaviour of different MCDM methodologies. If the methodology fails to converge consistently to the most preferred solution according to this idealized structure, when the responses of the DM do approximate this same preference structure (albeit with some perturbations), then this indicates a severe problem with the methodology which requires correction. (Of course, we are only providing in this way a necessary, and not a sufficient test of methodologies.) In order to do this, we need to specify both the idealized preference structure itself, and the manner in which a DM striving towards this ideal might respond to the inputs required of a particular MCDM methodology. These issues have been discussed at some length elsewhere (Stewart [1996a]), but let us briefly highlight some key features of the models we have used:

Idealized Preference Structures: There has been much debate in the decision analysis community regarding the role of axioms of "rational" choice. Certainly they are generally not descriptively valid, and their assumption in a normative sense has also been questioned. It nevertheless seems useful to invoke the standard assumptions for purposes of simulation, for at least two reasons: (a) It provides at least a clearly defined benchmark against which consistencies and biases of MCDM methodologies can be assessed; and (b) it is difficult to envisage a DM who would not accept that the basic principles of completeness, transitivity and continuity of preferences are an ideal towards which to strive in the process of constructing preferences.

The assumptions of completeness, transitivity and continuity of preferences, together with preferential independence of criteria, imply the existence of an additive value function in terms of some idealized set of criteria. This has thus been used as the basis for our simulations, together with the assumption that the marginal value functions are either concave (some earlier work), or more generally S-shaped, i.e. convex below some (perhaps unconscious) reference level and concave above it.

Simulation of ideal DM Responses: Even if a complete preference structure is assumed to be given, it is still not self-evident as how these will be translated into responses to the questions posed in implementing a specific MCDM methodology. The idealized preference structure model translates unambigously into a preference ordering of real or hypothetical alternatives (i.e. responses to: "Is A preferred to B"?), but not so self-evidently into the setting of goal or aspiration levels. Once more, our approach to this issue is documented elsewhere (Stewart [1996a, 1996b]), but we will return to this point later in the paper.

Non-idealities imposed by the problem structuring: The structure placed on the problem by the analyst or facilitator may distort the idealized preferences. Three key problems may be (a) poor definitions of criteria leading to violation of preferential independence; (b) the omission of criteria; and (c) impacts on the reference levels caused by framing of the problem. All three of these effects are easily incorporated into a simulation. Non-preferentially independent criteria are created by defining new criteria as linear functions of the ideal criteria, while omission of criteria and shifts of reference points can be imposed directly.

Discrepancies between perceived and idealized preferences: In simulating responses from a real DM, we also have to take into account deviations between the DM's own perceived preferences at the time of the analysis and the idealized preferences. In our simulations, we have modelled this by adding a "random noise" term to the calculated idealized value function.

Using the above framework, we can in principle study the extent to which an MCDM methodology tends to guide the DM towards the best option according to the idealized preferences, and/or tends to bias results towards particular types of solutions. The effects of the magnitude of non-idealities on these issues can also be investigated. In this paper we review the use of this approach in evaluating interactive methods ("progressive articulation of preferences") of the utility type (which can be viewed as a form of generalization of the Zionts-Wallenius procedure) and of the goal programming or reference point type. These results are largely based on the work in Stewart [1993, 1996b].

For the purposes of the ensuing discussion, we shall suppose that we are dealing with a discrete choice problem, in which choice has to be made between N alternatives (labelled by $r = 1, 2, \ldots, N$). The idealized preference structure is assumed to be based on q preferentially independent criteria (or attributes, which we take to be equivalent concepts for the purposes of this paper). Let x_i^r be the relevant measure of performance of alternative r in terms of idealized criterion $i(= 1, 2, \ldots, q)$, assumed to be defined in an increasing sense. Let $\mathbf{x}^r$ be the q-vector with elements x_i^r. Standard assumptions of completeness, continuity and transitivity imply the existence of a value function $V(\mathbf{x})$, such that alternative r is preferred to alternative s (according to the idealized preference structure) if and only if $V(\mathbf{x}^r) > V(\mathbf{x}^s)$. The assumption of preferential independence implies that $V(\mathbf{z})$ is expressible in the form:

$$V(\mathbf{x}) = \sum_{i=1}^{q} v_i(x_i)$$

where the $v_i(x_i)$ will be termed marginal value functions.

In the simulation studies to be described below, responses of the DM to any preference questions posed will be assumed to be generated by the idealized preference structure as described above plus some degree of random "error". The MCDM techniques will, however, be modelled as based on a set of $p \leq q$ criteria, and associated attribute vectors $\mathbf{z}$, in which one or more criteria may have been omitted, and the remainder may consist of mixtures of the original criteria (thus destroying preferential independence). The criteria represented by $\mathbf{z}$ will be termed the "*modelled criteria*". (For a full description and rationale for this approach, see Stewart [1996a].)

3 Utility-Based Procedures

In this approach to MCDM problems, the existence of an underlying value function, say $U(\mathbf{z})$, based on the modelled criteria, is assumed. Typically, additional properties such as pseudo-concavity will also be assumed for $U(\mathbf{z})$. Initially, the class of permissible value functions will be as large as possible, but this class is increasingly restricted in the light of DM preference statements, expressed either in the form of a strict preference, or as an indifference or trade-off. The procedure continues until the range of feasible decision alternatives that are still potentially optimal within the restricted class of value functions is sufficiently small (possibly even giving a unique solution in the case of discrete choice problems, or a uniquely optimal basis in multiobjective linear programming). Certainly the procedures of Zionts and Wallenius [1976, 1983] (which were however developed for multiple objective linear programming problems), as well as later variations (Korhonen et al. [1984]) fall into this category, as do the UTA method (Siskos [1980] and Stewart [1987]), and those of Musselman and Talavage [1980] and Rosinger [1981].

A generic form of this class of approaches to MCDM problems is described in Stewart [1993], for the case of discrete alternatives. This generic form is characterized by the following features:

- The value function $U(\mathbf{z})$ is assumed to be additive.
- The marginal value functions are assumed to be concave, and are approximated in piecewise linear form (in which the number of piecewise segments is a direct measure of the degree of non-linearity modelled).
- The DM is presented with pairs of attribute vectors, which may or may not correspond to actual decision alternatives; artificial alternatives may well be constructed, which differ on less than p attributes at a time. The DM may either express simple preferences (e.g. "$\mathbf{z}^r$ is preferred to $\mathbf{z}^s$"), or may adjust one of the two alternatives to obtain indifference between them. In the latter case, if only two attributes vary at a time, this is equivalent to the stating of tradeoffs.
- Each strict preference statement of the form $\mathbf{z}^r \succ \mathbf{z}^s$ induces an inequality constraint on the parameters of the value function, through relationships of the form:

$$U(\mathbf{z}^r) - U(\mathbf{z}^s) > 0$$

while the corresponding indifference induces the equality. Instead of trying to determine a feasible set of parameters for the value function, however, we seek rather a least infeasible set of parameters, by minimizing the maximum deviation from satisfying the constraints (where deviations on inequality constraints may be weighted differently to those on inequality constraints). This can be obtained by solving a linear programming problem, and generates a "middle-most" set of parameters (cf. Zionts and Wallenius [1983]).

A series of simulation studies was undertaken. In each simulation, a fixed number (N) of alternatives was randomly generated on a p-dimensional hypersurface (where p was the number of attributes), which was either concave or convex to the origin. Each run of the simulation was characterized by the following:

Problem structure defined by N, p, the shape of the hypersurface, and the form of the "true" value function. For the studies reported in Stewart [1993], the idealized and modelled criteria, i.e. the vectors $\mathbf{z}$ and $\mathbf{x}$ were taken to be identical.

Form of interaction with the DM defined by the numbers of attributes varied at a time (in comparing hypothetical alternatives), whether or not the DM was asked to identify indifference pairs (as opposed to simply expressing preferences), and the extent of judgemental errors made by the DM. If only two attributes are varied at a time, the identification of indifference pairs is simply a statement of trade-offs. For more than two attributes, weighted linear combinations of two hypothetical alternatives were considered, varying the weights until indifference to a third alternative was adjudged to have been achieved.

Degree of non-linearity modelled, *viz.* the number of piecewise segments used in estimating the marginal value functions.

Performance under each set of conditions was measured in terms of measures such as the mean rank (according to the "true", i.e. idealized, preference ordering) of the alternative selected, and the mean position of the true most preferred alternative in the estimated rank order, after 3 and 7 iterations of the procedure. For N between 50 and 200, and up to 10 or more attributes, results tended to stabilize after 7 iterations. Full results are described in the original paper, but the key findings can be summarized as follows:

(1) The search for pairs of (hypothetical) alternatives between which the DM is approximately indifferent, even when this assessment is subject to substantial inaccuracies (errors), results in substantially better rates of convergence to the best solutions (according to the true preference ordering) than is obtainable from simple preference statements alone. This effect was the largest and most significant found.

(2) The next most important, and still very substantial, effect was the degree of non-linearity modelled. Performance of the algorithm when value functions were approximated by linear functions (i.e. one piecewise segment) was dramatically worse than using even three or five piecewise segments. For example, use of a linear form only led to the true best alternative being ranked 4 to 5 places lower (out of a total of 50 alternatives) than obtained with 3 or 5 piecewise segments. This would take the true best alternative well outside of the top 3–5 alternatives identified, leading to potentially very unsatisfactory choices. The cause of the problem is evident. A linear value function is more prone to select extreme alternatives, i.e. performing very well on some criteria and very poorly on others, and will never select convex dominated alternatives. On the other hand, true preference may be more balanced, seeking good performance on all criteria even if this means selecting convex dominated (but of course not dominated) solutions. Thus restriction to linear value functions introduces a clear bias.

(3) There was some evidence that the use of trade-offs was preferable to comparisons involving larger numbers of attributes varying at a time (and we conjecture that the specification of trade-offs may also be the simpler cognitive task).

Stewart [1993] also demonstrates that, in terms of quality of solution and numbers of interactions with the DM, the interactive procedures and direct (*a priori*) assessment of value functions (as, for example, the approach of Keeney and Raiffa [1976], or SMART as described in von Winterfeldt and Edwards [1986]) perform essentially equivalently. We have elsewhere (Stewart [1996a]) described further and more detailed simulation studies for the case of direct assessment of value functions *a priori*. Although beyond the intended scope of the present paper, it is worth recording that these studies (a) re-inforced the conclusion that analyses based on linearized marginal value functions can be highly misleading, and (b) demonstrated that results are very sensitive to violations of preferential independence in the modelled attributes. The studies in Stewart [1993] did not address the issue of preferential or additive independence of criteria (since $\mathbf{z}$ was taken to be identical to $\mathbf{x}$), but the above results suggest strongly that insufficient care on this count can lead to highly spurious recommendations being generated by the procedures.

4 Interactive Goal Programming or Reference Point Methods

This represents the other main stream of interactive methods, and can be seen to include STEM (Benayoun *et al.* [1971]), the reference point method

(Wierzbicki [1980]) and interactive forms of goal programming (e.g. Masud and Hwang [1981]). Our "AIM" procedure (Lotfi et al. [1992]) can also be seen to draw heavily on this approach. A generic description of this class of MCDM method can be expressed in the following terms:

(1) Start with intitial "aspiration levels" for each criterion (the reference point or goals), a_i for $i = 1, 2, \ldots, p$: These may be set by the algorithm or the analyst at some arbitrary point such as the ideals, or directly by the DM.

(2) Find the feasible solution which best approaches these aspiration levels in some sense, by minimizing some measure of deviations (goal programming) or a "scalarizing function" (the reference point approach of Wierzbicki [1980])

(3) Present the resulting solution to the DM. If the DM is "satisfied" with this solution, then the procedure terminates; otherwise the aspiration levels (a_i) are adjusted by the DM and the procedure returns to the previous step. The adjustment of aspiration levels may by done either according to some prescribed systematic process, as in STEM, or in an *ad hoc* manner as in other forms of goal programming.

Simulation of the performance of interactive goal programming and related methods is made difficult by the absence of any formal model of how people set and modify goals. In recent work (Stewart [1996b]), however, we have argued that the axioms of completeness and transitivity remain normatively desirable as a model of the ultimate aim towards which the DM is striving. This still implies the existence of an underlying value function representing these idealized preferences, which will be additive if preferential or additive independence of the criteria is also assumed. For the purposes of simulating the results of using such procedures, we have conjectured that when the DM attempts to adjust goals, aspiration or reference levels, in the light of a given feasible solution, something like the following occurs:

- The DM perceives a direction of idealized maximum improvement: This would be in a direction in which all criteria are improving (the DM would not *desire* a deterioration in any criterion), but in which relatively larger improvements would be desired in some criteria than in others. This direction is presumably related to the gradient of the idealized value function described above.

- The DM perceives that the idealized direction is unachievable, and thus modifies it to be more realistic, in the sense that reductions in achievement levels on one or more criteria are conceded: This can be viewed as a projection of the idealized direction on to a *perceived* efficient frontier. This is perhaps one of the weaknesses of reference point or goal programming methods, as the DM's perception of what is achievable may differ substatially from what is actually achievable.

This process is relatively easily implemented in a simulation model, and some results are presented in Stewart[1996b]. The simulations produced very erratic results, with termination at sometimes severely sub-optimal solutions (relative to the simulated idealized preferences generating the simulated responses). The average rank order (in the idealized preferences) of the alternative selected, even under relatively ideal conditions, was around 15th out of 100. The feature of the simulation causing these results, but which we conjecture to be a feature of practice as well, is that directions of improvement may be stated which are infeasible, and thus no improvement is found (and the procedure terminates). There may, however, exist quite different directions in which more modest, but still meaningful gains, are possible. It is, in fact, quite simple to generate two-dimensional examples in which this can occur. For non-trivial problems, in which the DM needs aid or support, the DM may very well have very little feel for the efficient frontier, and any decision support needs to provide for a substantial degree of learning. These observations from the simulation results appear to confirm empirical evidence (Stewart [1988]) to the effect that simple approaches of this nature tend to terminate too early at poor solutions, before the DM has had time to explore the efficient frontier fully.

We thus conducted more detailed simulation studies on a modified version of Wierzbicki's reference point approach (as applied to discrete choice problems). The modifications were as follows:

(1) As each new solution is presented, the DM is asked whether this alternative is better than the previous one seen. If there is no improvement, then the aspiration levels are automatically adjusted, by rotating the previous vector of aspirations (i.e. the direction from the previous solution to the last chosen set of aspiration levels) towards the direction of improvement implied by the comparison. This process is repeated until improvement is found, or until the DM decides not to continue (simulated by a maximum number of iterations without improvement).

(2) In order to prevent early termination, minimization of the scalarizing function is carried out over the set of alternatives *excluding* the most recent one seen by the DM. There is then no automatic termination of the procedure, and the DM must decide when to stop (as per the previous step).

Detailed simulations of the effectiveness of this modified approach are reported in Stewart[1996b]. As for the simulations of value function based methods, different runs of the simulation examined the effects of:

Problem structure defined now primarily by the distribution of alternatives in the attribute space, and the form of the "true" value function (in which the marginal value functions were now allowed to be S-shaped as well as concave).

Form of interaction with the DM defined by parameters representing the DM's willingness to compromise on improvements, and the DM's persistence in continuing in the light of no improvement, as well as by the extent of judgemental errors.

Modelling issues, primarily effects of omitting criteria and/or defining criteria in forms which are not preferentially or additively independent.

The results of these simulations indicated that the modified reference point approach generated results which were largely comparable with those obtained by using value function methods, with the following two important exceptions:

(1) The reference point methods tended to perform less well than value function methods, when the marginal value functions were strongly "S-shaped" (i.e. strongly concave above a psychological reference level, and strongly convex below it, where this reference level falls well within the range of values obtained across the set of alternatives); of course, in this case, value function methods themselves only perform well if not constrained to be concave.

(2) The reference point methods are almost entirely insensitive to violations of preferential or additive independence amongst the attributes.

The modified reference point approach was also found to be sensitive to the number of iterations without improvement which were allowed before termination. Comparable results to the value function methods required that the DM persevere with at least 3 or 4 iterations without improvements, in order to ensure good convergence. This may not be entirely under the control of the analyst, but does emphasize the need for the analyst to encourage the DM to be persevering in exploring the decision space, and decision support software should be designed with this in mind.

5 Conclusions

The various simulation studies summarized above provide considerable evidence that interactive MCDM methods can help to guide the DM towards his/her most preferred solution (according to an idealized preference structure towards which he/she is striving), without imposing biasses, *provided that* due care is taken in the implementation of the procedures. "Due care" in this context means:

For value function based methods: that sufficient scope for non-linearity is built into the procedure, and that care is taken to ensure that the attributes are essentially preferentially independent;

For reference point or goal programming methods: that care is taken to ensure that the search proceeds in an improving direction and that premature termination is avoided, by implementation of modifications such as those proposed above.

Without such "due care", interactive methods can impose substantial biasses into the analysis (i.e. favouring certain types of solution over others, independently of DM preferences). With such due care, there is little to choose on technical grounds between value function based methods or interactive goal programming and reference point methods. Choice between these can be based, for example, on which mode of interaction is most comfortable for DM and decision aider. The only exception to this may be if there is serious concern about violation of additive independence (favouring goal programming or reference point methods), or concerning the possibility of severely "S-shaped" marginal value functions (favouring value function methods with sufficiently flexible functional forms).

References

A.S. Masud and C.L. Hwang (1981). "Interactive sequential goal programming" *Journal of the Operational Research Society*, Vol 32, 391-400

R. Benayoun, J. de Montgolfier, J. Tergny and O. Larichev (1971). "Linear programming with multiple objective functions: Step method (STEM)" *Mathematical Programming*, Vol 1, 366-375

R.L. Keeney and H. Raiffa (1976). *Decisions with Multiple Objectives*, Wiley, New York

P. Korhonen, J. Wallenius and S. Zionts (1984). "Solving the discrete multiple criteria problem using convex cones", *Management Science*, Vol 30, 1336-1345

V. Lotfi, T.J. Stewart and S. Zionts (1992). "An aspiration-level interactive model for multiple criteria decision making", *Computers and Operations Research*, Vol 19, 671-681

K. Musselman and J. Talavage (1980). "A tradeoff cut approach to multiple objective optimization", *Operations Research*, Vol 28, 1424-1435

E.E. Rosinger (1981). "Interactive algorithm for multiobjective optimization", *Journal of Optimization Theory and Applications*, Vol 35, 339-365

B. Roy (1993). "Decision science or decision-aid science", *European Journal*

of Operational Research, Vol 66, 184-203

J. Siskos (1980). "Comment modéliser les préférences au moyen de fonctions d'utilité additives", *RAIRO Recherche Opérationnelle/Operations Research*, Vol 14, 53-82

T.J. Stewart (1987). "Pruning of decision alternatives in multiple criteria decision making, based on the UTA method for estimating utilities", *European Journal of Operational Research*, Vol 28, 79-88

T.J. Stewart (1988). "Experience with prototype multicriteria decision support systems for pelagic fish quota determination." *Naval Research Logistics*, Vol 35, 719-731

T.J. Stewart (1993). "Use of piecewise linear value functions in interactive multicriteria decision support: A Monte Carlo study". *Management Science*, Vol 39, 1369-1381

T.J. Stewart (1996a). "Robustness of additive value function methods in MCDM." *Journal of Multi-Criteria Decision Analysis* (to appear)

T.J. Stewart (1996b). "Evaluation of interactive goal programming and reference point methods in MCDM: a simulation study" (In preparation)

D. von Winterfeldt and W. Edwards (1986). *Decision Analysis and Behavioral Research*, Cambridge University Press, Cambridge

A.P. Wierzbicki (1980). "The use of reference objectives in multiobjective optimization", in: *Multiple Criteria Decision Making Theory and Practice* (Editors: G Fandel and T Gal), Springer, Berlin

S. Zionts and J. Wallenius (1976). "An interactive programming method for solving the multiple criteria problem", *Management Science*, Vol 22, 652-663

S. Zionts and J. Wallenius (1983)."An interactive multiple objective linear programming method for a class of underlying nonlinear utility functions", *Management Science*, Vol 29, 519-529

Convergence of Interactive Procedures of Multiobjective Optimization and Decision Support

Andrzej P. Wierzbicki

Institute of Control and Computation Engineering,
Warsaw University of Technology, Nowowiejska 15/19, 00-665 Warsaw, Poland
and the Institute of Telecommunications, Szachowa 1, 04-894 Warsaw, Poland
The research reported in this paper was partly supported by the grant No. No. 3P40301806 of the Committee for Scientific Research of Poland.

Abstract. The paper presents an overview of issues of convergence of interactive procedures in multiobjective optimization and decision support. The issue of convergence itself depends on assumptions concerning the behavior of the decision maker – who, more specifically, is understood as a user of a decision support system. Known procedures with guaranteed convergence under classic assumptions are reviewed. Some effective procedures of accelerated practical convergence but without precise convergence proofs are recalled. An alternative approach to convergence based on an indifference threshold for increases of value functions or on outranking relations is proposed and illustrated by a new procedure called Outranking Trials.
Keywords: multiobjective optimization, decision support, interactive procedures, convergence

1 Introduction

An old and basic problem in interactive approaches to multiobjective optimization and decision support is the question how to organize interactive procedures which are convergent in some sense – or how to characterize and prove the convergence of an interactive procedure. A classic approach to this problem consists in emulating convergence analysis of mathematical programming algorithms – that is, in assuming that *the decision maker can be characterized by a value function* and the procedure helps her/him to maximize this function. Several approaches of this type are well known - such as the procedures of Geoffrion-Dyer-Feinberg (1972), of Zionts-Wallenius (1976), or of Korhonen-Laakso (1985).

However, other approaches assume that the essence of an interactive procedure is related to *learning, also by mistakes, thus changing value functions*. Therefore, an interactive procedure should concentrate more on exploring the set of attainable decision outcomes than on convergence. Again, there are several well known approaches of this type - starting with STEM (Benayoun *et al*, 1971), through *goal programming* (Charnes and Cooper, 1977), *reference point methods* (see *e.g.* Wierzbicki, 1980) and others.

Several questions arise from such a comparison. First, can we combine somehow an approach of the second type with one of the first type, to preserve the advantages of both types? The Korhonen-Laakso procedure can be viewed as such a combination; however, it will be shown here that some other combined procedures are also possible.

Secondly, how to account for uncertainty in such an interactive procedure? And what shall be meant by uncertainty in such a case? Full probabilistic description of decisions with uncertain outcomes, as in expected utility approaches, takes care of only one dimension of uncertainty. What if the decision maker is uncertain about her/his preferences – can we at all define the convergence of an interactive procedure in such a case?

Thirdly, what are possibly weakest assumptions necessary to define the convergence of an interactive procedure? Is the concept of a value or utility function necessary for this purpose, or can we use some weaker concepts of representing preferences?

The paper, after a short review of known approaches to the convergence of interactive procedures, addresses some aspects of the questions outlined above. However, before approaching these questions, we have to outline some general background.

We shall speak further about a decision maker in the sense of a *user* of a decision support system. The most frequent users of such systems are *analysts* that take advantage of computerized models to prepare analysis of various possible options for final political decision makers; or *modelers* that prepare and analyze computerized models describing a part of reality; or *designers* which have to construct a physical system – say, a computer network – while using the accumulated knowledge about such systems which can be converted into computerized models. When we reflect on the needs of analysts, modelers, designers, it becomes clear that a fast convergence toward the most preferred solution is of dubious value for them – they would rather like to be creative and thus do not know how to define the most preferred solution. Therefore, it is essential that a decision support system has properties supporting the exploration of possible outcomes, such as in reference point approaches.

After an exploration, the user might be interested in a convergent procedure of decision support; but we should not expect that she/he complies with the classical assumptions about the preferences of a decision maker. In order to define convergence, we might assume that the user has a value function – although we should not be surprised if she/he suddenly changes preferences and declares a finely tuned solution useless. Moreover, a typical user will instinctively evaluate objectives and value functions in ratio scales. Theoretically, this implies that his value function will be typically nonlinear and nonseparable – we can assume that value functions are typically linear or at least separable, but only if we consider these functions expressed in ordinal scale, see *e.g.* Rios, (1994).

Neither we can expect that the typical user will be interested in very small improvements of value. A preference structure with a threshold of indifference is best described by one of outranking relations of Roy, see *e.g.* Roy and Vincke (1981). In order to characterize convergence of various interactive procedures we shall assume that, after some time of exploration, the preferences of the user can be characterized by a value function; but we shall consider also the implications of an indifference threshold for this value function and shall also present a convergent procedure based on reference point exploration and on an outranking relation.

2 Preliminaries

2.1 A Multiobjective Analysis Problem

We shall discuss in the paper a general form of a multiobjective problem as follows. A *set of admissible decisions* X_0 in the decision space X is given – described by some inequality or equality constraints or other relations. We shall assume that the space X is topological and the set X_0 is compact. A space of decision outcomes or objectives (attributes, criteria) Y is defined; we shall assume $Y = \boldsymbol{R}^m$ *i.e.* limit our attention to the case of m criteria. Moreover, an *outcome mapping* $\mathbf{f} : X_0 \to Y$ is given; we shall assume that this mapping is continuous and thus the *set of attainable outcomes* $Y_0 = \mathbf{f}(X_0)$ is also compact.

The mathematical relations describing the set X_0 and the mapping $\mathbf{f}$ are called *the substantive model of a decision situation*; they describe the best knowledge about this situation we can rely on. The outcome relation is often defined *with uncertainty, i.e.* represented by a probabilistic or a fuzzy set model; but we shall not go into details of substantive models with uncertainty in this paper. Even if the substantive model might represent uncertainty, we assume that it is *structured, i.e.* mathematically well defined. If it is not, we call such a decision problem *unstructured*; although there exist techniques of dealing with unstructured information and knowledge (*e.g.* automatic learning by neural nets techniques), they are until now being developed and we shall not consider them here. In practice, most of decision problems are *semistructured:* out of the vast heritage of human knowledge, we must select such parts that are most pertinent to describe the model of given decision situation, then perform additional experiments *etc.*; the *art of model building,* necessary in all semistructured cases, relies on certain techniques but also on expert intuition, see *e.g.* Wierzbicki (1984, 1992a). Finally, a substantive model of a decision situation might have a *logical form* – with corresponding knowledge bases, inference engines *etc.* – or an *analytical form*; we shall limit our attention here to models of analytical form.

While a substantive model describes rather objective aspects and knowledge, each decision problem contains also much more subjective aspects, particularly concerning the preferences of a decision maker. The second part of a multiobjective analysis problem is thus a *preferential model* describing such preferences. A preferential model might have the form of a value function, *i.e.* a mapping $v : \boldsymbol{R}^m \rightarrow \boldsymbol{R}^1$, the higher values the better; or a utility function – a special form of a value function over the probabilistic distributions of decision outcomes – in the case of probabilistic representation of uncertainty. However, as discussed in the introduction, such a form of a preferential model is rather restrictive; we might use it in order to analyze the main subject of this paper, the convergence of interactive procedures, but it is better to consider also preferential models based on weaker assumptions.

One of them, basic for multiobjective analysis, is to represent the preferences of a decision maker by a partial order of Pareto type. We shall recall here some details of such representation. Such an order might be defined by selecting a positive (convex, proper, closed) cone D in outcome space and assuming the preference $\mathbf{y}^a \succ \mathbf{y}^b$ ($\mathbf{y}^a$ is preferred to $\mathbf{y}^b$) to be defined by the cone $\tilde{D} = D \setminus \{0\}$ as follows:

$$\begin{aligned} \mathbf{y}^a \succ \mathbf{y}^b \quad &\Leftrightarrow \quad \mathbf{y}^a \in (\mathbf{y}^b + \tilde{D}) = (\mathbf{y}^b + D \setminus \{0\}) \\ \mathbf{y}^a \succeq \mathbf{y}^b \quad &\Leftrightarrow \quad \mathbf{y}^a \in \mathbf{y}^b + D \end{aligned} \tag{1}$$

We shall consider here only the simplest case when $D = \boldsymbol{R}^m_+$ which amounts to maximizing all objectives, but should note that an appropriate modification of the cone D takes care not only of the case when some objectives are minimized, but also of the case when some objectives are *stabilized, i.e.* maximized when below and minimized when above certain reference level – see *e.g.* Lewandowski and Wierzbicki (1989).

Given a positive cone D, the set of efficient decision outcomes relative to this cone (called Pareto-optimal, if $D = \boldsymbol{R}^m_+$) is defined as follows:

$$\hat{Y}_0 = \{\hat{\mathbf{y}} \in Y_0 : Y_0 \cap (\hat{\mathbf{y}} + \tilde{D}) = \emptyset\} \tag{2}$$

The efficient decisions are defined as such that their outcomes are efficient, $\hat{\mathbf{x}} \in \hat{X}_0 \Leftrightarrow \mathbf{f}(\hat{\mathbf{x}}) \in \hat{Y}_0$. Beside the set of efficient decision outcomes, we shall use also the set of *weakly efficient* outcomes $\hat{Y}_0^w$ (and corresponding decisions) defined as in (2) but with $\tilde{D}$ substituted by the interior $IntD$, as well as the set of outcomes *properly efficient with a prior bound on trade-off coefficients* $\hat{Y}_0^{p\varepsilon}$ defined as in (2) but with $\tilde{D}$ substituted by the interior $IntD_\varepsilon$, where D_ε can be defined in various ways, see Wierzbicki (1986, 1992b), while the most useful definition is:

$$D_\varepsilon = \{\mathbf{y} \in \boldsymbol{R}^m : \max_{1 \le i \le m} y_i + \varepsilon \sum_{i=0}^{m} y_i \ge 0\} \tag{3}$$

where we assume that all objectives are dimensionless or have the same physical dimension (we shall later show how to weaken this assumption). While weakly efficient decisions and their outcomes are simpler to analyze mathematically but nonpractical, properly efficient decisions and outcomes with a prior bound on trade-off coefficients are eminently practical (in practice, decision makers do not distinguish cases of weakly efficient objectives and of objectives with very large trade-offs). It can be shown – see Wierzbicki (1992b), Kaliszewski (1995) – that, if $\hat{\mathbf{y}} \in \hat{Y}_0^{p\varepsilon}$, then the trade-off coefficients between objectives are bounded by the prior bound $M = 1 + \frac{1}{\varepsilon}$.

Maxima of a value function $v(\mathbf{y})$ correspond to efficient decision outcomes (or weakly efficient, or properly efficient with a prior bound) if this function is strictly monotone with respect to the cone $\tilde{D}$ (or $IntD$, or $IntD_\varepsilon$):

$$\mathbf{y}^a \in (\mathbf{y}^b + \tilde{D}) \Rightarrow v(\mathbf{y}^a) > v(\mathbf{y}^b) \tag{4}$$

When postulating that the decision maker has a value function, we shall assume such consistency of this function with the given partial order. We shall also use other functions – so called *scalarizing functions* $s(\mathbf{y}, \lambda)$ or $\sigma(\mathbf{y}, \bar{\mathbf{y}})$, where λ, $\bar{\mathbf{y}}$ are additional parameters, see e.g. Wierzbicki (1986) – which have also strict monotonicity property. Scalarizing functions might be considered as proxy value functions, but their actual role is to produce (when maximized) various efficient decisions and their outcomes and to control the selection of efficient outcomes by changing the parameters λ or $\bar{\mathbf{y}}$.

Another preferential model based on weaker assumptions than a value function is an outranking relation of the type proposed by Roy, see *e.g.* Vincke (1989). We assume then that the decision maker can tell, whether two decision outcomes $\mathbf{y}^a$ and $\mathbf{y}^b$ should be classified as incomparable, or one (strongly or weakly) preferred to another, or belonging to the same indifference class – while the most essential differences to the value function model are that not all decisions are necessarily comparable and and that there is a finite indifference threshold. We shall use later an outranking relation of this type together with a value function – thus, we shall assume that all decisions are comparable, but the finite indifference threshold implies that the decision maker does not distinguish small differences of value. We could thus define convergence of an interactive procedure only for the weaker model, an outranking relation, but to compare it with more classic approaches we must postulate that the outranking relation is consistent with a value function and an indifference threshold.

After this preliminaries, we can define now more precisely what is a *multiobjective analysis problem.* The word *problem* has here diverse meaning. First, given a *decision problem* (a *problem of real world*), we construct a model of the problem, consisting of two parts: substantive and preferential. Then, given a substantive model of a decision situation, we have a problem how to analyze the model – compared with the original decision problem, we might call this *analytical problem instance,* but we shall call it a *multiobjective analysis problem.* If we use a general preferential model based on Pareto-type order, multiobjective analysis means the analysis of the entire set of efficient outcomes and decisions. Analyzing a complicated, multidimensional set is itself a difficult issue: we must have at least tools for generating various specific, representative elements of this set. If we postulate a more detailed preferential model – an outranking relation, a value function – we might be interested either in finding the (or one of) most preferred decision and its outcomes, or again in generating a selected set of representative decisions and outcomes, possibly ranked and including the most preferred ones.

Practical experience in decision support shows that it is always good to use the more general, Pareto-type preferential model, possibly supplemented with more detailed ones, because the latter might easily change during the process of learning, inherent in decision support. Clearly, the more detailed types of preferential models should be consistent with the general one in the sense of monotonicity[1]. When we use a more detailed type of preferential model, we assume that the decision maker (the user of a decision support system) has already learned enough and thus the search for a most preferred decision is meaningful. The convergence of an interactive procedure of decision support is understood as the convergence of such search.

At this point, we could apply also the techniques used for *discrete* multiobjective (or multi-attribute) problems, when the set of admissible decisions X_0 is a discrete set, usually finite. Convergence in such a case is also finite. Thus, we might apply corresponding techniques of multi-attribute decision selection, say, between the ELECTRE, the PROMETHEE, the AHP (Analytical Hierarchy Process), see Roy and Vincke (1981), Brans and Vincke (1985), Saaty (1980), or any other known techniques – but not to produce a final answer, only in the sense of an iteration of an interactive process. If we actually have a *continuous* multiobjective problem, the set X_0 has infinite number of elements and continuous power, we could reduce the problem to a repetitive generation of a discrete, finite set of options and then to an application of some known technique of multiple attribute choice. In fact, some very attractive procedures of interactive decision support use similar ideas.

2.2 Neutral Compromise Solutions

Normative concepts of compromise solutions are usually presented, in a textbook of multiobjective optimization, as a counterpart of interactive procedures in multiobjective optimization. However, compromise solutions can be also viewed as a basis of interactive procedures, because:

- they might be used as starting points for interactive procedure – and the choice of a good starting point is important (especially in general, nonconvex cases);
- they can be parametrically modified, which can serve as a basis of interactive procedures.

[1]In practice, inconsistencies often occur. However, if the user of a decision support system discovers that an objective should no longer be maximized, the type of this objective can be changed to stabilized – hence the importance of dealing both with maximized or minimized and stabilized objectives.

Recall that, by a compromise solution, we understand typically in multiobjective analysis a decision with outcomes located somewhere in the middle of the efficient set; a more precise meaning of "somewhere in the middle" specifies the type of a compromise solution. This concept was investigated in detail first by Zeleny, see *e.g.* (1973), (1974), who has shown that a compromise solution might be defined by:

- selecting a concept of a norm in the objective space, together with weighting coefficients that might be used to modify this concept;
- selecting a reference point such that the minima of the distance from the reference point (defined by the selected norm) are efficient.

Suppose we maximize all objectives, with efficiency equivalent to Pareto-optimality and $D = \boldsymbol{R}^m_+$. By an ***utopia*** or ***ideal*** point we understand a point $\mathbf{y}_{uto}$ with components obtained by separate maximization of each objective; more generally, it is the lowest point (in the partial order generated by the cone D) such that $\hat{Y}_0 \subset \mathbf{y}_{uto} - D$. We shall use also an approximation of a ***nadir*** point $\mathbf{y}_{nad}$ – the highest point such that $\hat{Y}_0 \subset \mathbf{y}_{nad} + D$ – with the note of caution that the nadir point is much more difficult to precisely compute than the utopia point; fortunately, we usually need only some approximation of it.

Zeleny (1974) has shown that, in order to be sure of efficiency of solutions minimizing the distance even if the set Q_0 is not convex, the reference point $\bar{\mathbf{y}}$ for a scalarizing function $s(\mathbf{q}, \bar{\mathbf{y}}) = \| \mathbf{y} - \bar{\mathbf{y}} \|$ should be taken at the utopia point, $\bar{\mathbf{y}} = \mathbf{y}_{uto}$, or "above to the north-east" of this point, at a "displaced ideal" or simply upper bound $\bar{\mathbf{y}} = \mathbf{y}_{up} \in \mathbf{y}_{uto} + D$. Then, when minimizing a distance related to a l_p norm with $1 \leq p < \infty$, properly efficient (Pareto-optimal) compromise solutions are obtained. The Chebyshev (l_∞) norm results in only weakly efficient solutions. Dinkelbach and Isermann (1973) have shown that an augmented Chebyshev norm results in properly efficient solutions.

In order to use a norm we must be sure that all objective components are of the same dimension or dimension-free. Thus, we should anyway establish some ranges of objectives to rescale their increments – that is, to estimate lower bounds $y_{i,lo}$, *e.g.* approximations from below of nadir point components, and upper bounds $y_{i,up}$, *e.g.* approximations from above of utopia point components. Then we rescale increments from $| y_i - y_{i,up} |$ to $\frac{|y_i - y_{i,up}|}{|y_{i,up} - y_{i,lo}|}$. After such rescaling, when fixing the reference point at the upper bound point, we can define **neutral compromise solutions** (actually, their outcomes) as:

$$\hat{\mathbf{y}}^{(p)}_{neu} = \underset{\mathbf{y} \in Q_0}{\operatorname{argmin}} \left(\sum_{i=1}^{m} \frac{| y_{i,up} - y_i |^p}{| y_{i,up} - y_{i,lo} |^p} \right)^{1/p}, \quad 1 \leq p < \infty$$

$$\hat{\mathbf{y}}^{(\infty)}_{neu} = \underset{\mathbf{y} \in Q_0}{\operatorname{argmin}} \left(\max_{1 \leq i \leq m} \frac{| y_{i,up} - y_i |}{| y_{i,up} - y_{i,lo} |} \right),$$

$$\hat{\mathbf{y}}_{neu}^{(1,\infty)} = \underset{\mathbf{y}\in Q_0}{\text{argmin}} \left(\max_{1\le i\le m} \frac{|\, y_{i,up} - y_i \,|}{|\, y_{i,up} - y_{i,lo} \,|} + \varepsilon \sum_{i=1}^{m} \frac{|\, y_{i,up} - y_i \,|}{|\, y_{i,up} - y_{i,lo} \,|} \right) \tag{5}$$

the last one with some small $\varepsilon > 0$. While most neutral solutions are properly efficient, $\hat{\mathbf{y}}_{neu}^{(\infty)}$ might be only weakly efficient – and not uniquely defined in such a case; but we can select then an efficient solution by an additional lexicographic optimization, see *e.g.* Ogryczak *et al.* (1989). If $p = 1$, then we do not even need to use $\mathbf{y}_{up}$ as a reference point, because the neutral solution $\hat{\mathbf{y}}_{neu}^{(1)}$ can be obtained as well by maximizing the sum of rescaled objectives:

$$\hat{\mathbf{y}}_{neu}^{(1)} = \underset{\mathbf{y}\in Q_0}{\text{argmax}} \sum_{i=1}^{m} \frac{y_i}{|\, y_{i,up} - y_{i,lo} \,|} \tag{6}$$

However, depending on the shape of $\hat{Y}_0$, the solution $\hat{\mathbf{y}}_{neu}^{(1)}$ might be lokated quite far away from "the middle" of this set, while the solutions $\hat{\mathbf{y}}_{neu}^{(\infty)}$ and $\hat{\mathbf{y}}_{neu}^{(1,\infty)}$ are as close to "the middle" as possible. The neutral solution $\hat{\mathbf{y}}_{neu}^{(1,\infty)}$ can be also obtained equivalently by maximizing a different function. Note that we could define a reference point precisely in the middle of estimated objective ranges, appropriately redefine the rescaled objective components:

$$\bar{y}_{i,mid} = \frac{y_{i,up} + y_{i,lo}}{2}, \quad \frac{y_{i,up} - y_i}{y_{i,up} - y_{i,lo}} = \frac{1}{2} - \frac{y_i - \bar{y}_{i,mid}}{y_{i,up} - y_{i,lo}} \tag{7}$$

and use the equivalence:

$$\begin{aligned} \max_{1\le i\le m} \frac{|\, y_{i,up} - y_i \,|}{|\, y_{i,up} - y_{i,lo} \,|} + \varepsilon \sum_{i=1}^{m} \frac{|\, y_{i,up} - y_i \,|}{|\, y_{i,up} - y_{i,lo} \,|} &= \\ -\min_{1\le i\le m} \frac{y_i - \bar{y}_{i,mid}}{y_{i,up} - y_{i,lo}} - \varepsilon \sum_{i=1}^{m} \frac{y_i - \bar{y}_{i,mid}}{y_{i,up} - y_{i,lo}} &+ \frac{1}{2}(1 + \varepsilon m) \end{aligned} \tag{8}$$

Hence, minimizing the distance (induced by the augmented Chebyshev norm) from the upper bound point is equivalent to maximizing the following *order-consistent achievement function*:

$$\sigma(\mathbf{y}, \bar{\mathbf{y}}_{mid}) = \min_{1\le i\le m} \frac{y_i - \bar{y}_{i,mid}}{y_{i,up} - y_{i,lo}} + \varepsilon \sum_{i=1}^{m} \frac{y_i - \bar{y}_{i,mid}}{y_{i,up} - y_{i,lo}}, \quad \bar{y}_{i,mid} = \frac{y_{i,up} + y_{i,lo}}{2} \tag{9}$$

while such achievement scalarizing functions are used – instead of norms – in *reference point* approaches. The advantage of order-consistent achievement scalarizing functions over norms is that they are monotone and thus their maximal points are efficient – in fact, even properly efficient with a prior bound on trade-off coefficients – even if we use arbitrary (in particular, attainable or not) reference points instead of using only points higher than the utopia point. In *goal programming* approaches, see e.g. Charnes and Cooper (1977), we also use norms and arbitrary reference points, but the efficiency of obtained solutions cannot be guaranteed; thus, reference point approaches are sometimes called *generalized goal programming.*

2.3 Weighted Compromise Solutions

Neutral solutions as discussed above might serve only as starting points for interaction with the decision maker. On the other hand, the interaction can be organized when using the concept of weighted compromise solutions. We ask first the decision maker to state what is the relative importance of criteria, then we express this relative importance in terms of weighting coefficients $\lambda_i > 0$, $\forall i = 1, \dots m$, $\sum_{i=1}^{m} \lambda_i = 1$. Usually, the responses of the decision maker can be interpreted in terms of the ratios $\omega_{ij} = \lambda_i/\lambda_j$, and we must apply a method of converting these responses into the vector λ. A broadly applied method, the AHP (Analytical Hierarchy Process) approach, is to exploit the properties of eigenvalues of rank-one matrices in order to shorten computations needed for such conversion, see Saaty (1980). A best approximation of the vector λ assuming some random errors in the responses $\omega_{i,j}$ of the decision maker would require more complicated computations.

Once the weighting coefficients λ are determined, the weighted compromise solutions $\hat{\mathbf{y}}_\lambda^{(p)}$ are defined by:

$$
\begin{aligned}
\hat{\mathbf{y}}_\lambda^{(p)} &= \operatorname*{argmin}_{\mathbf{y}\in Q_0} \left(\sum_{i=1}^{m} \lambda_i \frac{| y_{i,up} - y_i |^p}{| y_{i,up} - y_{i,lo} |^p} \right)^{1/p}, \quad 1 \le p < \infty \\
\hat{\mathbf{y}}_\lambda^{(\infty)} &= \operatorname*{argmin}_{\mathbf{y}\in Q_0} \left(\max_{1\le i\le m} \lambda_i \frac{| y_{i,up} - y_i |}{| y_{i,up} - y_{i,lo} |} \right), \\
\hat{\mathbf{y}}_\lambda^{(1,\infty)} &= \operatorname*{argmin}_{\mathbf{y}\in Q_0} \left(\max_{1\le i\le m} \lambda_i \frac{| y_{i,up} - y_i |}{| y_{i,up} - y_{i,lo} |} + \varepsilon \sum_{i=1}^{m} \lambda_i \frac{| y_{i,up} - y_i |}{| y_{i,up} - y_{i,lo} |} \right)
\end{aligned}
\tag{10}
$$

While the concept of weighted compromise is sufficient for efficiency – if $\mathbf{y}_{up} \in \mathbf{y}_{uto} + D$, all weighted compromise solutions $\hat{\mathbf{y}}_\lambda^{(p)}$ for $1 \le p < \infty$ are properly efficient (except $\hat{\mathbf{y}}_\lambda^{(\infty)}$ which might be only weakly efficient) – it is not sufficient for organizing a convergent interactive procedure. Given a properly efficient outcome $\hat{\mathbf{y}} \in \hat{Q}_0^p$ that might be the maximum of a value function, for a given $p < \infty$ we cannot be sure that we find such a vector of weighting coefficients $\hat{\lambda} \in \operatorname{Int} \boldsymbol{R}_+^m$ that $\hat{\mathbf{y}}_{\hat{\lambda}}^{(p)} = \hat{\mathbf{y}}$. It can be only shown – see *e.g.* Sawaragi *et al.* (1985) – that we can find $\hat{\mathbf{y}}_\lambda^{(p)}$ in some neighborhood of the given $\hat{\mathbf{y}}$ by jointly changing λ and p. Thus, for a given $p < \infty$, we cannot say that we can obtain any properly efficient outcome desired by the decision maker by changing weighting coefficients.

Moreover, the character of *the dependence* of $\hat{\mathbf{y}}_\lambda^{(p)}$ on λ *is not easy to interpret.* In some applications, this might be a procedural advantage. If a group of high-level decision makers cannot agree on selecting a decision, it might be advantageous to propose to them a rational procedure without clearly predictable outcomes. Then they should, say, agree first by voting on the comparative importance of criteria, then the weighting coefficients will be determined as an outcome of the vote and a weighted compromise solution would be computed and presented to the group. However, in the case of a single decision maker – particularly, if she/he is an analyst or a modeler – the lack of a clear interpretation of this dependence is disadvantageous.

There is, fortunately, one case in which the dependence of a weighted compromise solution on weighting coefficients is easy to interpret, namely the case of Chebyshev norms. Let us consider the augmented Chebyshev norm and $\hat{\mathbf{y}}_\lambda^{(1,\infty)}$. Suppose we choose a weighting coefficient vector λ with $\lambda_i > 0$, $\sum_{i=1}^m \lambda_i = 1$, and a scalar coefficient $\eta > 1/\lambda_i \; \forall i = 1, \dots m$. Then we can assign an aspiration level $\bar{y}_i$ to each λ_i:

$$\bar{y}_i = y_{i,up} - \frac{y_{i,up} - y_{i,lo}}{\eta \lambda_i} \tag{11}$$

while $y_{i,lo} \le \bar{y}_i < y_{i,up} \; \forall i = 1, \dots m$ because $\eta > 1/\lambda_i$ – although the aspiration levels $\bar{y}_i$ might change with η, in which case equation (11) describes a line segment in $\boldsymbol{R}^m$ ending at $\mathbf{y}_{up}$ as $\eta \to \infty$. Equivalently, for any aspiration point $\bar{\mathbf{y}}$ such that $y_{i,lo} \le \bar{y}_i < y_{i,up} \; \forall i = 1, \dots m$ we can set:

$$\begin{aligned} \lambda_i &= \frac{y_{i,up} - y_{i,lo}}{y_{i,up} - \bar{y}_i} \Bigg/ \sum_{j=1}^m \frac{y_{j,up} - y_{j,lo}}{y_{j,up} - \bar{y}_j} \\ \eta &= \sum_{j=1}^m \frac{y_{j,up} - y_{j,lo}}{y_{j,up} - \bar{y}_j} \end{aligned} \tag{12}$$

which defines the inverse to the transformation (11). We can interpret this inverse transformation in the following way: the ratios $\omega_{ij} = \lambda_i / \lambda_j$ of importance of criteria are defined by selected aspiration levels as an inverse ratio of their relative distances from upper bound levels:

$$\omega_{ij} = \frac{\lambda_i}{\lambda_j} = \frac{y_{j,up} - \bar{y}_j}{y_{i,up} - \bar{y}_i} \tag{13}$$

The transformation (12) has been implicitly used by Steuer and Choo (1983) in a procedure using the augmented Chebyshev norm and $\hat{\mathbf{y}}_\lambda^{(1,\infty)}$, but controlled interactively by the decision maker who specified aspiration levels that were used to define the weighting coefficients. The outcomes of such a procedure are properly efficient; Steuer and Choo show that any properly efficient outcome can be obtained by this procedure for convex sets Q_0.

However, we can show more: under transformations (11), (12), the weighted compromise solution $\hat{\mathbf{y}}_\lambda^{(1,\infty)}$ can be equivalently obtained by the maximization of an order-consistent achievement scalarizing function with such aspiration levels used as a reference point. This is because we have:

$$\lambda_i \frac{y_{i,up} - y_i}{y_{i,up} - y_{i,lo}} = \left(\frac{y_{i,up} - y_i}{y_{i,up} - \bar{y}_i}\right) \Big/ \eta = \left(1 - \frac{y_i - \bar{y}_i}{y_{i,up} - \bar{y}_i}\right) \Big/ \eta \tag{14}$$

therefore, since $y_{i,up} \geq y_i \geq y_{i,lo}$:

$$\max_{1 \leq i \leq m} \frac{| y_{i,up} - y_i |}{| y_{i,up} - y_{i,lo} |} + \varepsilon \sum_{i=1}^{m} \frac{| y_{i,up} - y_i |}{| y_{i,up} - y_{i,lo} |} = \tag{15}$$

$$(- \min_{1 \leq i \leq m} \frac{y_i - \bar{y}_i}{y_{i,up} - \bar{y}_i} - \varepsilon \sum_{i=1}^{m} \frac{y_i - \bar{y}_i}{y_{i,up} - \bar{y}_i} + 1 + \varepsilon m)/\eta$$

Hence, minimizing the weighted distance (induced by the augmented Chebyshev norm) from the upper bound point is equivalent to maximizing the following order-consistent achievement function:

$$\sigma(\mathbf{y}, \bar{\mathbf{y}}) = \min_{1 \leq i \leq m} \frac{y_i - \bar{y}_i}{y_{i,up} - \bar{y}_i} + \varepsilon \sum_{i=1}^{m} \frac{y_i - \bar{y}_i}{y_{i,up} - \bar{y}_i} \tag{16}$$

with the reference point $\bar{\mathbf{y}}$ defined as the transformation (11) of the vector of weighting coefficients λ with any (sufficiently large) parameter η.

Moreover, even if the set Q_0 is not convex (or even if it is a discrete set), we can take any properly efficient outcome $\hat{\mathbf{y}}$ with trade-off coefficients scaled down by the deviations from the upper levels and bounded by:

$$t_{ij}(\hat{\mathbf{y}}) \leq (1 + 1/\varepsilon) \frac{y_{i,up} - \hat{y}_i}{y_{j,up} - \hat{y}_j} \tag{17}$$

At any such point, we can define a reference point and weighting coefficients – by taking precisely $\bar{\mathbf{y}} = \hat{\mathbf{y}}$ and applying the transformation (12) – in such a way that the maximal point of $\sigma(\mathbf{y}, \bar{\mathbf{y}})$, equal to the weighted compromise solution $\hat{\mathbf{y}}_\lambda^{(1,\infty)}$, coincides with $\hat{\mathbf{y}}$, see e.g. Wierzbicki (1992b). Equivalently, for any properly efficient $\hat{\mathbf{y}}$ we can choose a sufficiently small ε such that (17) is satisfied; then we can choose an aspiration point (equal *e.g.* to $\hat{\mathbf{y}}$, but any point on the line segment joining $\mathbf{y}_{up}$ and $\hat{\mathbf{y}}$ would do) together with the corresponding weighting coefficients such that the point $\hat{\mathbf{y}}$ is obtained either by maximizing the achievement function or by minimizing the weighted and augmented Chebyshev distance from the upper bound point[2].

---

[2]We must stress, however, that ***this equivalence does not mean that the order-consistent achievement function is a distance function.*** The order-consistent achievement function compares points $\mathbf{y}$ and $\bar{\mathbf{y}}$, not $\mathbf{y}$ and $\mathbf{y}_{up}$, and cannot be a distance between $\mathbf{y}$ and $\bar{\mathbf{y}}$, because it must preserve monotonicity at $\mathbf{y} = \bar{\mathbf{y}}$. In fact, it is a more complicated function of these two variables than a distance function. The equivalence means only that this more complicated function can be constructed while utilizing the distance between other points, $\mathbf{y}$ and $\mathbf{y}_{up}$.

While using a similar equivalence as discussed above, Kok and Lootsma (1985) proposed a technique similar to the AHP for determining the weighting coefficients and then combining them with the Chebyshev or augmented Chebyshev distance minimization in order to obtain a weighted compromise solution. This has clearly an advantage when compared to the use of other norms, since we can interpret the ratios of weighting coefficients as in (13). We can thus inform the decision maker that, if he specifies some objective j to be, say, 5 times more important than objective i, then he is likely to obtain the relative deviation of the objective j from its best possible level – measured in the percentage of entire range of changes of this objective – 5 times smaller than the relative deviation of the objective i. In this statement, "likely to obtain" means "precisely, independently of the convexity of the set of outcomes" if the utopia point is used as the upper bound and a point with such deviation ratios is contained in the set of properly efficient outcomes (with an appropriate bound on trade-off coefficients).

However, if the decision maker is a modeler interested in multi-objective model analysis, she/he does not need to specify such importance ratios. The modeler might as well specify directly aspiration levels and use the optimization of an achievement scalarizing function, while knowing that if the aspiration level for the i-th objective deviates from the upper level 5 times more (in the relative sense) than the one for the j-th objective, this proportion is likely to be preserved in the efficient outcomes.

Lootsma *et al.* (1995) proposed also another way of defining a weighted compromise solution: by maximizing an aggregated improvement of objectives as compared with their lower bound – e.g. the nadir point component. A suggested way of aggregation is to use a weighted geometric mean *e.g.* of the form:

$$s_g(\mathbf{y}, \mathbf{y}_{lo}) = \prod_{i=1}^{m} \left(\frac{y_i - y_{i,lo}}{y_{i,up} - y_{i,lo}} \right)^{\gamma_i}, \quad \gamma_i > 0 \; \forall i = 1, \ldots m, \; \sum_{i=m}^{m} \gamma_i = 1 \qquad (18)$$

Lootsma *et al.* recall an interesting property of the weighted geometric mean $s_g(\mathbf{y}, \mathbf{y}_{lo})$ which was actually known before, see *e.g.* Sawaragi *et al.* (1995): at a given point $\mathbf{y}$, the proportion of partial derivatives of the geometric mean, scaled by the inverse ratio of the deviations from the lower bounds, is equal to the ratio of weighting coefficients:

$$\left(\frac{\partial s_g}{\partial y_i} \Big/ \frac{\partial s_g}{\partial y_j} \right) \frac{y_i - y_{i,lo}}{y_j - y_{j,lo}} = \frac{\gamma_i}{\gamma_j} \qquad (19)$$

The ratio γ_i/γ_j can be again interpreted as a measure of relative importance of objective i over objective j and identified with the help of *multiplicative* AHP proposed by Lootsma, see *e.g.* (1993). If $\mathbf{y}$ is efficient, then the ratio on the left-hand side of (19) can be interpreted as a trade-off coefficient scaled down by the ratio of improvements $(y_i - y_{i,lo})/(y_j - y_{j,lo})$. The weighted geometric mean of improvements $s_g(\mathbf{y}, \mathbf{y}_{lo})$ is a strictly monotone function of $\mathbf{y}$, hence its maxima over $\mathbf{y} \in Q_0$ are efficient, even properly efficient because of the boundedness of (19).

The procedures based on weighted compromise solutions described above can and have been used as interactive procedures of multiobjective decision support. However, they typically do not have convergence proofs – because proving convergence based on selecting weighting coefficients is not an easy task.

3 Interactive Procedures With Proven Convergence

3.1 Geoffrion-Dyer-Feinberg Procedure

The first of interactive procedures with guaranteed convergence was given by Geoffrion, Dyer and Feinberg (1972). For a detailed discussion of this procedure, see *e.g.* Steuer (1986); here we shall give only short comments. The procedure is applicable for general, nonlinear but differentiable substantive models; it is assumed that the decision maker has also nonlinear but differentiable value function[3]. The procedure is an extension of the Frank-Wolfe nonlinear optimization algorithm, where the interaction with human decision maker replaces the modules of function and gradient determination. The determination of the gradient of a value function is a complicated task, requiring many pairwise comparison questions to be answered by the decision maker. The procedure, though theoretically elegant, did not find many applications in multiobjective decision support. We have thus learned that, in order to be effective, an interactive procedure must concentrate more on the details of interaction with the decision maker, less on imitating mathematical programming algorithms.

However, this procedure is a prototype of other convergent procedures that exploit the concepts of convergence of mathematical programming algorithms. As it is known – see *e.g.* Luenberger (1973) – an unconstrained nonlinear programming algorithm based on directional searches converges under two general assumptions:

- the directions of search should give uniform improvement, that is the angle between these directions and the gradient should be uniformly (in the case of maximization) acute;
- the step-size coefficients determining the length of steps along the directions should not converge to zero too fast – which can be specified by various tests (of Goldstein, of Wolfe) or expressed by the assumption that, if they converge to zero, then the sequence of points generated by the algorithm converges to the maximum or at least a stationary point of the function.

[3]Note that the differentiability of a value function implies that it is not an ordinal value function, but is measured on a stronger scale.

The Frank-Wolfe algorithm produces directions of uniform improvement, and the second assumption sufficient for convergence is fulfilled in the Geoffrion-Dyer-Feinberg procedure because of the details of a specific line search used in the procedure.

3.2 Zionts-Wallenius Procedure

The procedure of Zionts and Wallenius (1976) was designed for a much more limited class of models, but more attention was paid to the details of interaction with the decision maker. Substantive models were restricted to multiobjective linear programming models, where the set Y_0 is polyhedral; as a preferential model, the linear value function:

$$s(\mathbf{y}, \lambda) = \lambda^T \mathbf{y} = \sum_{i=1}^{m} \lambda_i y_i \tag{20}$$

was chosen, but only as an approximation of any concave value function.

The procedure starts at any efficient extreme point of Y_0 – for example, from $\hat{\mathbf{y}}_{neu}^{(1)}$ corresponding to $\lambda_i = \frac{1}{m}, \forall i = 1, \ldots m$. The decision maker is asked whether she/he prefers:

- other extreme points of Y_0 that are adjacent to the current point but distinctly different from it, with a given difference threshold;
- trade-off vectors generated in an multiobjective linear programming algorithm and corresponding to such edges that lead to adjacent extreme points which are too close, below given difference threshold.

The adjacent extreme points are compared to the current one, but the trade-off vectors are evaluated in more general terms of liking them or not. Later experience has shown that decision makers might feel uncomfortable when answering such general questions. Since we actually ask the decision maker whether she/he likes improvements in a given direction, thus – if the adjacent extreme points are too close – an alternative way of asking the question is to present to her/him further, distinctly different points on the same straight line, not necessarily belonging to Y_0.

After such round of questions either a preference for an adjacent point or for a direction is established or all edges leading to adjacent extreme points of the current point are evaluated as nonprofitable. Conceptually, this would suffice to construct a convergent interactive procedure. Indeed, moving to the preferred extreme point results in an increase of the value function. Since the procedure moves among finite number of extreme points only, then the convergence (assuming compact Y_0) is actually finite – precise for linear value functions and only approximate for concave value functions that might have maxima on the facets, not at some extreme points of Y_0.

However, in the actual procedure of Zionts and Wallenius, the data obtained from the answers of the decision maker way are used additionally for updating the set of weighting coefficients $\Lambda^{(k)}$, where the upper index (k) denotes the iteration number. At the starting point, $\Lambda^{(0)} = \{\lambda \in \boldsymbol{R}^m : \lambda_i \geq 0, \sum_{i=1}^m \lambda_i = 1\}$ is assumed. Each increase of preference along a given direction, $\mathbf{y} \succ \mathbf{y}^{(k)}$, indicates that the set $\Lambda^{(k+1)}$ might be constructed from $\Lambda^{(k)}$ by adding the constraint $\lambda^T(\mathbf{y} - \mathbf{y}^{(k)}) \geq \delta$, where $\delta > 0$ is an accuracy threshold. Tradeoff vectors can be treated similarly as directions $\mathbf{y} - \mathbf{y}^{(k)}$. Errors in interaction might be corrected by deleting the oldest remembered constraints on λ.

Given an updated set $\Lambda^{(k+1)}$, the next extreme efficient point $\mathbf{y}^{(k+1)}$ is generated not as the preferred adjacent extreme point, but as a solution to the problem:

$$\max_{\mathbf{y} \in Y_0} \lambda^{(k+1)T} \mathbf{y} \tag{21}$$

where $\lambda^{(k+1)}$ is an (arbitrary) element of $\Lambda^{(k+1)}$.

Actually, a reasonable modification of Zionts-Wallenius procedure would be to let the decision maker choose between such $\mathbf{y}^{(k+1)}$ and the preferred adjacent extreme point. The original Zionts-Wallenius procedure, though it pays much attention to the interaction with the decision maker, is rather complicated in details; many further modifications, often simplifying this procedure, were proposed, see *e.g.* Zionts and Wallenius (1983). The main advantage of this procedure, as seen from todays perspective, was to use the properties of the set Y_0 and the thresholds of accuracy in interaction with the decision maker to produce a procedure with finite convergence.

3.3 Korhonen-Laakso Procedure and Pareto Race

Korhonen and Laakso (1985) proposed a visual interactive method for controlling the selection of efficient outcomes and decisions, incorporated in a software tool called Pareto Race.

The essence of the Pareto Race method is a directional search of improvements of the satisfaction of the decision maker with the outcomes of efficient decisions (or of a value function). Starting from an efficient objective outcome point $\hat{\mathbf{y}}^{(k)}$, $k = 1$, the search is a repetition of following three phases:

- the determination of a direction $\mathbf{d}^{(k)} \in \boldsymbol{R}^m$ such that it would point towards some improvements of the satisfaction of the decision maker;
- the computation of several efficient objective outcomes and decisions corresponding to several reference points distributed along the direction;
- the choice of the step-length in the direction, *i.e.* a decision and objective outcome determined in previous phase that provides best improvement of the satisfaction of the decision maker; the resulting point becomes the next $\hat{\mathbf{y}}^{(k+1)}$.

Thus, the general idea of Korhonen-Laakso procedure is in a sense similar to that of Geoffrion-Dyer-Feinberg procedure, with some essential differences.

The direction $\mathbf{d}^{(k)}$ is chosen in Pareto Race by the decision maker, but she/he might do it either arbitrarily (and take full responsibility for the convergence of the procedure) or be supported by an appropriate procedure. Such a procedure might consist in asking the decision maker, for each objective component y_i, how much this component should be – as compared to $\hat{y}_i$ – increased or decreased in relative terms (in % of the known range $[y_{i,lo}, y_{i,up}]$). This can be done *e.g.* in a prepared questionnaire form, asking the decision maker how much decreased or increased should be given objective y_i.

Such a direction $\mathbf{d}^{(k)}$ can be interpreted as a *reference direction*. Several reference points can be chosen along the direction *e.g.* as follows:

$$\begin{aligned} \bar{\mathbf{y}}^{(k,j)} &= \hat{\mathbf{y}}^{(k)} + \tau_j \mathbf{d}^{(k)}, \\ \tau_j &= 2^j, \; j = -3, -2, -1, 0, 1, 2, 3 \end{aligned} \tag{22}$$

which can be additionally projected on the ranges $[y_{i,lo}, y_{i,up}]$, if necessary. The efficient objective outcomes that are presented to the decision maker for evaluation are obtained in the Pareto Race method by a maximization of an order-consistent achievement function *e.g.* of the form (16):

$$\hat{\mathbf{y}}^{(k,j)} = \underset{\mathbf{y} \in Q_0}{\operatorname{argmax}} \, \sigma(\mathbf{y}, \bar{\mathbf{y}}^{(k,j)}) \tag{23}$$

These points can be considered as *projections*[4] of the directional changing reference points $\hat{\mathbf{y}}^{(k,j)}$ *on the properly efficient frontier* $\hat{Y}_0^{p\varepsilon}$. Thus, another essential difference to the Geoffrion-Dyer-Feinberg procedure – and similarity to the Zionts-Wallenius procedure – is that Pareto Race moves among efficient points.

The choice of a step-length in the direction $\mathbf{d}^{(k)}$ can be prepared by displaying graphically these projections as profiles of changes of component objectives $\hat{y}_i^{(k,j)}$ along changing j. An important feature of the Pareto Race tool is the graphically supported interaction with the decision maker which uses such profiles in order to select some j for which the improvement of all objective components is best balanced. This choice constitutes the third stage of the method and determines next starting point for a new iteration:

$$\hat{\mathbf{y}}^{(k+1)} = \hat{\mathbf{y}}^{(k,j)} \text{ for selected } j \tag{24}$$

If the decision maker states that she/he is not satisfied with any j – since neither of $\hat{\mathbf{y}}^{(k,j)}$ is for her/him better than $\hat{\mathbf{y}}^{(k)}$ – the decision support system should ask her/him to specify another direction $\mathbf{d}^{(k)}$, hopefully leading to an improvement, in which case the decision maker takes the responsibility for the

[4]It should be stressed, however, that these projections can be interpreted as results of a distance minimization only if $\mathbf{y}^{(\bar{k},j)}$ are "above to the north-east" of $\hat{Y}_0^{p\varepsilon}$. If $\bar{\mathbf{y}}^{(k,j)} \in Y_0$ (which might occur *e.g.* in non-convex cases), then these projections represent rather a uniform improvement of all objective components than distance minimization, since an order-consistent achievement function is generally not a norm.

convergence. The search stops if the decision maker states that she/he cannot suggest a direction of improvement. Korhonen and Laakso (1985) presented a convergence proof of this procedure under appropriate assumptions. Further details of convergence of such reference point or reference direction procedures were discussed by Bogetoft *et al.* (1988).

3.4 Stochastic Approximation and Quasi-gradient Procedures

The interactive procedures described above have some corrective properties in cases of errors made by the decision makers, but do not take into account such errors systematically. There exist, however, a way of obtaining slow but certain convergence for directional search methods even if the evaluations of the decision maker are subject to random errors. This way is known from the old technique of *stochastic approximation,* or from the more general[5] technique of *stochastic quasi-gradient* of Ermolev – see e.g. Ermolev and Wets (1988). It consists in simply applying a harmonic sequence of fixed step-length coefficients for each iteration (k):

$$\begin{aligned} \tau_{(k)} &= \tau_{(1)}/k; \quad k = 1, 2 \ldots; \\ \bar{\mathbf{y}}^{(k+1)} &= \hat{\mathbf{y}}^{(k)} + \tau_{(k)}\mathbf{d}^{(k)}; \\ \hat{\mathbf{y}}^{(k+1)} &= \underset{\mathbf{y}\in Q_0}{\operatorname{argmax}}\, \sigma(\mathbf{y}, \bar{\mathbf{y}}^{(k+1)}) \end{aligned} \tag{25}$$

The step to $\hat{\mathbf{y}}^{(k+1)}$ might be either made always (which results in a not monotone but still convergent variant of the procedure) or only if the decision maker accepts $\hat{\mathbf{y}}^{(k+1)}$ as better than $\hat{\mathbf{y}}^{(k)}$; if the step is not made, she/he has just to redefine the direction $\mathbf{d}^{(k)}$. Due to the fixed but harmonically decreasing coefficients, the procedure converges under the assumption that the directions specified by the decision maker approximate (with random errors but without bias) the stochastic quasi-gradient of her/his value function, see Ermolev and Wets (1988). Michalevich (1986) has used such a technique to propose an interactive procedure of decision support with proven convergence.

The essence of the convergence of such techniques consists in sufficiently slow convergence such that the stochastic errors of evaluations can be averaged. This accounts for the uncertainty about the preferences of the decision maker, if this uncertainty can be expressed in probabilistic terms. This idea is very attractive and powerful theoretically; however, practical experiments show that decision makers usually find such convergence too slow.

---

[5]Taking into account also nondifferentiable cases.

4 Procedures With Accelerated Convergence

In order to account for the value of the time used up by decision makers, different principles of interaction must be used. It is true that pairwise comparison is a relatively simple cognitive task and thus might be reliably used in interactive procedures; but most decision makers are able and would prefer to perform more complicated cognitive tasks if the time of procedure could be shorten this way. In a sense, decision makers are willing to work in a *parallel* fashion (as we can use today computers of remarkable processing power – and if this power is too small, we can always use computers with scalably parallel architecture).

There exists several approaches aimed at accelerating convergence of interactive procedures, all based on presenting to the decision maker a larger sample of decision options and their outcomes. We assume then that the decision maker is able to point to the most preferred outcome of such a sample. Theoretically, it is a difficult cognitive task; moreover, no convergence proofs were given for such procedures. Practically, however, such procedures save time. Therefore, we outline here briefly some of such procedures; then we shall propose a new procedure of accelerated convergence, together with convergence proof, indicating how the convergence of such approaches could be investigated.

4.1 Reference Ball

One of the first ideas in reference point methods (see Wierzbicki, 1980) was to use – as an approximation of the preferred decision outcome – a central point $\hat{\mathbf{y}}^{(k)}$, which at $k = 1$ could be selected as a neutral point, and the most natural way of scattering the efficient points around it:

$$
\begin{aligned}
\delta_{(k)} &= 0.1/k, \quad k = 1, 2, \ldots \\
\bar{\mathbf{y}}^{(k,j)} &= \hat{\mathbf{y}}^{(k)} + \delta_{(k)}\mathbf{e}^{(j)}, \ \mathbf{e}^{(j)} = (0\,0 \ldots 1_{(j)} \ldots 0\,0)^T \\
\hat{\mathbf{y}}^{(k,j)} &= \operatorname*{argmax}_{\mathbf{y}\in Q_0} \sigma(\mathbf{y}, \bar{\mathbf{y}}^{(k,j)})
\end{aligned}
\tag{26}
$$

The $m + 1$ efficient points $\hat{\mathbf{y}}^{(k)}$, $\hat{\mathbf{y}}^{(k,j)}$, $j = 1, \ldots m$ are presented then to the decision maker who selects the most preferred one as the next central point $\hat{\mathbf{y}}^{(k+1)}$. Because of the harmonic decrease of the radius of scattering, we can expect that the procedure might be convergent, even in the presence of random evaluation errors. However, no formal proof of convergence was given; moreover, practical trials have shown that this procedure is too slow for a typical decision maker.

On the other hand, this procedure illustrates the idea of a *reference ball:* given a current point $\hat{\mathbf{y}}^{(k)}$, scatter a given number of additional points in a ball of radius $\rho^{(k)}$ centered at the current point and use the additional points as reference points. If we take the Chebyshev norm to define the ball, it becomes just a box centered at $\hat{\mathbf{y}}^{(k)}$:

$$\tilde{B}(\hat{\mathbf{y}}^{(k)}, \rho^{(k)}) = \{\bar{\mathbf{y}} \in \boldsymbol{R}^m : |\bar{y}_i - \hat{y}_i^{(k)}| \leq \rho^{(k)}(y_{i,up} - y_{i,lo})\}, \ \ \rho^{(k)} \in (0, 1) \quad (27)$$

where the symbol $\tilde{}$ in the denotation $\tilde{B}(\hat{\mathbf{y}}^{(k)}, \rho^{(k)})$ stresses that $\rho^{(k)}$ is only a relative radius of the box or ball. Suppose we choose some number P of randomly generated points $\bar{\mathbf{y}}^{(l,j)}$, $j = 1, \dots P$ either on the boundary of this box or first in the box and then filter those who are most distant from the box center. For example, if we want to generate the points on the box boundary, it is sufficient to select randomly the index of the component i and then to generate randomly – *e.g.* with an uniform distribution – an aspiration component:

$$\bar{y}_i \in [\hat{y}_i^{(k)} - \rho^{(k)}(y_{i,up} - y_{i,lo}),\ \hat{y}_i^{(k)} + \rho^{(k)}(y_{i,up} - y_{i,lo})] \quad (28)$$

then project this component on the box boundary:

$$\begin{aligned} \bar{y}_i^{(k,j)} &= \hat{y}_i^{(k)} + \rho^{(k)}(y_{i,up} - y_{i,lo}) \text{ if } \bar{y}_i \geq \hat{y}_i^{(k)} \\ \bar{y}_i^{(k,j)} &= \hat{y}_i^{(k)} - \rho^{(k)}(y_{i,up} - y_{i,lo}) \text{ if } \bar{y}_i < \hat{y}_i^{(k)} \end{aligned} \quad (29)$$

For other components $i+1, \dots, m, 1, \dots i-1$ of the vector $\bar{\mathbf{y}}^{(k,j)}$, we repeat the generation as in (28), but not necessarily the projection as in (29) – if we repeat the projection, we obtain obtain a random sequence of corner points of the box. Having generated the points $\bar{\mathbf{y}}^{(k,j)}$, $j = 1, \dots P$, we can filter them additionally: *e.g.* if we do not produce only corner points, we can select a half of the points most distant from the box center in the sense of the Chebyshev metric, relative to the objective ranges.

Then we proceed to compute the corresponding properly efficient outcome points by repeated maximization of an order-consistent achievement function:

$$\hat{\mathbf{y}}^{(k,j)} = \underset{\mathbf{y} \in Q_0}{\operatorname{argmax}}\, \sigma(\mathbf{y}, \bar{\mathbf{y}}^{(k,j)}),\ j = 1, \dots P \quad (30)$$

The resulting properly efficient outcomes $\hat{\mathbf{y}}^{(l,j)}$ will be not necessarily the most distant from $\hat{\mathbf{y}}^{(l)}$, even if we choose randomly aspiration points only in the box corners. Therefore, an additional filtering of these outcomes might be performed (if there are too many of them to be presented to the decision maker). Thus selected number of spread points $\hat{\mathbf{y}}^{(k,j)}$, $j = 1, \dots P'$ and the center $\hat{\mathbf{y}}^{(k)}$ are presented to the decision maker who is supposed to select the most preferred one as the next $\hat{\mathbf{y}}^{(k+1)}$.

4.2 Contracted Cone

A similar idea is employed in the *contracted cone* procedure of Steuer and Choo (1983). The original procedure was to define a contracting set of weighting coefficients which, due to the transformation (11), can be interpreted as a cone of directions in objective space, starting at the upper bound or utopia point, going through the reference points and "illuminating" a part of the efficient surface. The decision maker would choose a point in the "illuminated" part, around which a new but contracted – "narrower" cone would be centered. The procedure was defined originally in terms of weighting coefficients and the Chebyshev norm defining the distance from the utopia point and interpreted in terms of contracting directional cones. Note that, due to the transformations (11), (12), we can equivalently define this procedure in terms of a ball of reference points with a contracting radius and use the maximization of an order-consistent achievement function.

The convergence of such a procedure depends critically on the choice of the sequence $\rho^{(k)}$. We can choose either an exponential or a harmonic sequence:

$$\rho^{(k)} = 1/2^k \text{ or } \rho^{(k)} = 1/2k, \; k = 1, \ldots \infty \tag{31}$$

Steuer and Choo proposed the exponential sequence. Although their procedure has found many successful applications, exponential sequences might result in a lack of convergence in special cases[6]. If we choose the harmonic sequence instead, it could be proved – as indicated earlier – that the procedure converges (in the expected value sense) to the maximum of the value function of the decision maker, constrained to Q_0, even if the decision maker makes occasional errors in selecting the most preferred point. However, the convergence with the harmonic sequence is rather slow.

4.3 Light Beam Search

Another similar idea was applied by Jaszkiewicz and Słowiński, see *e.g.* (1994), when illuminating a part of efficient surface in their Light Beam Search method. They noted, however, that the use of an order-consistent achievement scalarizing function instead of the Chebyshev metric method makes it possible to further modify the ideas of Steuer and Choo. They locate then "the light source" at any non-attainable reference point ("to the North-East" of efficient frontier, not necessary at the utopia point) and generate "the cone of light" with the help of additional weighting coefficients.

We should add here a further comment that, when using both aspiration and reservation levels instead of one reference point in an order-consistent achievement function, see *e.g.* Granat *et al.* (1994), we can use an aspiration point to locate the "source of light" and appropriate changes in reservation levels to generate the "cone of light". For example, if an aspiration point $\bar{\mathbf{y}}^{(k)}$ and a reservation point $\bar{\bar{\mathbf{y}}}^{(k)}$ were used to obtain an efficient point $\hat{\mathbf{y}}^{(k)}$ by maximizing an order-consistent achievement function $\sigma(\mathbf{y}, \bar{\mathbf{y}}^{(k)}, \bar{\bar{\mathbf{y}}}^{(k)}$, then a spread of reservation points around $\hat{\mathbf{y}}^{(k)})$ such as described above by (28), (29) will generate an appropriate "cone of light".

[6]Consider the following single-dimensional example: maximize $y = -(x-2)^2$ starting with $x^{(1)} = 0$ and with the sequence of limited steps $x^{(k+1)} - x^{(k)} = 1/2^k$, $k = 1, \ldots \infty$. Obviously, $\lim_{k \to \infty} x^{(k)} = 1$ which falls short of $\hat{x} = 2$.

Another important aspect of the Light Beam Search method of Jaszkiewicz and Słowiński was the construction of an outranking relation in the sense of Roy, see *e.g.* Vincke (1989), in the illuminated part of efficient frontier. We shall not present here this feature of Light Beam Search method in detail; instead, we shall propose in next section another method which combines the construction of an outranking relation directly with the specification of reference levels. The proof of convergence of this method indicates how to approach convergence of methods such as Light Beam Search or other interactive procedures with accelerated convergence that, until now, lack convergence proofs.

5 Outranking Trials

The proposed method is as follows. Suppose the decision maker (the user of a decision support system) has performed as many experiments of multiobjective optimization with the substantive model as she/he wished, but finally selected a preferential model in the form of a set of objectives with a specified partial order; suppose a properly efficient outcome $\hat{\mathbf{y}}$ (say, summarizing the experience gained in the initial experiments) is also given.

Starting from this outcome, the decision maker asks for help in checking whether a more satisfactory decision outcome can be found and in finding the most satisfactory outcome. We assume that, by asking for help in convergence, the decision maker indicates that she/he has learned enough about the substantive model and has formed a preferential model. Although this preferential model might be expressed by a value function, we assume that the decision maker does not care about small differences of value and does not want to compare decisions which result in too widely changing objectives. Thus, we shall use another preferential model of an outranking relation. Both the value function and the outranking relation are assumed to be consistent with each other (which will be discussed in detail later) and with a partial order of objectives specified by the decision maker. For simplicity of description, we assume that the partial order corresponds to the maximization of all objectives, although the method can be easily generalized for minimized objectives and, with some additional details, also for stabilized objectives.

Therefore, at this point of procedure, the decision maker is asked to specify some parameters of her/his outranking relation. We could use here *e.g.* the outranking relation of the type used in ELECTRE 3 or 4, see Vincke (1989). However, we shall modify these outranking relations to exploit the full possibilities of reference point methods. The outranking relation proposed here is constructed as follows. For each component objective y_i, while knowing the range of changes of this objective and the point $\hat{\mathbf{y}}$, the decision maker is asked to specify four incremental threshold values – two obligatory, other two optional. If a range of reference levels (an aspiration level $y_{i,asp}$ and a reservation point $y_{i,res}$ such that $y_{i,res} < \hat{y}_i < y_{i,asp}$)) is given, then these thresholds can be also defined as corresponding parts of this range:

- the *veto threshold* $\Delta y_{av,i} > 0$, *e.g.* $\Delta y_{av,i} = \beta(\hat{y}_i - \bar{y}_{i,res})$, $\beta \in (0,1]$;
- the *negative indifference threshold* $\Delta y_{ni,i} > 0$, $\Delta y_{ni,i} < \Delta y_{av,i}$ or $\Delta y_{ni,i} = \xi \Delta y_{av,i}$, $\xi \in (0,1)$;
- the *component outranking threshold* $\Delta y_{co,i} > 0$, *e.g.* $\Delta y_{co,i} = \beta(\bar{y}_{i,asp} - \hat{y}_i)$, $\beta \in (0,1]$;
- the *positive indifference threshold* $\Delta y_{pi,i} > 0$, $\Delta y_{pi,i} < \Delta y_{co,i}$ or $\Delta y_{pi,i} = \xi \Delta y_{co,i}$, $\xi \in (0,1)$;

The number of objective components increased beyond positive indifference threshold will be compared with the number of components decreased beyond negative indifference threshold in order to determine specific outranking relations. All above threshold values might depend on the current point, but we shall not indicate this dependence explicitly, only let the decision maker correct them if needed.

We assume that all component objectives are measured on ratio scales; since the decision maker already learned about their ranges of change $y_{i,up} - y_{i,lo}$, all above thresholds might be also expressed in percentage terms of these ranges. We do not assume that any objective is much more important than another one, although moderate ratios of importance of objectives might be expressed implicitly by the selection of thresholds and their ratios. We assume here that the decision maker might specify separate negative and positive indifference threshold (although in ELECTRE the thresholds $\Delta y_{ni,i}$ and $\Delta y_{pi,i}$ are usually equal) because of known psychological studies – see *e.g.* Kahneman and Tversky (1982) – showing that we usually value differently a loss and a gain.

The specification of four outranking thresholds for each objective component might require too much effort from the decision maker. Therefore, we suggest that she/he specifies obligatory only two of them: the veto threshold $\Delta y_{av,i}$ and the component outranking threshold $\Delta y_{co,i}$ for each i; their suggested values might be just given percentage (*e.g.* 5%) of the objective ranges, or a given part β of the distances to the reservation and aspiration levels, as indicated above. If the decision maker does not like to define specific values for the other (indifference) thresholds $\Delta y_{ni,i}$ and $\Delta y_{pi,i}$, she/he might define a number $\xi \in (0,1)$ which defines them generally, with the suggested value $\xi = 1/m$.

For any given pair of $\hat{\mathbf{y}}$ (which is interpreted as the current efficient point in objective space) and $\mathbf{y}$ (a compared point in objective space), the specified outranking thresholds might be used to define the following five index sets:

$$\begin{array}{lcl} I_+ & = & \{i \in \{1, \ldots m\} : y_i > \hat{y}_i + \Delta y_{pi,i}\} \\ I_- & = & \{i \in \{1, \ldots m\} : y_i < \hat{y}_i - \Delta y_{ni,i}\} \\ I_0 & = & \{1, \ldots m\} \setminus (I_+ \cup I_-) \\ I_c & = & \{i \in \{1, \ldots m\} : y_i > \hat{y}_i + \Delta y_{co,i}\} \\ I_v & = & \{i \in \{1, \ldots m\} : y_i < \hat{y}_i - \Delta y_{av,i}\} \end{array} \tag{32}$$

Furthermore, the decision maker is asked to define an integer number m_0 in comparing the numbers of objectives increased and decreased beyond indifference thresholds, with suggested value *e.g.*: the smallest integer greater than $m/6$. The *component $\mathcal{C}$, normal $\mathcal{P}$ and weak $\mathcal{Q}$ outranking relations* between $\mathbf{y}$ and $\hat{\mathbf{y}}$ can be defined then as follows:

$$\begin{aligned} \mathbf{y}\,\mathcal{C}\,\hat{\mathbf{y}} &\Leftrightarrow (I_c \neq \emptyset) \wedge (I_- = \emptyset) \\ \mathbf{y}\,\mathcal{P}\,\hat{\mathbf{y}} &\Leftrightarrow (|I_+| - |I_-| \geq m_0) \wedge (I_v = \emptyset) \\ \mathbf{y}\,\mathcal{Q}\,\hat{\mathbf{y}} &\Leftrightarrow (0 \leq |I_+| - |I_-| < m_0) \wedge (I_v = \emptyset) \end{aligned} \tag{33}$$

where $|I_+|$, $|I_-|$ denote the numbers of elements of sets I_+, I_-. Thus, the normal outranking of $\hat{\mathbf{y}}$ by $\mathbf{y}$ means that there is no veto and that the number of objectives "in favor of $\mathbf{y}$" exceeds the number of objectives "in favor of $\hat{\mathbf{y}}$" at least by m_0 which is a number threshold defined also by the decision maker. For small m, a natural choice is $m_0 = 1$. The weak outranking relation means that there is no veto and that the number of objectives "in favor of $\mathbf{y}$" is at least equal but does not exceed by m_0 the number of objectives "in favor of $\hat{\mathbf{y}}$". This weak outranking is interpreted, similarly as in ELECTRE, as a case when the decision maker might hesitate.

For example, consider the case $m = 2$, $m_0 = 1$; there might be hesitation if one component improves beyond its positive indifference threshold and the other one deteriorates beyond its negative indifference threshold but not beyond the veto threshold. Normal outranking means, for $m = 2$, that one component improves beyond its positive indifference threshold but the other one does not deteriorate beyond its negative indifference threshold. The component outranking is stronger than the normal one: one component improves beyond its component outranking threshold, greater than the positive indifference threshold, and the other one does not deteriorate beyond its negative indifference threshold. It is easy to consider similar cases, say, for $m = 3$, $m_0 = 1$.

It is convenient to consider both the outranking relations $\mathcal{P}$ and $\mathcal{Q}$ as represented by one *parametric outranking relation* $\mathcal{PQ}_{m_1}$ with a given parameter $m_1 \in \{0, \dots m-1\}$:

$$\mathbf{y}\,\mathcal{PQ}_{m_1}\,\hat{\mathbf{y}} \Leftrightarrow (|I_+| - |I_-| = m_1) \wedge (I_v = \emptyset) \tag{34}$$

The outranking relations considered above are slightly more sensitive than those used in ELECTRE, because we will use them to produce decision options and their outcomes that *are suspected of outranking* the current decision and its outcome; *the final choice, whether the decision option really outranks the current one, is reserved for the decision maker* who plays a sovereign role in this procedure.

The essence of this procedure is the use of reference point methodology to produce points suspected of outranking the current one. This is based on a basic property of an order-consistent achievement scalarizing function of the type (16):

$$\bar{\mathbf{y}} \in Y_0 \Rightarrow \left\{ \begin{array}{rl} \max_{\mathbf{y}\in Y_0} \sigma(\mathbf{y},\bar{\mathbf{y}}) & \geq 0; \\ \hat{\mathbf{y}} = \operatorname{argmax}_{\mathbf{y}\in Y_0} \sigma(\mathbf{y},\bar{\mathbf{y}}) & \geq \bar{\mathbf{y}} \end{array} \right\} \tag{35}$$

Thus, if $\max_{\mathbf{y}\in Y_0} \sigma(\mathbf{y},\bar{\mathbf{y}}) < 0$, then $\bar{\mathbf{y}} \notin Y_0$. Therefore, in order to test for $\mathbf{y}\,\mathcal{C}\,\hat{\mathbf{y}}$, it is sufficient to test the attainability of the reference points $\bar{\mathbf{y}}^{(j)}$ with component reference levels:

$$\bar{y}_i^{(j)} = \left\{ \begin{array}{ll} \hat{y}_i + \Delta y_{co,i}, & i = j \\ \hat{y}_i - \Delta y_{ni,i} & i \neq j \end{array} \right\}, \; j \in \{1, \dots m\} \tag{36}$$

If the maximum of $\sigma(\mathbf{y},\bar{\mathbf{y}}^{(j)})$ is nonnegative, we take the point maximizing this function as suspected for component outranking and present it to the decision maker for acceptance. If the maximum is negative, no component outranking point exists for the objective j. In this case, and also in the case when the decision maker does not accept a suggested point, we just check for next $j \in \{1, \dots m\}$; we must thus perform m maximizations of the order-consistent achievement function (16).

The testing for normal and weak outranking is more complicated. We shall actually test for parametric outranking $\mathcal{PQ}_{\Updownarrow\infty}$ with m_1 decreasing from $m-1$ to 0. In order to do this, we define sets $I_+^{(j)}$, $I_-^{(j)}$, $I_0^{(j)}$ such that:

$$|I_+^{(j)}| - |I_-^{(j)}| = m_1 \tag{37}$$

where (j) denotes one of all possible subdivisions of the set $\{1, \dots, m\}$ into such sets. The number of such possible subdivisions, denoted here by $p(m, m_1)$, can be generated in a combinatorial way. This number denotes all possible partitions of the set $I = \{1, \dots m\}$ into three sets $I_+^{(j)}$, $I_-^{(j)}$, $I_0^{(j)}$, including the cases in which the sets $I_-^{(j)}$, $I_0^{(j)}$ are empty.

For example, if $m = 3$, $m_1 = 1$, then $p(m, m_1) = 6$, since there are the following partitions:

$$\begin{array}{lcllllll} j & = & 1 & 2 & 3 & 4 & 5 & 6 \\ I_+^{(j)} & = & \{1,2\} & \{1,3\} & \{2,3\} & \{1\} & \{2\} & \{3\} \\ I_-^{(j)} & = & \{3\} & \{2\} & \{1\} & \emptyset & \emptyset & \emptyset \\ I_0^{(j)} & = & \emptyset & \emptyset & \emptyset & \{2,3\} & \{1,3\} & \{1,2\} \end{array} \tag{38}$$

If $m = 4$, $m_1 = 1$, then $p(m, m_1) = 12$, *etc.*; it is not difficult to write a computer program generating all partitions $I_+^{(j)}$, $I_-^{(j)}$, $I_0^{(j)}$, $j = 1, \dots p(m, m_1)$ for any m and m_1. For each of such partitions we can define a specific reference point (for each m_1, we start the count $j = 1, \dots p(m, m_1)$ anew):

$$\bar{y}_i^{(j)} = \left\{ \begin{array}{ll} \hat{y}_i + \Delta y_{pi,i}, & i \in I_+^{(j)} \\ \hat{y}_i - \Delta y_{av,i} & i \in I_-^{(j)} \\ \hat{y}_i - \Delta y_{ni,i} & i \in I_0^{(j)} \end{array} \right\}, \quad \begin{array}{l} i = 1, \dots m, \\ j = 1, \dots p(m, m_1) \end{array} \tag{39}$$

For each $j = 1, \dots p(m, m_1)$, we have now again to maximize the order-consistent achievement function $\sigma(\mathbf{y}, \bar{\mathbf{y}}^{(j)})$. Using again the basic property of such functions we note that if any of these maxima is greater or equal zero, then there might exist an attainable outcome outranking the point $\hat{\mathbf{y}}^{(k)}$ and such outcome can be chosen as the maximal point of this function. Conversely, if all these maxima are smaller than zero, then the reference points $\bar{\mathbf{y}}^{(j)}$) are not attainable for all $j = 1, \dots p(m, m_1)$ and there are no attainable points outranking the point $\hat{\mathbf{y}}$.

If such points suspected of outrankimg exist, we present all of them (for a given m_1) to the decision maker. The sum of numbers $p(m, m_1)$ has an upper bound 2^m and thus the total number of points to be tested in Outranking Trials might be as many as 2^m, if the decision maker wants to check all possibilities for all $m_1 = m-1,\ m-2, \dots,\ 1, 0$; but she/he has the right and motivation to stop earlier, if the stronger outranking tests give negative results. On the other hand, solving such a large number (say, $2^m = 128$ if $m = 7$) of maximization problems is not a strong restriction today, because – if necessary – we can apply scalably parallel computers and exploit special parallel computation algorithms, or use computing power distributed in computer networks.

We can turn now to the full procedure of Outranking Trials which can be organized as follows:

1. At a given current outcome point $\hat{\mathbf{y}}^{(k)}$, where (k) denotes iteration number, check whether the decision maker would not like to *redefine the outranking thresholds.* If yes, correct these thresholds.

2. *Check for component outranking:* using reference point methodology, check whether there exist an admissible decision with outcomes $\mathbf{y}$ that component outrank the current point. If yes, present it for acceptance to the decision maker. If $\mathbf{y}$ is accepted, shift current outcome point $\hat{\mathbf{y}}^{(k+1)} = \mathbf{y}$ and start a new iteration. If $\mathbf{y}$ is not accepted, again (for other possible case of different objective components) check whether there exist another admissible decision with outcomes that component outrank the current point and, if yes, present it to the decision maker. Repeat until either a new point is accepted and new iteration started, or all possible cases of component outranking are tested negatively.

3. *Check for normal outranking:* proceed as above, but testing the existence and presenting to the decision maker possible cases of normal outranking by testing for parametric outranking with m_1 decreasing from $m-1$ to m_0, until a new iteration is started or all possible normal outranking cases are tested negatively.

4. *Check for weak outranking:* proceed as above for m_1 decreasing from $m_0 - 1$ to 0, testing the existence and presenting to the decision maker possible cases of weak outranking, until a new iteration is started or all possible weak outranking cases are tested negatively.

5. If all cases were tested negatively (either outranking decisions do not exist or the decision maker does not accept them as outranking), *stop;* the current point is most satisfactory according to the accepted outranking relation.

The use of three different outranking relations simplifies computations in the initial phases of the procedure: since finding one decision option that is accepted as outranking by the decision maker is sufficient to start a new iteration, we start with checking the strongest relation and proceed to a weaker one first when the results of testing the stronger one are negative. When we come down to the weak outranking, we just test for decisions that might be only suspected of outranking. Thus, if there are too many of such suggestions, the decision maker has the right to stop the procedure at any point (on the understanding that further trials will produce only results that are less strongly outranking than former trials).

In order to accelerate convergence, we might also modify the above procedure of Outranking Trials as follows: when checking for outranking of a given type, points suspected of outranking are presented to the decision maker not separately, but jointly – for example, in the case of component outranking, for all $j \in \{1, \dots m\}$ such that the maxima of $\sigma(\mathbf{y}, \bar{y}_i^{(j)})$ are nonnegative. The decision maker can either select any of these points or reject them all.

This modification is especially important when testing for normal and weak outranking, because there might be many such tests.It can be seen that one iteration of Outranking Trials might be long, particularly if the decision maker wants to check all possible cases of outranking. However, the advantage of Outranking Trials is their assured and finite convergence.

In order to investigate convergence, we shall first note that the decision maker might decline to accept a possibly outranking option because she/he believes that this option does not increase her/his value function sufficiently. Thus, we can use the concept of a value function in order to analyze the convergence of Outranking Trials; we must only assume that there is a finite indifference threshold for a meaningful increase of the value function. Observe that the decision maker, because she/he decides sovereignly about all parameters and details of the procedure, can act consistently with her/his value function and with the indifference threshold, by defining sufficiently small outranking thresholds to produce all options suspected of outranking and accepting only those from such options that result in a meaningful increase of the value function, above the indifference threshold. Thus, we define the following condition of consistence of an outranking with a value function:

Outranking-value consistence. Outranking relations defined by (34) are consistent with a continuous, monotone value function $v(\mathbf{y})$ if there is a finite, positive indifference threshold Δv such that, if $v(\mathbf{y}) - v(\hat{\mathbf{y}}^{(k)}) \geq \Delta v$, then $\mathbf{y}$ outranks $\hat{\mathbf{y}}^{(k)}$ at least in the weak sense.

Assuming outranking-value consistence, the proof of the convergence of Outranking Trials is immediate. If the set Q_0 is compact, the continuous value function can only finitely increase in this set; since each iteration in which we find an outranking point increases this function at least by Δv, the procedure must stop after a finite number of iterations at a final point at which no outranking point can be found. We have shown earlier that if no such point can be found, then no attainable outranking point exists, due to the basic property of order-consistent achievement function used for seeking such points. Because of outranking-value consistence, no point exists which would increase the value function more than Δv when compared to the final point of the procedure.

We note that the use of a value function and of the outranking-value consistence property is necessary to investigate the convergence in classic terms; without such assumptions, we can naturally deduce the convergence in the more general terms of outranking relations.

We shall also note that the concept of convergence of Outranking Trials might be only *local*, since the use of veto thresholds implies a refusal to compare decisions with widely different objective components. However, if the assumed value function is concave and the set Y_0 is convex, local convergence is equivalent to the global one.

While it remains to be tested practically whether the Outranking Trials procedure is acceptable for a typical decision maker, this procedure provides at least an alternative theoretical approach to the issue of convergence of some interactive procedures of multiobjective optimization and decision support. For example, the convergence of Contracted Cone or Light Beam Search procedures might probably be investigated while using similar concepts as applied for Outranking Trials.

6 Conclusions

We have presented in this paper an overview of issues related to the investigation of convergence of interactive procedures of multiobjective optimization and decision support. While this subject has a long tradition, we have learned in many applications of decision support systems that the human decision maker cannot be treated as an automaton. It is the complexity of human behavior that necessitates alternative approaches to procedures of interactive decision support and their convergence.

References

[1] Bogetoft, P., Å. Hallefjord and M. Kok (1988) On the convergence of reference point methods in multiobjective programming. *European Journal of Operations Research* **34**, 56-68.

[2] Brans, J.P. and Ph. Vincke (1985) A preference ranking organization method. *Management Science* **31**, 647-656.

[3] Charnes, A. and W.W. Cooper (1977) Goal programming and multiple objective optimization. *J. Oper. Res. Soc.* **1** 39-54.

[4] Dinkelbach, W. and H. Isermann (1973) On decision making under multiple criteria and under incomplete information. In J.L. Cochrane, M. Zeleny (eds.) *Multiple Criteria Decision Making*. University of South Carolina Press, Columbia SC.

[5] Ermolev, Yu., and R. J-B. Wets (1988) *Numerical Techniques for Stochastic Optimization*. Springer Verlag, Berlin-Heidelberg.

[6] Geoffrion, A.M., J.S. Dyer and A. Feinberg (1972) An Interactive Approach for Multicriterion Optimization, with an Application to the Operation of an Academic Department. *Management Science* **19**, 357-368.

[7] Granat, J and A.P. Wierzbicki (1994) Interactive Specification of DSS User Preferences in Terms of Fuzzy Sets. Working Paper of the International Institute for Applied Systems Analysis, WP-94-29, Laxenburg, Austria.

[8] Granat, J., T. Kręglewski, J. Paczyński and A. Stachurski (1994) IAC-DIDAS-N++ Modular Modeling and Optimization System. Part I: Theoretical Foundations, Part II: Users Guide. Report of the Institute of Automatic Control, Warsaw University of Technology, March 1994, Warsaw, Poland.

[9] Jaszkiewicz, A. and R. Słowiński (1985) The LBS package - a microcomputer implementation of the Light Beam Search method for multi-objective nonlinear mathematical programming. Working Paper of the International Institute for Applied Systems Analysis, WP-94-07, Laxenburg, Austria.

[10] Kahneman, D. and A. Tversky (1982) The Psychology of Preferences. *Scientific American* **246** 160-173.

[11] Kaliszewski, I. (1994) *Quantitative Pareto Analysis by Cone Separation Techniques*. Kluwer Academic Publishers, Dordrecht.

[12] Kok, M. and F.A. Lootsma (1985) Pairwise comparison methods in multiple objective programming. *European Journal of Operational Research* **22**, 44-55.

[13] Korhonen, P. and J. Laakso (1985) A Visual Interactive Method for Solving the Multiple Criteria Problem. *European Journal of Operational Research* **24** 277-287.

[14] Lewandowski, A. and A.P. Wierzbicki (1989), eds. *Aspiration Based Decision Support Systems*. Lecture Notes in Economics and Mathematical Systems **331**, Springer-Verlag, Berlin-Heidelberg.

[15] Lootsma, F.A. (1993) Scale sensitivity in multiplicative AHP and SMART. *Journal of Multi-Criteria Decision Analysis* **2**, 87-110.

[16] Lootsma, F.A., T.W. Athan and P.Y. Papalambros (1985) Controlling the Search for a Compromise Solution in Multi-Objective Optimization. To appear in

[17] Luenberger, D.G. (1973) Introduction to Linear and Nonlinear Programming. Addison-Wesley, Reading, Ma.

[18] Michalevich, M.V. (1986) Stochastic approaches to interactive multicriteria optimization problems. WP-86-10, IIASA, Laxenburg.

[19] Ogryczak, W., K. Studzinski and K. Zorychta (1989) A generalized reference point approach to multiobjective transshipment problem with facility location. In A. Lewandowski and A. Wierzbicki, eds.: *Aspiration Based Decision Support Systems* Lecture Notes in Economics and Mathematical Systems, 331, Springer-Verlag, Berlin-Heidelberg.

[20] Rios, S. (1994) *Decision Theory and Decision Analysis: Trends and Challenges*. Kluwer Academic Publishers, Boston-Dordrecht.

[21] Roy, B. and Ph. Vincke (1981) Multicriteria Analysis: Survey and New Directions. *European Journal of Operational Research* **8**, 207-218.

[22] Saaty, T. (1980) *The Analytical Hierarchy Process*. McGraw-Hill, New York.

[23] Sawaragi, Y., H. Nakayama and T. Tanino (1985) *Theory of Multiobjective Optimization.* Academic Press, New York.

[24] Steuer, R.E. and E.V. Choo (1983) An interactive weighted Tchebycheff procedure for multiple objective programming. *Mathematical Programming* **26**, 326-344.

[25] Steuer, R.E. (1986) *Multiple Criteria Optimization: Theory, Computation, and Application.* J.Wiley & Sons, New York.

[26] Vincke, Ph. (1992) *Multicriteria Decision Aid.* J.Wiley, Chichester-New York.

[27] Wierzbicki, A.P. (1980) The use of reference objectives in multiobjective optimization. In G. Fandel, T. Gal (eds.): *Multiple Criteria Decision Making; Theory and Applications,* Lecture Notes in Economic and Mathematical Systems **177**, 468-486, Springer-Verlag, Berlin-Heidelberg.

[28] Wierzbicki, A.P. (1984) *Models and Sensitivity of Control Systems.* Elsevier-WNT, Amsterdam.

[29] Wierzbicki, A.P. (1986) On the completeness and constructiveness of parametric characterizations to vector optimization problems. *OR-Spektrum,* **8** 73-87.

[30] Wierzbicki, A.P. (1992a) The Role of Intuition and Creativity in Decision Making. WP-92-078, IIASA, Laxenburg.

[31] Wierzbicki, A.P. (1992b) Multiple Criteria Games: Theory and Applications. WP-92-079, IIASA, Laxenburg.

[32] Zeleny, M. (1974) A concept of compromise solutions and the method of the displaced ideal. *Comput. Oper. Res.* Vol **1**, 479-496.

[33] Zeleny, M. (1973) Compromise Programming. In J.L. Cochrane and M. Zeleny (eds.) *Multiple Criteria Decision Making.* University of Carolina Press, Columbia, SC.

[34] Zionts, S. and J. Wallenius (1976) An Interactive Programming Method for Solving the Multicriteria Problem. *Management Science* **22** 653-663.

[35] Zionts, S. and J. Wallenius (1983) An Interactive Multiple Objective Linear Programming Method for a Class of Underlying Nonlinear Utility Functions. *Management Science* **29** 519-529.

A HYBRID INTERACTIVE TECHNIQUE FOR THE MCDM PROBLEMS

Ignacy Kaliszewski[1] **Wojtek Michalowski**[2] **Gregory Kersten**[3]

[1]Systems Research Institute, Polish Academy of Sciences, ul. Newelska 6, 01-447 Warszawa, Poland.
[2]School of Business, Carleton University, Ottawa, Ontario, Canada K1S 5B6.
[3]School of Business, Carleton University, Ottawa, Ontario, Canada K1S 5B6.

Abstract. An interactive technique proposed in this paper allows the decision maker to use different search principles depending on his/her perception of the achieved values of the objectives and trade-offs. While an analysis of values of the objectives may guide the initial search for a satisfactory solution, it can be replaced by trade-off evaluations at some later stages of interactive decision making.

Keywords: vector optimization, interactive decision making, trade-offs, efficiency

1 Introduction

A classical multiple criteria decision making problem (Bell,Raiffa,Tversky [1988]; French [1986]) in its most general formulation may take the form of the following vague statement:

$$\text{"given a set of decisions, choose the best one according to decision making circumstances"} \tag{1}$$

Problem (1) is seldom restrained by requiring that a choice be made from the efficient decisions. This condition, however, appears in a following normative formulation of the multiple criteria decision making problem (1):

$$\text{"max"} f(x) \text{ s.t. } x \in S, \tag{2}$$

where $f : \mathcal{R}^n \to \mathcal{R}^k$, $f = (f_1, f_2, \dots, f_k)$, is a vector of the objective functions f_i, S is the set of feasible solutions (admissible decisions), and "max" stands for the operator of determining all efficient solutions of S . With the notation $f(x) = y$, $f(S) = Z$, we call y an *outcome* and Z an *outcome set*. Thus, problem (1) is translated into a problem of the generation of efficient solutions according to a predetermined set of objectives. Such a formulation allows one to identify "the best" decision only if decision making circumstances, in terms of relationships between the outcomes in (2), are fully specified.

Information about these relationships can be acquired and processed in an interactive fashion (Wierzbicki [1980]) Chankong,Haimes [1983]; Yu [1985]; Steuer [1986]). However, within the framework of interactive decision making, detailed information (such as provided by the trade-offs) and deep insights into mutual relationships between solutions of (2), if not correctly structured, may cause the unstable choice behaviour of the decision maker (DM). He/she may reverse previously expressed preferences, may become inclined towards moving in the direction of high relative gains in certain objectives at the expense of moderate losses in other, neglecting the previously accepted levels of the objective function values, or may change his/her risk profile. These aspects of the DM's behaviour need to be considered while modelling and supporting decision processes.

There are two distinct entities of decision process modelling and support: the DM, and the solver (Kersten,Noronha [1996]). The solver comprises of models and solution procedures; they are used to represent important aspects of the decision problem and the solution process. However, in the interactive decision systems there are elements of a decision problem that are not modelled and that require direct input from the DM. Thus, the DM "fills-in" the gaps that are embedded in the solver either by design or by lack of some information. In either case, the solver cannot determine a solution without an active DM's intervention.

The elements of the decision problem and the problem solving process that are embedded in the solver may be quantitative and qualitative. Studies in the decision making indicate that all (or at best most) of the elements that need to be externalized by the DM are inherently qualitative. Moreover, studies in user-system interactions show the relevance and expressiveness of the graphical interfaces and symbolic (linguistic) variables in comparison with the numerical information. Hence, the DM requires a solver that communicates on his/her level of comprehension, that is using qualitative attributes, and that allows for qualitative reasoning.

The research presented in this paper is a step towards building such a solver. Whilst formal quantitative models and algorithms are at the heart of the interactive technique, its informational requirements are very similar to those of the qualitative process theory (D'Ambrosio [1989]) and qualitative reasoning (Werther [1994]). Scope of interaction and the DM's inputs are captured through the specification of three qualitative values such that "+" stands for improvement, "-" stands for indifference, and "0" stands for maintenance of current values of the objective functions, and the specification of the landmark values (Kuipers [1994]) expressed as the bounds on the trade-offs.

These qualitative values are used in the assessment of the numerical values of the objective functions, and the DM's postulates with respect to their change, while the landmark values allow to "break" a set of outcomes into qualitatively distinct regions. The expectation that the DM can specify an exact value of a required trade-off or even state that obtained value is satisfactory, is rarely reasonable. However, much weaker requirement that he/she can give an upper (or lower) bound for a trade-off may often be met.

The rationale for the introduction to the interactive decision making of the elements used in qualitative reasoning is two-fold. First, it takes into account that the decision process is a learning process, it involves the DM's realization of his/her preferences and trade-offs. That is qualitative reasoning which has been developed to model problems with incomplete knowledge can be applied here to search for "the best" decision and to uncover reasons behind preferences, trade-offs, and choices. Secondly, it allows to weaken the information requirements of the interactive decision making while maintaining the rigour of the formal methodology. The hybrid interactive technique asks for less precise information from the DM than the other solvers, and still provides the DM with significant and structured support.

The paper is organised as follows. In the next section we give necessary definitions of trade-offs and efficiecy. In Section 3 we review the literature pertinent to the interactive decision making. In Section 4 we operationalize the hybrid interactive technique, and in Section 5 we give theoretical foundations of the proposed methodology. The paper concludes with a discussion in Section 6.

2 Trade-offs and Efficiency

Throughtout this section, for the sake of simplicity, we present all the results in terms of elements y of an outcome set $Z = f(S)$. However, since Z is rarely given explicitly, usually computations are to be made in terms of the solutions x, as defined in (2).

We start with the necessary notation and formal definitions.

Let $\bar{y} \in Z$, $Z \subseteq \mathcal{R}^k$. For $i = 1, ..., k$, we denote:

$$Z_i^<(\bar{y}) = \{y \in Z \mid y_i < \bar{y}_i \,,\; y_l \geq \bar{y}_l \,,\; l = 1, ..., k \,,\; l \neq i\} \,,$$

Definition 2.1 *Let* $\bar{y} \in Z$. **Trade-off** $T_{ij}^G(\bar{y})$ *involving objectives* i *and* j, $i, j = 1, ..., k$, $i \neq j$, *is defined as*

$$\sup_{y \in Z_j^<(\bar{y})} \frac{y_i - \bar{y}_i}{\bar{y}_j - y_j} \,.$$

We adopt a convention that if $Z_j^<(\bar{y}) = \emptyset$, then $T_{ij}^G(\bar{y}) = \infty$, $i = 1, ..., k$, $i \neq j$. This means that we do not distinguish between the cases when the corresponding supremum is is not bounded and when it is undefined.

The following are commonly accepted definitions of various types of efficiency.

Definition 2.2 *The outcome* $\bar{y} \in Z$ *is* **weakly efficient** *if there is no* y , $y \in Z$, *such that* $y_i > \bar{y}_i$, $i = 1, ..., k$.

Definition 2.3 *The outcome* $\bar{y} \in Z$ *is* **efficient** *if* $y_i \geq \bar{y}_i$, $i = 1, ..., k$, $y \in Z$, *and this implies that* $y = \bar{y}$.

Definition 2.4 (Geoffrion [1968]) *The outcome* $\bar{y} \in Z$ *is* **properly efficient** *if it is efficient and there exists a finite number* $M > 0$ *such that for each* i *we have*

$$\frac{y_i - \bar{y}_i}{\bar{y}_j - y_j} \leq M$$

for some j *such that* $y_j < \bar{y}_j$, *whenever* $y \in Z$ *and* $y_i > \bar{y}_i$.

It is well known that an efficient outcome is also weakly efficient, but the opposite is not true. Contrary to other definitions of trade-offs, we do not require outcome $\bar{y}$ for which a trade-off is being calculated, to be efficient. When Z is convex and $\bar{y}$ is not weakly efficient, then it is easy to show that a trade-off does not exist. However, for nonconvex Z, trade-offs can exist for the outcomes which are not weakly efficient, as demonstrated by the case when Z is finite.

3 Literature Review

Interactive decision making shares a common feature of a dialog scenario linking alternative stages of converting problem (2) into a single criterion optimization problem and solving it (Gardiner, Steuer [1994]). A rough taxonomy of dialogs in interactive decision making involves: *distinctions among outcomes*, *treatment of trade-off*, and *manipulation of a reference point*. An overview and evaluation of interactive methods may be found in Evans [1984]; Goicoechea et al. [1982]; Michalowski [1987]; Szidarovszky et al. [1986]; and Steuer [1986], among others.

Distinctions among outcomes were considered earlier in lexicographic programming (Ignizio [1982]), and this concept has been introduced into interactive decision making through the partition of outcomes. In the method of Chankong,Haimes [1983] information about the partition of outcomes is utilised to construct a proxy utility function which is applied to the selection of the most preferred solution. In the method of Benayoun et al. [1971], an optimisation problem is solved in order to identify a candidate for the most preferred solution. The function being optimised is built using weights derived from the information about partition of outcomes. In the methods proposed in Michalowski [1988] and Michalowski,Szapiro [1989,1992] the same information helps to displace points of reference and to guide the direction of the search for the most preferred solution. The classical method by Geoffrion et al. [1972] also belongs to this group of approaches.

In other approaches, the implicit partition of the outcomes is replaced by an analysis of trade-offs between them. Point-to-point trade-off evaluation was also used in Zionts,Wallenius [1983] to control a review of the adjacent extreme points of the (polyhedral) set of feasible solutions. In principle, all methods which make use of comparison of two or more solutions (via comparing corresponding outcomes) can be viewed as an indirect application of point-to-point trade-off information.

Manipulation of points of reference is another form of a dialog scenario used in interactive decision making. A change of reference may be accomplished through the displacement of an ideal outcome (Zeleny [1982]), or some other points of reference (Lewandowski,Wierzbicki [1989]; Michalowski [1988]; Michalowski,Szapiro [1992]; Zeleny [1982]). A point of reference may be specified by a model (Nakayama,Sawaragi [1984]), or it may be elicited from the DM (Korhonen,Laakso [1986]; Lewandowski, Wierzbicki [1989]; Michalowski,Szapiro [1989,1992]).

A dialog scenario of an interactive decision making usually is structured around either evaluation of the outcomes or trade-offs. So far there were not attempts to explore a complementary character of these two evaluations and to incorporate them into a single dialog scenario framework. A hybrid interactive technique described in the following section is an attempt to create such a framework.

4 A Hybrid Interactive Decision Making Technique

A hybrid interactive decision making approach presented in this section allows the DM to apply different search principles depending on his/her preception of the achieved values of the objectives and the trade-offs. One possibility is to search for satisfactory outcomes paying no attention to the values of trade-offs. Another possibility is to search for solutions with acceptable values of trade-offs regardless of the values of the objective functions. It seems that none of these possibilities is consistently followed by the DM. While an analysis of values of the objectives may guide the initial search for a satisfactory solution to (2), it can be replaced by trade-off evaluations at some later stages of interactive decision making. Thus, the DM initially discriminates among efficient solutions of (2) with the help of outcomes, and terminates his/her search by identifying some satisfactory efficient solution with a help of trade-off information. The operation of a hybrid interactive technique which enables switching between search principles is given below.

The VTB (*Value and Trade-offs Bounding*) Technique

1. Find an efficient solution $\bar{x}$ to (2).
2. Ask the DM to evaluate the candidate solution $\bar{x}$ in terms of objective function values and trade-offs.
At this point the DM is supposed to specify objectives which he/she would like to improve (+), to maintain (0) at the current level, and where he/she is indifferent (-) to changes.
The DM is also given an opportunity to set bounds on trade-offs.
3. Express the DM's preferences with respect to change of objective function values by constraining the feasible set.
4. Express DM's preferences with respect to upper bounds on trade-offs by setting trade-off control parameters.
5. Determine an efficient solution x satisfying requirements specified in Step 3 and calculate trade-offs.
Substitute this solution for $\bar{x}$.
6. If trade-offs of $\bar{x}$ satisfy the specified bounds, go to Step 2, otherwise go to Step 7.
7. Determine a subset of solutions with acceptable trade-offs which were elicited in Step 4.
Find an element of this subset which is "closest" to $\bar{x}$.
Substitute this solution for $\bar{x}$.
Go to Step 2.

Application of the VTB technique does not impose normative priority on the importance of the analysis of values of the objective functions and the evaluation of trade-offs. However, it may be expected that in a search for a satisfactory solution the DM first explores solutions with acceptable outcomes and later discriminates among them by looking for satisfactory trade-offs.

The seven steps of the VTB technique are quite natural in the context of decision making and do not require specific justification, with the possible exception of Step 7. Taking the "closest" (measured in some subjective fashion) solution to $\bar{x}$ is justified by the fact that it cannot be guaranteed that there exists a solution which satisfies the DM's preferences with respect to the values of objective functions *and* the trade-offs.

Additional necessary specifications for the VTB technique are:
Stopping rule: The operation of VTB terminates in Step 2 whenever the DM is satisfied with the current solution $\bar{x}$.

In Step 2: Very large values for bounds on trade-offs can be used as a default.

In Step 4: Trade-off control parameters are explained in the next section.

In Step 5: A general method of calculating trade-offs is given in Kaliszew -ski [1994].

In Step 7: A methodology to generate a subset of properly efficient solutions to (2) satisfying preimposed bounds on trade-offs is discussed in the next section.

5 Theoretical Background for the VTB Technique

This section provides a methodological foundation (Theorem 5.5) for generation of weakly efficient solutions with a *common* upper bound on trade-offs for a *selected* subset of all possible $n(n-1)/2$ trade-offs. Further refinements of that foundation (Kaliszewski, Michalowski [1996], see Theorem 5.6) allow to bound groups of trade-offs by *different* upper bounds, and to generate properly efficient solutions with the required properties of their trade-offs (Theorem 5.8).

Some theorems presented in this section draw from other research. Theorem 5.1 is a well known result dealing with generating properly efficient elements of convex outcome sets. Theorem 5.2 shows how weighting coefficients in a linear scalarizing function are related to bounds on values of trade-offs. Theorem 5.3 recalls an earlier result on scalarizing problem (2) by the modified Tchebycheff metric, and Theorem 5.4 identifies a relationship between a parameter ρ of this metric and bounds on trade-offs. Theorem 5.5 shows how to preimpose a bound on a selected trade-off of weakly efficient elements of an outcome set Z. Finally, with Theorem 5.6 we arrive at a generalization (Theorem 5.7) of Theorem 5.4.

Theorem 5.1 (Geoffrion [1968]) *Assume that Z is convex. An outcome $\bar{y} \in Z$ is properly efficient if and only if there exists a vector λ such that $\bar{y}$ solves the problem*

$$\max_{i} \sum_{i} \lambda_i y_i \qquad (3)$$

for some $\lambda > 0$.

Theorem 5.2 (Kaliszewski [1994]) *Let $\bar{y}$ solves problem (3). Then*

$$T_{ij}^{G}(\bar{y}) \leq \max_{i \neq j} \frac{\lambda_j}{\lambda_i}$$

for all $i, j = 1, ..., k$, $i \neq j$.

Let y^* be an element in $\mathcal{R}^k$ such that $Z \subseteq y^* - int(R^k_+)$, where R^k_+ denotes the nonegative orthant of $\mathcal{R}^k$ and *int* denotes the interior of a set.

Theorem 5.3 (Choo,Atkins [1983]; Wierzbicki [1986]; Kaliszewski [1987,1994]) *An outcome $\bar{y} \in Z$ is properly efficient if and only if there exists a vector $\lambda > 0$ and a number $\rho > 0$ such that $\bar{y}$ solves*

$$\min_{y \in Z} \max_i \lambda_i((y_i^* - y_i) + \rho e^k(y^* - y)), \tag{4}$$

where e^k is a k-dimensional row vector whose all components are equal to 1.

The following theorems provide an operational base for the VTB technique.

Theorem 5.4 (Wierzbicki [1990]; Kaliszewski [1994]) *Suppose $\bar{y}$ solves problem (4) for some $\lambda > 0$ and $\rho > 0$. Then,*

$$T_{ij}^G(\bar{y}) \leq (1+\rho)\rho^{-1}$$

for all $i, j = 1, ..., k$, $i \neq j$.

Theorem 5.5 (Kaliszewski,Michalowski [1995]) *An outcome $\bar{y} \in Z$ is weakly efficient and $T_{ji}^G(\bar{y}) \leq (1+\rho)\rho^{-1}$ if and only if for some vector $\lambda > 0$ and number $\rho > 0$ it solves the following problem*

$$\min_{y \in Z} \max(\lambda_i((1+\rho)(y_i^* - y_i) + \rho(y_j^* - y_j)), \max_{l \neq i} \lambda_l(y_l^* - y_l)), \tag{5}$$

where $l = 1, ..., k$.

Let $I = \{1, ..., k\}$ and $I_1 \subseteq I$, $I_2 = I \setminus I_1$.

Theorem 5.6 (Kaliszewski,Michalowski [1996]) *An outcome $\bar{y} \in Z$ is weakly efficient and for each $i \in I_1$ and each $t \in I_1$, $t \neq i$, $T_{ti}^G(\bar{y})$ is bounded from above by a positive finite number if and only if there exists $\lambda_i > 0$ and $\rho_i > 0$, $i \in I_1$, such that $\bar{y}$ solves*

$$\min_{y \in Z} \max(\max_{i \in I_1} \lambda_i((y_i^* - y_i) + \sum_{t \in I_1} \rho_t(y_t^* - y_t)), \max_{i \in I_2} \lambda_i(y_i^* - y_i)). \tag{6}$$

Any solution to the above problem satisfies $T_{ti}^G(\bar{y}) \leq (1+\rho_i)\rho_t^{-1}$, $i \in I_1$, $t \in I_1$, $t \neq i$.

The following result generalizes Theorem 5.4.

Theorem 5.7 *Suppose $\bar{y}$ solves the following problem*

$$\min_{y\in Z}\max_i \lambda_i((y_i^* - y_i) + \sum_{t\in I}\rho_t(y_t^* - y_t)), \tag{7}$$

where $\lambda_i > 0$ and $\rho_i > 0$ for each i. Then,

$$T_{ti}^G(\bar{y}) \leq (1+\rho_i)\rho_i^{-1}$$

for each $i, t \in I$, $i \neq t$.

Proof The proof follows immediately from Theorem 5.6. Indeed, putting in this theorem $I_1 = I$, $I_2 = \emptyset$ we immediately get (7). Theorem 5.6 states that in this case $T_{ti}^G(\bar{y}) \leq (1+\rho_i)\rho_i^{-1}$ for all i and all $t \neq i$, which is the required result. □

The next theorem generalizes Theorem 5.3.

Theorem 5.8 (Kaliszewski,Michalowski [1996]) *An outcome $\bar{y} \in Z$ is properly efficient if and only if there exists a vector λ, $\lambda > 0$, and numbers ρ_i, $\rho_i > 0$, $i = 1, ..., k$, such that $\bar{y}$ solves*

$$\min_{y\in Z}\max_i \lambda_i((y_i^* - y_i) + \rho_i e^k(y^* - y)).$$

The theorems presented in this section establish theoretical foundations for the VTB technique. The determination of a subset of properly efficient solutions to (2) with acceptable trade-offs is made possible due to the results of Theorem 5.7 and Theorem 5.8 which allow to operationalize the VTB technique. However, in order to demonstrate a close link of these results to other research in the field, we included also background findings (Theorems 5.1 to 5.6). Such treatment of the theoretical foundations of the VTB contributes to their self-contained presentation and puts our results in a broader perspective.

6 Conclusions

In this paper we have presented a new approach to interactive decision making. The novelty of the VTB technique lies in the premises of varying principles of searching for a satisfactory solution. In contrast to the existing interactive methods, our technique gives the DM an opportunity to focus, on any stage of the decision process, on "the speed of changes" (trade-offs) of criteria, instead of on the values of criteria. Following Simon (Simon [1956]), as satisfactory (in the sense of Simon's aspiration levels) solutions are reached, values of criteria should no longer be improved. Our technique allows the DM to explore further these solutions using another search principle based on trade-off evaluations.

The methodology presented in this paper is flexible because it is not contracted to a particular class of normative decision problems. Using the principles of the VTB technique, it is possible to analyze both continuous and discrete decision making situations, the only limitation being the possibility of solving problems (4) and (5). Moreover, the only formally required and not restrictive assumption is the postulate for existence of an outcome y^*. We believe that with the VTB technique we open new directions in interactive decision making research, where known behavioral biases of the DM can be addressed and accomodated regardless of the type of an underlying decision model. In that sense our hybrid approach is truly prescriptive decision making technique.

In a near future we plan to extend the VTB technique and consider mechanisms of the qualitative operators (+, -, 0) used in the representation of the DM's preferences with respect to the values of the objective functions. This in turn should allow to enhance solver's expressiveness and to use the DM's inputs in a non-uniform manner. Such a treatment should contribute to one of the goals of qualitative reasoning which is a discovery of underlying quantitative functions. A challenging task will be then to adapt the VTB in such a way that it will guide the DM while about learning his/her preferences.

References

D'Ambrosio, B. (1989). *Qualitative Process Theory Using Linguistic Variables*, Springer Verlag, New York.

Bell, D., Raiffa, H., Tversky, A. (1988). *Decision Making: Descriptive, Normative, and Prescriptive Interactions*, Cambridge University Press, Cambridge.

Benayoun, R., De Montgolfier, J., Tergny, J., Laritchev, O. (1971). "Linear programming with multiple objective functions: Step Method (STEM)", *Mathematical Programming*, 8, 366-375.

Chankong, V., Haimes, Y.Y. (1983). *Multiobjective Decision Making. Theory and Methodology*, Elsevier, New York.

Choo, E.U., Atkins, D.R. (1983). "Proper efficiency in nonconvex programming", *Mathematics of Operations Research*, 8, 467-470.

Evans, G.W. (1984). "An overview of techniques for solving multiobjective mathematical programs", *Management Science*, 30, 1268-1282.

French, S. (1986). *Decision Theory: An Introduction to the Mathematics of Rationality*, Ellis Horwood , Chichester.

Gardiner, L.R., Steuer, R.E. (1994). "Unified interactive multiple objective programming", *European Journal of Operational Research*, 74, 391-406.

Geoffrion, A.M. (1968). "Proper efficiency and the theory of vector maximization", *Journal of Mathematical Analysis and its Applications*, 22,

618-630.

Geoffrion, A.M., Dyer, J.F., Feinberg, A. (1972). "An interactive approach for multi-criterion optimization, with an application to the operation of the academic department", *Management Science*, 19, 357-368.

Goicoechea, A., Hansen, D.R., Duckstein, L. (1982). *Multiobjective Decision Analysis with Engineering and Business Applications*, John Wiley, New York.

Ignizio, J.P. (1982). *Linear Programming in Single and Multiple Objective Systems*, Prentice Hall, Englewood Cliffs.

Kaliszewski, I. (1987). "A modified weighted Tchebycheff metric for multiple objective programming", *Computers and Operations Research*, 14, 315-323.

Kaliszewski, I. (1994). *Quantitative Pareto Analysis by Cone Separation Technique*, Kluwer Academic Publishers, Dordrecht.

Kaliszewski, I., Michalowski, W. (1995). "Generation of outcomes with selectively bounded trade-offs", *Foundations of Computing and Decision Sciences*, 20, 113-122.

Kaliszewski, I., Michalowski, W. (1996). "Efficient solutions and bounds on trade-offs", accepted for publication in Journal of Optimization Theory and Applications.

Kersten, G., Noronha, S. (1996). "The goodness of decision making: in search of the universal measure", *Journal of Multi-Criteria Decision Analysis*, 5, 12-15.

Korhonen, P., Laakso, J. (1986). "A visual interactive method for solving the multiple criteria problem", *European Journal of Operational Research*, 24, 277-287.

Kuipers, B. (1994). *Qualitative Reasoning. Modelling and Simulation with Incomplete Knowledge*, MIT Press, Cambridge.

Lewandowski, A., Wierzbicki, A.P. (eds) (1989). *Aspiration Based Decision Support Systems, Theory, Software, Applications*, Lecture Notes in Economics and Mathematical Systems, 331, Springer Verlag, Berlin.

Michalowski, W. (1987). "Evaluation of multiple criteria interactive programming approach: an experiment", *INFOR*, 25, 165-173.

Michalowski, W. (1988). "Use of the displaced worst compromise in interactive multiobjective programming", *IEEE Trans. Syst. Man Cybernet.*, 18, 472-477.

Michalowski, W., Szapiro, T. (1989). "A procedure for worst outcomes displacement in multiple criteria decision making", *Computers and Operations Research*, 16, 195-206.

Michalowski, W., Szapiro, T. (1992). "A bi-reference procedure for interactive multiple criteria programming", *Operations Research*, 40, 247-258.

Nakayama, H., Sawaragi, Y. (1984). "Satisficing trade-off method for multiobjective programming", *Interactive Decision Analysis*, (eds M.Grauer and A.P.Wierzbicki), Lecture Notes in Economics and Mathematical

Systems, 229, 113-122, Springer Verlag, Berlin.

Simon, H. (1956). "Rational choice and the structure of the environment", *Psychological Review*, 63, 129-138.

Steuer, R.E. (1986). *Multiple Criteria Optimization: Theory, Computation and Application*, John Wiley, New York.

Szidarovszky, F., Gershon, M.E., Duckstein, L. (1986). *Techniques for Multiobjective Decision Theory*, Elsevier, Amsterdam.

Werther, H. (1994). *Qualitative Reasoning. Modelling and the Generation of Behavior*, Springer Verlag, New York.

Wierzbicki, A.P. (1980). "The use of reference objectives in multiobjective optimization", *Multiple Criteria Decision Making; Theory and Applications*, (eds G.Fandel and T.Gal), Lecture Notes in Economics and Mathematical Systems, 177, 468-486, Springer Verlag, Berlin.

Yu, P.L. (1985). *Multiple Criteria Decision Making: Concepts, Techniques and Extensions*, Plenum Press, New York.

Zeleny, M. (1982). *Multiple Criteria Decision Making*, McGraw-Hill, New York.

Zionts, S., Wallenius J. (1983). "An interactive multiple objective linear programming method for a class of underlying nonlinear utility functions", *Management Science*, 29, 519-529.

A Comparison of Aspiration Level Interactive Method (AIM) and Conjoint Analysis in Multiple Criteria Decision Making

Madhukar Angur [1] and Vahid Lotfi[2]

Abstract

The predictive validity of the two methods, Aspiration-level interactive method (AIM) and conjoint analysis, used for solving decision problems involving discrete alternatives are compared. An empirical analysis based on subjects' preferences for a multiattribute product (buying a house) and a service (selecting an MBA program for study) indicated that consumer preferences derived from AIM may be more valid than the preferences derived from the full-profile conjoint analysis method.

Introduction

The areas of consumers' preferences for multiattribute products and the models for solving multiple criteria decision making (MCDM) have emerged as an integrated body of knowledge over the past two decades [e.g., Dyer, Fishburn, and Steuer 1992]. A consumers' choice process, for example, can be seen as a case in multicriteria or multiattribute decision making. For managers and marketers, understanding consumer preferences is extremely valuable in designing and promoting new products as well as in repositioning an existing product.

Since the early 1970s, conjoint analysis has been regarded as a major technique for measuring consumers' preferences for multiattribute

---

[1] Madhukar Angur, Assistant Professor of Marketing, University of Michigan-Flint, School of Management, Flint, Michigan 48502-2186

[2] Vahid Lotfi, Professor of Management Science, University of Michigan-Flint, School of Management, Flint, Michigan 48502-2186

products and services; consequently, since that time researchers and practitioners have paid considerable attention to conjoint analysis. Based on the original work by Luce and Tukey [1964], conjoint analysis was first adapted to the marketing field by Green and Rao [1971]. Since then many new developments in conjoint analysis and related methods have been reported in the literature. Green and Srinivasan [1990] provide a comprehensive review of the developments in conjoint analysis. The conjoint measurement technique uses an individual's ranking or rating of deliberately manipulated attributes of a product or service to determine his/her utilities for various levels of those attributes. Readers new to conjoint analysis are encouraged to read Green and Wind [1975].

Relatively recently, Lotfi, Stewart, and Zionts [1992] introduced Aspiration-level Interactive Method (AIM) to solve discrete alternative multiple criteria problems. AIM is based on the concept of users' level of aspiration. The method is interactive and enables the user to explore the efficient frontier by adjusting his/her levels of aspiration. In an experimental application, Lotfi, Stewart, and Zionts [1992] reported that AIM performed significantly better than the Expert Choice method or the manual approach in predicting final choice of subjects.

2. Research Objective

The objective of this research was to compare the predictive validity of the two methods, AIM and conjoint analysis, as evidenced by decision makers' (DM) preferences involving a product and a service, each with multiple attributes. Results of empirical studies comparing the validity and reliability of methods, models and techniques are periodically reported in the literature [e.g., Agarwal 1988; Finkbeiner and Platz 1986; Lotfi, Stewart, and Zionts 1992; Safizadeh 1989]. We hope that our assessment of the two methods will further contribute to this endeavor. Comparison of methods is important because researchers need to know whether one evaluation method will lead to the same conclusions as another method. If correlations among the evaluation measures are high, the researcher can choose any convenient method. If the correlations are low, however, the researcher can choose the better method, or use different methods under different circumstances.

3. General Exposition of the Two Methods

3.1 Aspiration-level Interactive Method

Aspiration-level interactive method (AIM) is a discrete alternative, multiple criteria approach to predicting preferences. AIM is an eclectic approach which is based on the concept of the level of aspiration. Alternatives in AIM are assumed to be deterministic. Each alternative is observed on p objectives. The objectives (criteria) can be maximizing, minimizing, or target type, where deviations from lower and upper thresholds are equally undesirable. For each of the three objective function types, and especially the third, there may be associated satisficing thresholds, such that the DM is effectively indifferent to values above or below the threshold (in the case of the first two objective types) or within the range of threshold values in the third case. The DM is asked to identify the stated thresholds explicitly prior to the initiation of the solution process.

The method begins by identifying and presenting to the DM the "basic information" determined from various problem characteristics. These include "ideal point," "nadir point," "next better," and "next worse." Other information such as the proportion of alternatives that are at least as good as the above values are also displayed. The DM is presented with the "median value" of each objective independently and an initial "nearest solution". AIM assumes a Tchebycheff utility function for the decision maker's preferences. Hence, the nearest solution is defined as an alternative that minimizes the maximum weighted distance between the current goals or aspiration levels and the efficient frontier. Initially, the current goals are set at the median values for each objective.

The DM can then adjust his/her levels of aspiration (current goals) and arrive at other non-dominated solutions. As the DM adjusts the current goals, the nearest solution together with all other basic information are updated. The DM also can scan all of the solutions that satisfy the current goals. These alternatives are ranked based on their degree of closeness to the current goal levels. In addition, the DM can request a set of "neighboring" solutions. AIM uses a simplified version of outranking as used in the ELECTRE methods to find “neighbors” of the nearest solution. The DM continues to experiment with the method and explore the efficient frontier until he/she finds an acceptable solution alternative.

3.2 Conjoint Analysis

Conjoint analysis comprises a set of methods used to predict DM (consumer) preferences for a multiattribute product. Green and Srinivasan [1978 p. 104] have defined conjoint analysis as, "any decompositional method that estimates the structure of a consumer's preferences (i.e., estimates preference parameters such as part-worths, importance weights, ideal points), given his or her overall evaluations of a set of alternatives that are prespecified in terms of levels of different attributes." In conjoint analysis, the DM is required to rate or rank order preferences of product profiles presented in one of two ways: either as a tradeoff matrix using two attributes at a time, or as a full profile using all of the selected attributes.

Combinations of attributes set at discrete levels are used in both approaches. In the tradeoff approach, DM is required to rank each combination of levels of two attributes, from the most preferred to the least preferred. However, in the full-profile approach, the DM is given stimulus cards that describe complete product or service configurations. Then the DM is asked to rate or rank the stimulus cards based on their preferences. Although researchers have used both data collection approaches in conjoint analysis, the full-profile approach is generally favored over the tradeoff (two attributes at a time) approach because: (a) the description of the product/service concepts are more realistic since all attributes are considered simultaneously; (b) a ranking or rating scale can be used for evaluating the task; and (c) the DM has to make fewer judgements (especially in the case of a fractional factorial design of the full-profile approach) than if the tradeoff approach is used. Wittink and Cattin [1989] report that almost 60 percent of studies undertaken in recent years used the full-profile approach, 10 percent used a combination of full-profile and tradeoff approaches, and only about 6 percent used the tradeoff approach.

In conjoint analysis, DM preferences are assumed to be modeled by adding the utilities associated with the various attribute levels. These utilities are estimated from data collected either through the full-profile or tradeoff approaches. Several studies have reported the reliability and validity of conjoint analysis. For example, Bateson, Reibstein, and Boulding [1987] provide a comprehensive review of reliability studies in conjoint analysis. Similarly, several studies [e.g., Anderson and Donthu 1988; Krishnamurthi 1988; Mohn 1990] have addressed validity issues in conjoint analysis. These studies include comparing predicted market share using a conjoint simulator with actual market share [e.g., Clarke 1987], and comparing individual-level preferences with actual choices at a later date [e.g., Srinivasan 1988]. Regarding conjoint analysis validity, Green and Srinivasan [1990, p. 13] comment that, "the empirical evidence points to the validity of conjoint analysis as a *predictive* technique."

4. Methodology

As discussed earlier, the primary purpose of this study was to assess the predictive validity of AIM compared to the predictive validity of conjoint analysis. Specifically, this study compares individuals' preferences obtained through AIM and the conjoint method. To do so, a product and a service were chosen based on their relevance to the subjects for the study. The subjects consisted of 40 students enrolled in the Master's of Business Administration (MBA) program at a leading university located in the Midwest. All of the students had a minimum of three years of full-time professional work experience.

4.1 Orientation

During a two-lecture presentation, subjects were introduced to discrete alternative MCDM problems, to AIM, and to the conjoint analysis method. A tutorial demonstration of both AIM and conjoint analysis was used. For the tutorial, the decision problem involved the purchase of a personal computer using five criteria: price, random access memory (RAM), storage capacity, speed, and type of display.

4.2 Test Problems

Two cases, one involving product selection (buying a house) and the other addressing the choice of a service (selecting an MBA program), were utilized. In the case of buying a house, the problem involved selecting a house from among a set of 50 houses obtained from multiple listings through a realtor agency. Each house was rated on the following attributes: number of bathrooms, number of bedrooms, square footage, list price, and lot size. The levels of the five attributes employed in the study are presented in Table 1. In the service selection case, subjects were asked to choose an MBA program from among the top 20 programs in the country published in *Business Week's Guide to the Best Business Schools* [Byrne 1990]. Each program was rated on the following attributes: tuition cost, recruiters' rank, size of the business school, and the cost of living index. Table 2 presents the levels of the four attributes used in this case.

4.3 Experimental Setting

The experiment was conducted in two phases. In the first phase, the subjects were told to assume that they were in the market to buy a house (product) as well as to select an MBA program for study (service). The subjects used the AIM method in a computer lab employing a hands-on approach; specifically, subjects used the AIM computer software program to identify their most preferred choice for each case -- buying a house and selecting an MBA program for study.

The same subjects were then asked to rank order according to desirability stimulus cards representing varying levels of the attributes selected for each case. The conjoint technique used for the study was a full-profile fractional factorial design [Green, Carroll and Carmone 1978] involving 49 combinations for the first case (buying a house) and 9 combinations for the second case (selecting an MBA program for study). Fractional factorial designs are used to reduce the number of choices presented to the DM since in such designs most, if not all, of the interactions are dismissed [Safizadeh 1989]. The rank-ordered profiles were then analyzed using conjoint analysis to generate each subject's most-preferred choices for both cases. It is important to note that a computer-interactive method was not utilized to collect preference rankings for the full-profile conjoint analysis. Adaptive Conjoint Analysis (ACA) developed by Johnson [1987] uses a computer-interactive mode to collect preference data from subjects. Johnson [1987] has explained that computer interaction increases subjects' interest and involvement with the task. However, the results of two studies that empirically compared the performance of the ACA method with the full-profile conjoint analysis method indicate that the full-profile "pencil and paper" method performed better than the ACA method. Consequently, Green and Srinivasan [1990, p. 11] state, "Overall, the empirical validation results to date do not seem to favor the ACA method." In view of this, the full-profile "pencil and paper" method was used in the study.

After two weeks, subjects were revisited as part of the second phase of the experiment. Each subject was presented with three outcomes (alternatives) for each of the two case problems. The three alternatives consisted of: (a) the preference indicated by the AIM method in the first phase; (b) the preference derived through the conjoint method in the first phase; and (c) a control choice selected at random. The source of the alternatives presented, however, was not revealed to the subjects. The subjects were required to rank order according to desirability the three alternatives for each case.

5. Results and Discussion

The results of the rankings are shown in Table 3. In the first case (buying a house), the AIM solution was ranked as their first preference by 28 subjects, as their second preference by 10 subjects and as their third preference by 2 subjects. Similarly, in the second case (selecting an MBA program for study), the AIM solution was ranked as their first preference by 28 subjects, as their second preference by 9 subjects, and as their third preference by 3 subjects. In both cases, it is apparent that a significant proportion of subjects (70 percent in each case) preferred the outcome from the AIM method, followed by the outcome from the conjoint method. As expected, the control choice, selected at random, was the least preferred.

In order to determine whether the rankings were significantly different from each other, Friedman's non-parametric F-test for several related samples [Conover 1980] was utilized. The results indicated that the three rankings for both cases were significantly different (at the 0.001 level). After rejecting the null hypothesis relating to the quality of treatments, Freidman's multiple comparisons t-test was used to compare the methods pairwise. For the first case (buying a house), results indicated that the AIM method was significantly better (at the 0.05 level) than the solutions provided by the other two methods. Further, the conjoint method was found to be significantly better than the random choice method. In the second case also (selecting an MBA program for study), results indicated that the AIM method was significantly better (at the 0.05 level) than the other two methods. However, the solution provided by the conjoint method was not significantly different from the random choice method. It is possible that the reason the results obtained by the conjoint method were not significantly different from those obtained by the random choice method in this case (selecting an MBA program) is that fewer levels of attributes were manipulated than in the first case (buying a house). Results clearly indicate that AIM performed significantly better in terms of predicting *intended* choice of the DM compared to conjoint analysis and the random choice method.

6. Conclusions

The purpose of this study was to compare the validity of two methods, AIM and the conjoint method, used for solving multiple criteria problems involving discrete alternatives. The study involved the use of subjects as decision makers in two different choice situations, *viz,* buying a house (product choice) and selecting an MBA program for study (service choice).

The participation of MBA students as experimental subjects is a potential limitation of this study. If the study's objective had been the actual identification of individuals' preferences for house selection or MBA program selection, the use of MBA students would raise legitimate questions about the generalizability of the findings. However, the objective was to evaluate the comparative performance of the two methods, and preferences for houses or programs serves only as an example. Thus, student subjects should not detract from the significance of the study. Further, the homogeneous educational background of the subjects extirpates a possible source of variation in the expected results. In addition, one of the major strengths of this study is that both a product and a service were used.

Future research should attempt to ascertain the validity of the AIM and the conjoint method as a function of product types, number of attributes involved, and a respondent's *a priori* knowledge of the attributes. Also, the validity of the methods must be evaluated by comparing results with actual choice behavior, rather than only the most preferred choice.

TABLE 1
Case 1: Buying a House--Attributes and Levels

Attribute	Attribute-levels
Square Feet	1000 or less
	1001-1300
	1301-1600
	1601-1900
	1901-2200
	2201-2500
	2501 or more
Number of Bathrooms	1.0
	1.5
	2.0
	2.5
	3.0 or more
Number of Bedrooms	2
	3
	4
	5 or more
Price	70,000 or less
	71,000-100,000
	101,000-130,000
	131,000-160,000
	161,000-190,000
	191,000-220,000
	221,000 or more
Lot Size	.25 acre or less
	.26-.50 acre
	.51-.75 acre
	.76-1.0 acre
	over 1 acre

TABLE 2
Case 2: Selecting an MBA Program for Study

Attribute	Attribute-levels
Recruiters' Rank	Above Median Rank Median Rank Below Median Rank
Tuition Cost	Above Median Cost Median Cost: $28,000 Below Median Cost
Size of Business School	Above Median Size Median Size: 710 students Below Median Size
Cost of Living Index	Above Median Cost of Living Index Median Cost of Living Index: 116 Below Median Cost of Living Index

Table 3
Results of Subjects' Rankings for the Two Cases

Case 1: Buying a Home

Outcome provided by	Number of Respondents		
	First Preference	Second Preference	Third Preference
AIM Method	28	10	2
Conjoint method	9	22	9
Random choice method	3	8	29
Total	40	40	40

Case 2: Selecting an MBA Program for Study

Outcome provided by	Number of Respondents		
	First Preference	Second Preference	Third Preference
AIM Method	28	9	3
Conjoint method	7	17	16
Random choice method	5	14	21
Total	40	40	40

REFERENCES

Agarwal, Manoj (1988), "An Empirical Comparison of Traditional Conjoint and Adaptive Conjoint Analysis," Working Paper No. 88-140, School of Management, State University of New York at Binghamton.

Anderson, James C. and **Naveen Donthu** (1988), "A Proximate Assessment of the External Validity of Conjoint Analysis," *1988 AMA Educators' Proceedings*, G. Frazier et al., eds. Series 54. Chicago: American Marketing Association, 87-91.

Bateson, John E., David J. Reibstein, and **William Boulding** (1987), "Conjoint Analysis Reliability and Validity: A Framework for Future Research," in Review of Marketing, michael J. Houston, ed. Chicago: American Marketing Association, 451-81.

Byrne, John A. (1990), *Business Week's Guide to the Best Business Schools*, McGraw-Hill Publishing Company, New York, N.Y.

Clarke, Darral G. (1987), *Marketing Analysis and Decision Making*. Redwood City, CA: The Scientific Press, 180-92.

Conover, W.J. (1980), *Practical Nonparametric Statistics*, John Wiley & Sons, New York.

Dyer, James S., Peter C. Fishburn, and **Ralph E. Steuer** (1992), "Multiple Criteria Decision Making, Multiattribute Utility Theory: The Next Ten Years," *Management Science*, 38 (May), 645-54.

Finkbeiner, Carl T. and **Patricia J. Platz** (1986), "Computerized Versus Paper and Pencil Methods: A Comparison Study," paper presented at the Association for Consumer Research Conference, Toronto (October).

Green Paul E, J. D. Carroll and **F. J. Carmone** (1978), "Some New Types of Fractional Factorial Designs for Marketing Experiments," In J. N. Sheth (Ed.), *Research in Marketing*, Vol. 1. Greenwich, CT:JAI Press.

Green, Paul E., and **A. M. Krieger** (1991), *"Segmenting Markets With Conjoint Analysis,"* Journal of Marketing, October, 20-31.

Green, Paul E., and **V. R. Rao** (1971), "Conjoint Measurement for Quantifying Judgmental Data," *Journal of Marketing Research*, 8, 355-63.

Green, Paul E., and **V. Srinivasan** (1990), "Conjoint Analysis in Marketing: New Developments With Implications for Research and Practice," *Journal of Marketing*, October, 3-19.

Green, Paul E. and **V. Srinivasan** (1978), "Conjoint Analysis in Consumer Research: Issues and Outlook," *Journal of Consumer Research*, 5 (September), 103-23.

Green, Paul E., and **Yoram Wind** (1975), *"New Way to Measure Consumers' Judgements,"* Harvard Business Review, 53 (July-August), 107-17.

Johnson, Richard M (1987), "Adaptive Conjoint Analysis," in *Sawtooth Software Conference on Perceptual Mapping, Conjoint Analysis, and Computer Interviewing*. Ketchum, ID: Sawtooth Software, 253-65.

Krishnamurthi, Lakshman (1988), *"Conjoint Models of Family Decision Making,"* International Journal of Research in Marketing, 5, 185-98.

Lotfi, Vahid, Theodor J. Stewart, and **Stanley Zionts** (1992), "An Aspiration-level Interactive Model for Multiple Criteria Decision Making," *Computers and Operations Research*, 19 (7), 671-681.

Luce, R.D., and **J. W. Tukey** (1964), "Simultaneous Conjoint Measurement: A New Type of Fundamental Measurement," *Journal of Mathematical Psychology*, 1, 1-27.

Mohn, N. Carroll (1990), Simulated Purchase 'Chip' Testing vs. Tradeoff (Conjoint) nalysis-Coca Cola's Experience," *Marketing Research*, 2 (March), 49-54.

Safizadeh, Hossein M. (1989), "The Internal Validity of the Trade-Off Method of Conjoint Analysis," *Decision Sciences*, Vol 20, 451-61.

Srinivasan, V. (1988), "A Conjunctive-Compensatory Approach to the Self-Explication of Multiattributed Preferences," *Decision Sciences*, 19 (Spring), 295-305.

Wittink, Dick R. and **Philippe Cattin** (1989), "Commercial Use of Conjoint Analysis: An Update," *Journal of Marketing*, 53 (July), 91-96.

Reference Direction Approach To Multiple Objective Linear Programming: Historical Overview

Pekka Korhonen, Department of Economics and Management Science, Helsinki School of Economics, Runeberginkatu 14-16, 00100 Helsinki, FINLAND

Abstract. The purpose of this paper is to provide a historical overview and state-of-art review of the development of the reference direction approach to multiple objective linear programming, originally proposed by Korhonen and Laakso [1986a]. We will discuss the original approach and several development phases. First, the basic idea was extended by allowing a dynamic specification of the reference direction. This extension, which enables a decision-maker to freely search nondominated solutions, is called Pareto Race by Korhonen and Wallenius [1988]. Pareto Race is an essential part of the decision support system VIG, developed by Korhonen [1987a], to support the solving of multiple objective linear programming problems. Furthermore, several other development phases are reviewed .

Keywords: Multiple Objective Programming, Achievement Scalarizing Function, Decision Support, Visual, Interactive, Computer Graphics.

1. Introduction

The solution of multiple criteria decision problems always requires the intervention of the decision-maker (DM), because the solution is not generally unique. Therefore, most procedures are interactive. Several dozen procedures and computer implementations have been developed during the last 25 years to address both *Multiple Criteria Evaluation* and *Design Problems*. The term "evaluation" refers to problems, in which the most preferred solution is chosen from a set of explicitly given alternatives. "Design" refers to problems, in which alternatives are defined by constraints. For an excellent review of several interactive multiple criteria procedures, see Steuer [1986].

The specifics of these procedures vary, but they have several common characteristics. For example, at each iteration, a solution, or a set of solutions, is generated for a DM's examination. As a result of the examination, the DM inputs information in the form of tradeoffs, pairwise comparisons, aspiration levels, etc. The responses are used to generate a presumably, improved solution. The ultimate goal is to find the so-called most preferred solution of the DM. Which search technique and termination rule is used is heavily dependent on the underlying assumptions postulated about the behavior of the DM and the way in which these assumptions are implemented. In the MCDM-research there is a growing interest in the behavioral realism of such assumptions.

Multiple Objective Linear Programming (MOLP) is the most commonly studied problem in *Multiple Criteria Decision Making* (MCDM). In early methods, a common feature was to operate with criterion weights, limiting consideration to efficient extreme points (see, e.g., Zionts and Wallenius [1976]). Today, many systems are based on the use of aspiration level projections, where the projection is performed using Chebyshev-type achievement scalarizing functions. These functions can be controlled either by varying weights (keeping aspiration levels fixed) or by varying the aspiration levels (keeping weights fixed). The same idea was originally proposed in a somewhat different form by Steuer and Choo [1983] and Wierzbicki [1980].

The *Achievement Scalarizing Function* is the main theoretical basis on which the *Reference Direction Approach* proposed by Korhonen and Laakso [1986a] also lies. The parameterizing of the achievement scalarizing function was one of the original ideas in the reference direction approach. When a direction is projected onto the nondominated frontier, a curve traversing across the nondominated frontier is obtained. Then an interactive line search is performed along this curve. The idea enables the DM to make a continuous search on the nondominated frontier. Actually, the idea to use a direction was adopted from the GDF method (Geoffrion, Dyer, and Feinberg [1972]). However, the Frank-Wolfe algorithm used by Geoffrion et al. also deals with dominated solutions. Hemming [1981] was the first to propose the projecting of the direction in the GDF method onto the nondominated frontier, but he suggested no method to implement it. Zionts and Wallenius [1983] proposed the use of the optimality conditions for extreme point solutions. That idea has been generalized for any nondominated solution in the reference direction approach. Thus the reference direction approach is a combination of the ideas of Geoffrion et al. [1972], Wierzbicki [1980], and Zionts and Wallenius [1976], flavored with some of our own ideas.

The original reference direction approach has been further developed into many directions. First, Korhonen and Wallenius [1988] improved upon the original procedure by making it dynamic. The dynamic version was called *Pareto Race*. In Pareto Race, the DM can freely move in any direction on the efficient frontier he/she likes, and no restrictive assumptions concerning the DM's behavior are made. Furthermore, the objectives and constraints are presented in a uniform manner. This idea, originally adopted from Ignizio [1983], was explicitly applied in Pareto Race, and further developed in Korhonen and Narula [1993] into a systematic method called *Evolutionary Approach*.

Pareto Race is only suitable to solving moderate size problems. When the size of the problem becomes large, computing time makes the interactive mode inconvenient. To solve large-scale problems Korhonen, Wallenius and Zionts [1992] proposed a method based on Pareto Race. An (interactive) free search is performed to find the most preferred direction. Based on the direction, an efficient curve can be generated in a batch mode if desired.

The use of the simple achievement scalarizing function in the original approach and further in Pareto Race guaranteed that the solution points are weakly efficient, but not always efficient. Because the appearance of these kinds of points was originally assumed to be a rare exception, the problem was not originally paid attention to. However, later on it was detected that solution points, which are only weakly efficient may actually be quite common in practical problems. The problem to avoid weakly efficient solution points was solved by means of lexicographic parametric programming developed by Korhonen and Halme [1995].

Moreover, the reference direction idea was also extended to the evaluation problems (Korhonen [1988]). However, the parametrization is more difficult to perform in a finite set of alternatives than, e.g., in multiple objective linear programming. Difficulties also appear when the reference direction approach is applied to solve *Multiple Objective Linear Integer Programming* problems (Karaivanova, Korhonen, Wallenius, and Vassilev [1995]). The achievement scalarizing function and the reference direction can also be used in the solution technique based on the use of the interior point method developed by Karmarkar [1984]. The search directions are generated by applying the *Interior Point Algorithm* for solving linear programming problems (Arbel and Korhonen [1996]).

The rest of the paper is organized as follows: In section 2, we will review the theory of the achievement scalarizing function. In section 3, we describe the original reference direction approach. The most important further developments are discussed in section 4. We conclude the paper with a few remarks in section 5.

2. Some Theory of the Achievement Scalarizing Function

The MOLP problem can be written in the general form as follows:

$$\begin{aligned} &\text{“}max\text{”} \quad q = Cx \\ &\text{s.t.} \\ &\qquad x \in X = \{x \mid Ax \geq b,\ x \geq 0\} \end{aligned} \tag{2.1}$$

where $x \in \Re^n$, $b \in \Re^m$, the constraint matrix $A \in \Re^{m \times n}$ is of full rank m, and the objective function matrix $C \in \Re^{k \times n}$.

In MOLP, the efficiency concept is defined using objective function vectors:

Definition 1. In (2.1), $x^* \in X$ is *efficient* iff there does not exist another $x \in X$ such that $Cx \geq Cx^*$ and $Cx \neq Cx^*$.

Definition 2. In (2.1), $x^* \in X$ is *weakly efficient* iff there does not exist another $x \in X$ such that $Cx > Cx^*$.

Let $Q = \{ q = Cx \mid x \in X\}$ be the set of *feasible* objective or criterion function vectors (i.e., the feasible region in the objective or criterion function space). Vectors $q \in Q$ corresponding to efficient solutions are called *nondominated* criterion vectors and vectors $q \in Q$ corresponding to weakly efficient solutions are called *weakly nondominated* criterion vectors. The set of all efficient solutions is called the *efficient set* (denoted E), and the set of all nondominated criterion vectors is called the *nondominated set* (denoted N). For weakly efficient solutions, we use E^W and N^W, respectively.

For characterizing the nondominated set, Wierzbicki [1980] suggested the use of *an achievement (scalarizing) function*. Consider the following mathematical programming problem:

$$min\ s(g, q, w, \rho) = min\ (\max_{i \in K} [(g_i - q_i)/w_i] + \rho \sum_{i=1}^{k} (g_i - q_i)/w_i), \qquad (2.2)$$

$$\text{s.t.} \qquad Q = \{ q = Cx \mid x \in X\},$$

where $w > 0$, $w \in \Re^k$, is a vector of weights, $\rho > 0$ is a small number, and $K = \{1, 2, ..., k\}$. Vector $g \in \Re^k$ is a given point in the objective function space, the components of which are called *aspiration levels*. Using the model (2.2), we may project any given (feasible or infeasible) point $g \in \Re^k$ into the set of nondominated solutions. The following two theorems provide the necessary theoretical basis.

Theorem 1. Let $w > 0$ be an arbitrary vector. Point $x^* \in X$ is efficient if $\exists\, g \in \Re^k$ and $\rho > 0$ such that $q^* = Cx^*$ is a solution of problem (2.2). If $x^* \in X$ is efficient, then $\exists\, \rho > 0$ such that $q^* = Cx^*$ is a solution of problem (2.2), when $g = q^*$. Then the optimal value of $s(g, q, w, \rho)$ is zero.

Proof. See, e.g., Wierzbicki [1986].

Theorem 2. Let $w > 0$ be an arbitrary vector. Point $x^* \in X$ is weakly efficient if $\exists\, g \in \Re^k$ and $\rho \geq 0$ such that $v^* = Cx^*$ is a solution of problem (2.2). If $x^* \in X$ is weakly efficient, then $q^* = Cx^*$ is a solution of problem (2.2), when $g = q^*$ and $\rho = 0$. Then the optimal value of $s(g, q, w, 0)$ is zero.

Proof. See, e.g., Wierzbicki [1986, Theorem 10].

Remark 1. If point $g \in \Re^k$ is feasible, then for a solution $q^* \in N^W$, we have $q^* \geq g$.

Remark 2. If the solution q^* is unique, then $q^* \in N$; otherwise it may be dominated (see, e.g., Korhonen and Halme [1995], Corollary 2.)

To illustrate the use of the achievement scalarizing function s(g, q, w, ρ), consider a two-criteria problem with a feasible region having three extreme points {(0,0), (8,0), (0,4)}, as shown in Fig. 1.

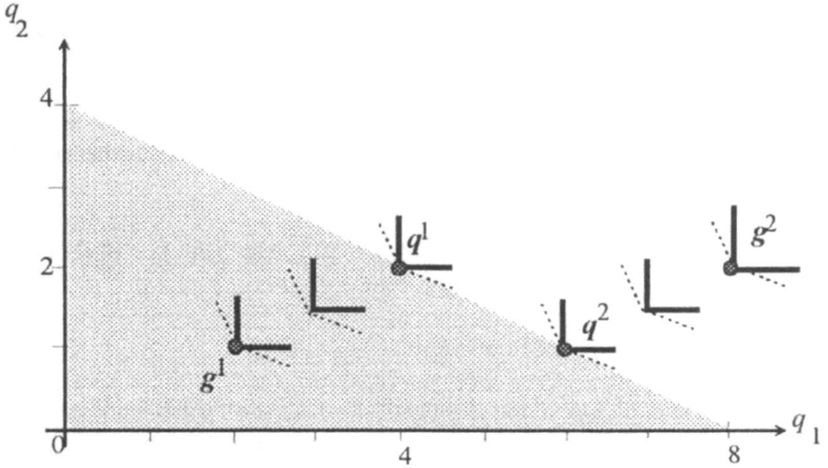

Figure 1. Illustrating the projection of a feasible and an infeasible aspiration level point onto the nondominated surface

Let us assume that the DM first specifies a feasible aspiration level point g^1 = (2,1). Using a weight vector $w = [2, 1]^T$, the minimum value of the achievement scalarizing function (-1) is reached at point q^1 = (4,2) (cf. Fig. 1). Correspondingly, if an aspiration level point is infeasible, say g^2 = (8,2), then the minimum of the achievement scalarizing function (+1) is reached at point q^2 = (6,1). When the aspiration level point is dominated by a feasible point, then the value of the scalarizing function is always nonpositive; otherwise it is nonnegative. It is zero, if an aspiration level point is weakly nondominated. The indifference curves of the achievement scalarizing function in case $\rho = 0$ is illustrated with the solid lines in Fig. 1, and the dotted lines stand for a case $\rho > 0$.

Given $g \in \Re^k$, the minimum of an achievement scalarizing function s(g, q, w, ρ) is found by solving the following LP-problem (see, e.g., Wierzbicki [1980]):

$$\min \quad \varepsilon + r \sum_{i=1}^{k} (g_i - q_i)/w_i$$

s.t. (2.3)

$$\varepsilon \geq (g_i - C_i x)/w_i, \; i = 1, 2, \; \ldots \;, p$$
$$x \in X$$
$$x \geq 0,$$

where C_i ($i = 1, 2, \ldots, p$) refers to the ith row of the objective function matrix C. The problem (2.3) can be further written as:

$$\min \quad \varepsilon + r \sum_{i=1}^{k} (g_i - q_i)/w_i$$

$$\text{s.t.} \qquad\qquad\qquad\qquad\qquad\qquad\qquad\qquad \text{(2.4)}$$

$$Cx + \varepsilon w - z = g$$
$$x \in X$$
$$x, z \geq 0.$$

Because $\varepsilon = z_i/w_i + (g_i - C_i x)/w_i = z_i/w_i + (g_i - q_i)/w_i$, $i = 1, 2, \ldots, k$, we may write the objective function of (2.4) in the form:

$$\varepsilon + r \sum_{i=1}^{k} (\varepsilon - z_i/w_i) = \varepsilon(1 + k\,r) - r \sum_{i=1}^{k} z_i/w_i = (1 + k\,r)[\varepsilon - \frac{r}{1 + k\,r} \sum_{i=1}^{k} z_i/w_i]. \quad (2.5)$$

By writing $\delta = \frac{r}{1 + k\,r}$, we may resort to the following LP-formulation to solve problem (2.2):

$$\min \quad \varepsilon - \delta 1^T y. \qquad\qquad \text{(2.6)}$$

s.t.

$$Cx + \varepsilon w - z = g$$
$$x \in X$$
$$x, z \geq 0,$$

where $y = (z_1/w_1, z_2/w_2, \ldots, z_p/w_p)$ and $\delta > 0$ is a small number.

The use of parameter $\delta > 0$ in computer algorithms is, however, problematic. If δ is too big, not all nondominated points can be diagnosed nondominated. On the other hand, due to too small δ some weakly nondominated points which are dominated will be diagnosed nondominated (because of the finite accuracy which has to be used in computer algorithms). To overcome this problem, some researchers (Sawaragi, Nakayama, and Tanino [1985, p. 276] and Steuer [1986, p. 445]) have proposed the use of the lexicographic formulation:

$$\text{lex min} \quad \{\varepsilon, -1^T z\} \qquad\qquad \text{(2.7)}$$

subject to the constraints in (2.6).

Notation '*lex min*' means the following. The objective function ε is first minimized and if the solution turns out not to be unique, the second objective function -1^Tz is minimized lexicographically by adding the constraint $\varepsilon \leq \varepsilon^*$, where ε^* is the minimum value of ε. Note that the lexicographic formulation is only needed, when the optimal solution of the first objective function (ε) is not unique. Otherwise, we may operate with the first objective function. Generally, the use of the lexicographic formulation does not require much extra work.

3 The Original Version of the Reference Direction Approach

In the following, we will give the outline of the original *Reference Direction Approach* in the form Korhonen and Laakso [1986a] proposed in their original article.

Step 0. Find an arbitrary point q^0 in the criterion space. Let $h := 1$.

Step 1. Specify a vector g^h and take vector $r^h := g^h - q^{h-1}$ as the new reference direction.

Step 2. Find the set Q^h of the efficient vectors q that solve the problem:

$$\min \quad s(q, q^{h-1} + tr^h, w, 0)$$

$$\text{s.t.} \quad q \in N,$$

as t is increased from zero to infinity.

Step 3. Find the most preferred solution q^h in Q^h.

Step 4. If $q^{h-1} \neq q^h$, let $h := h + 1$ and return to Step 1. Otherwise, check the optimality conditions. If the conditions are satisfied, stop; q^h is an optimal solution. If the conditions are not satisfied, let $h := h + 1$, and r^h = a new search direction identified by the optimality checking procedure, and return to Step 2.

The main idea of the reference direction approach is illustrated in Fig. 2. Vector q^0, in Step 1, does not need to be efficient, or even feasible, because it will be projected onto the efficient frontier in Step 2, in any case. However, an efficient solution was already used as an initial solution in the original implementation of the procedure.

In Step 2, the reference direction was specified by using vector g^h. This vector was chosen to be a vector of *aspiration levels* in the spirit of Wierzbicki's original reference point approach (Wierzbicki [1980]). Vector g^h and/or the reference direction r^h can be set up in some alternative ways, but the use of aspiration levels is quite natural and it also turned out to be the best in an experiment by Korhonen and Lantto [1986]. An alternative way is to use, e.g., the AHP (see, Korhonen [1987b]).

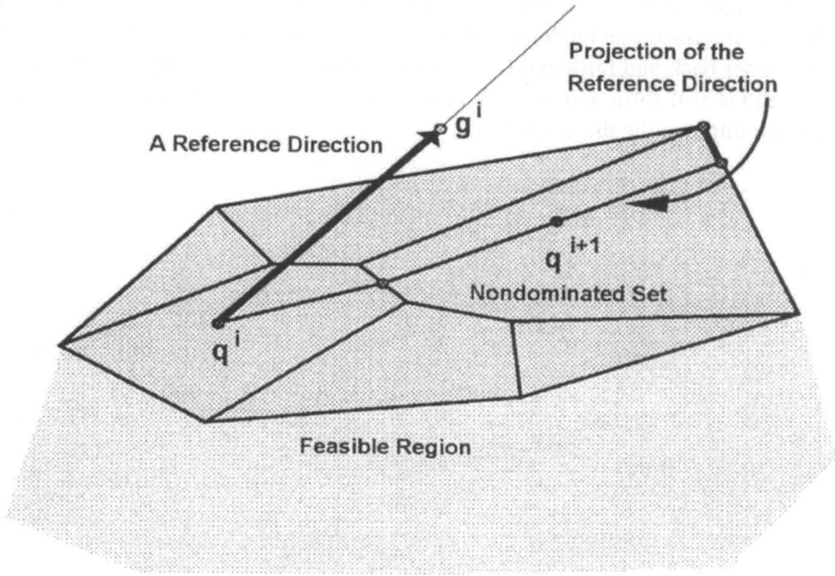

Figure 2. Illustration of the Reference Direction Approach

In the original approach, the achievement scalarizing function was used in its simple form ($\rho = 0$). Set N instead of Q was used in the procedure as a feasible set to guarantee that the solution is efficient, not only weakly efficient (see, Theorem 2). However, the procedure was implemented by using set Q as a feasible set, because it was initially assumed that weakly efficient solutions are rather an exception than a rule. Thus set Q^h was determined by solving the following parametric program:

$$\min \quad \varepsilon \tag{3.1}$$

s.t.

$$Cx + \varepsilon w - z = g + tr^h$$
$$x \in X$$
$$x, z \geq 0,$$

when $t: 0 \to \infty$.

In Step 3, the DM is asked to evaluate solutions on the efficient curve Q^h and to choose the most preferred one. The solutions are easy to present as the functions of the parameter t, and that's why the interface similar to the one proposed by Geoffrion et al. [1972] was a natural choice. The interface used in the original reference direction approach is illustrated in Fig. 3.

The main difference of the interface by Geoffrion et al. [1972] and ours was that the interface was implemented in an interactive fashion. The DM may move the cursor back and forth and he/she will see the values of the objective functions in a numerical form and visually in the line graph. The colors of the lines are used to improve the illustration.

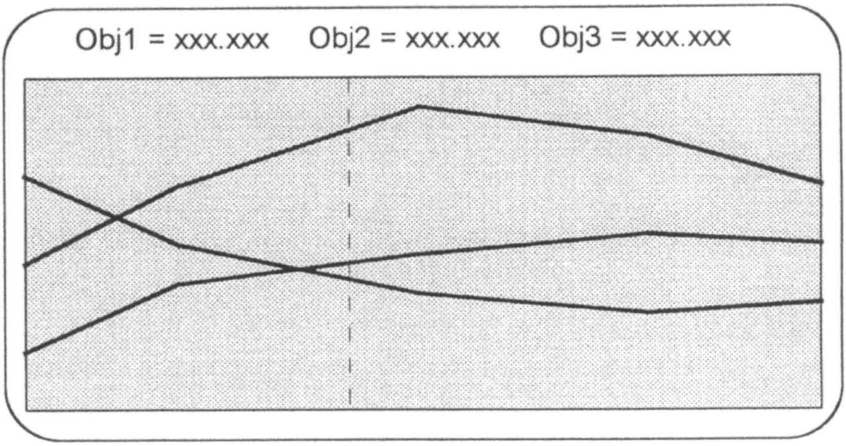

Figure 3. The Interface of the Original Reference Direction Approach

The purpose of Step 4 was to help the DM find the directions of improvement, when he/she is unable to find any. However, in our experience, the DMs do not seem to be very fond of this feature. When they are not able to find directions of improvement, they are not willing to consider other directions either. In addition, to implement optimality checking may be complicated in some special cases (see, Halme [1992]).

The reference direction approach has been further developed into many directions. We will overview some of them in the following section.

4. Some Further Developments

Pareto Race

In the original reference direction approach, the intervention of the DM is controlled. He/she is asked to specify the aspiration levels for objectives and then to search for the most preferred solution on the nondominated curve produced by the approach. However, the formula (3.1) provides sufficient tools to apply the search in a dynamic way. The parametric vector r^h may be redefined at any moment whatever value parameter t has. Based on this idea, Korhonen and Wallenius [1988] developed a dynamic version, in which the reference direction is dynamically varied during the process. This approach enables the DM to make a free search on the nondominated frontier. The idea is illustrated in Fig. 4:

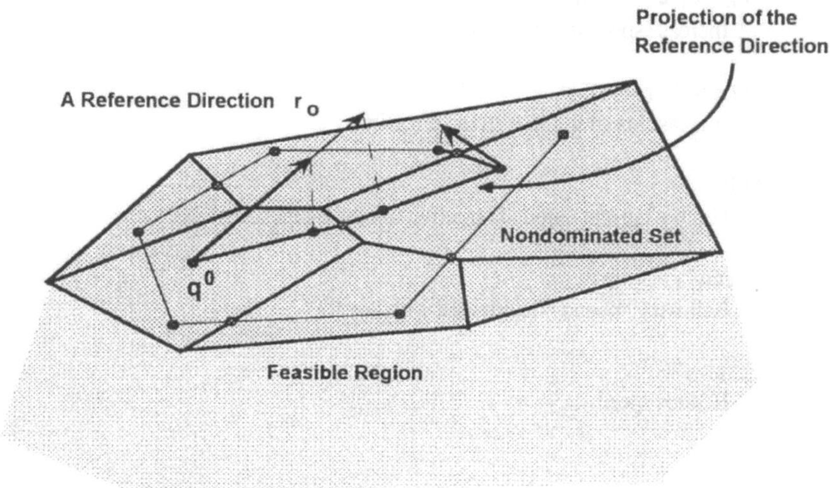

Figure 4: Illustration of the Dynamic Version of the Reference Direction Approach

The method and its implementation is called *Pareto Race*. Pareto Race is a visual, dynamic, search procedure for exploring the nondominated frontier of a multiple objective linear programming problem. In Pareto Race the user sees the objective function values on a display in numeric form and as bar graphs, as he/she travels along the nondominated frontier. The keyboard controls include an accelerator, gears, brakes, and a steering mechanism. The search on the nondominated frontier is like driving a car. The DM can, e.g., increase/decrease the speed, make a turn and brake at any moment he/she likes.

Technically, two parameters are used to control the motion on the nondominated frontier: 1) reference vector r (= *direction*) and 2) the scalar parameter t (= *step size or speed*). The mechanism to change those parameters is as follows:

- When the DM wants to improve some objectives, change r in such a way that the improvements can be seen in the values of the objectives in question.

- When the DM wants to "travel" faster, increase the step-size.

To implement those features, Pareto Race uses certain control mechanisms, which are controlled by the following keys:

(SPACE) BAR: An "Accelerator"

Proceed in the current direction at constant speed.

F1: "Gears (Backward)"

Increase speed in the backward direction.

F2: "Gears (Forward)"

Increase speed in the forward direction.

F3: "Fix"

Use the current value of objective *i* as the worst acceptable value.

F4: "Relax"

Relax the "bound" determined with key F3.

F5: "Brakes"

Reduce speed.

F10: "Exit"

num: "Turn"

Change the direction of motion by increasing the component of the reference direction corresponding to the goal's ordinal number $i \in [1, k]$ pressed by DM.

An example of the Pareto Race screen is given in Fig. 5 below.

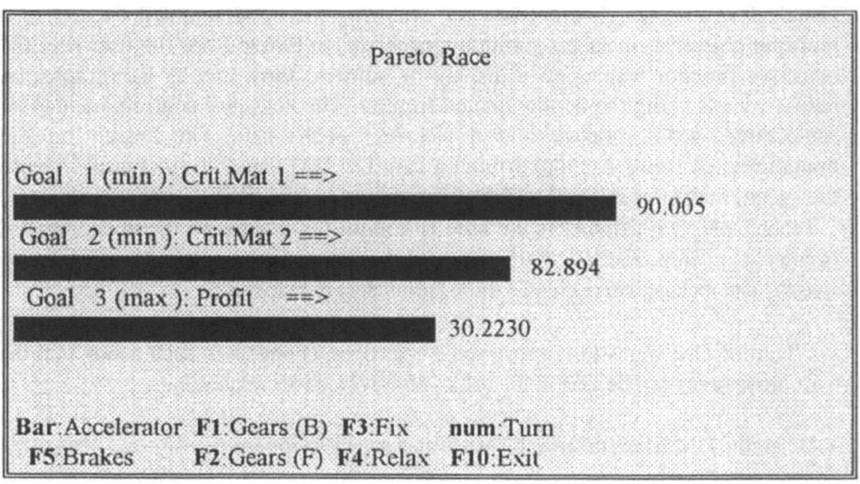

Figure 5: An Example of the Pareto Race Screen

Pareto Race is an essential part of the Multiple Criteria Decision Support System VIG (see, Korhonen [1987a]). A detailed description of the system is given in the booklet by Korhonen, Siljamäki, and Wallenius [1989].

Evolutionary Approach

Following Ignizio [1983], Korhonen and Laakso [1986b] formulated objectives and constraints in a uniform manner as goals. The constraints were called inflexible goals and the objectives flexible goals, correspondingly. According to this philosophy, formula (3.1) can be written as:

$$\min \quad \varepsilon \qquad (4.1)$$

$$\text{s.t.} \quad A^+x + \varepsilon w^+ - z^+ = g^+ + tr^+$$
$$x, z \geq 0,$$

when $t: 0 \rightarrow \infty$, where $A^+ = \begin{pmatrix} C \\ A \end{pmatrix} \in \Re^{(k+m)xn}$, $w^+ = \begin{pmatrix} w \\ 0 \end{pmatrix} \in \Re^{k+m}$, $z^+ = \begin{pmatrix} z \\ z_A \end{pmatrix} \in \Re^{k+m}$, $g^+ = \begin{pmatrix} g \\ b \end{pmatrix} \in \Re^{k+m}$, and $r^+ = \begin{pmatrix} r \\ 0 \end{pmatrix} \in \Re^{k+m}$. Notation z_A refers to the vector standing for the surplus variables associated with the inequalities defining a feasible set. The components of vectors w^+ and r^+ are defined as follows:

$$w_i^+ \quad \begin{cases} = 0, & \text{if } i \text{ refers to a constraint} \\ > 0, & \text{if } i \text{ refers to an objective} \end{cases}$$

and

$$r_i^+ \quad \begin{cases} = 0, & \text{if } i \text{ refers to a constraint} \\ \neq 0, & \text{if } i \text{ refers to an objective.} \end{cases}$$

Formulation (4.1) now provides a simple formulation to change the role of objectives and constraints. When a constraint is changed into an objective, a certain positive value is given to the corresponding component of the weight vector w^+, and a non-zero value is given to the corresponding component of vector r^+. When the change is made in the opposite direction, the corresponding components of vectors w^+ and r^+ are set zero.

Korhonen and Narula [1993] presented a systematic way to utilize this simple "role change" idea for problems, in which the DM cannot specify -- in the beginning of the search process -- which rows of the model are constraints and which objectives. The approach was called *Evolutionary Approach.* The Evolutionary Approach was applied to a media selection problem in Korhonen, Narula, and Wallenius [1989].

Large Scale Problems

Large scale MOLP problems cannot be solved in the same way interactively as small scale problems by moving from one efficient solution to another, and allowing the DM to change his/her mind frequently. Each iteration takes some time, and therefore the solution process has to be planned in such a way that the total number of iterations is minimized. At each iteration, the method should gather as much information about the DM's preferences as possible and use this information as effectively as possible. For example, we might ask the DM to carefully consider the environment of the current solution to make sure that a new search direction is really good. Using this information we may then generate an efficient path that starts from the current solution and passes through the entire efficient frontier, until a boundary of the feasible region is reached. The most time consuming phase is the generation of new efficient solutions. In the future, this phase should probably be handled on a mainframe computer. The user would still interface with the system using a microcomputer. Korhonen, Wallenius, and Zionts [1992] proposed a method, in which the DM will make an environmental search locally using Pareto Race. A new direction is searched in the environment of the current solution in a so-called covering cone. The covering cone is a tangent cone generated in the criterion space, and thus the dimensions of the problem considered in the environmental search are much smaller than those of the original one, making it possible to specify the direction interactively.

In large scale mathematical programming, technically speaking, pivoting is the critical operation in classical methods based on the use of the simplex method. Moving around on the efficient frontier requires a very large number of pivots. That's why in the above approach, the solution process is divided into an interactive and a batch mode. The generation of new solutions is made in the batch mode, because it is too time consuming to be practical to solve interactively. An interesting alternative approach is to approximate the efficient frontier from inside using the ideas of interior point methods developed by Karmarkar [1984]. Arbel and Korhonen [1996] have proposed a new type of approach, which is based on the interior point approach. By varying the precision of the approximation, we may control computing time, making it possible -- at least in principle -- to solve even large scale problems interactively. So far, we have only a prototypical program to solve small textbook examples. It is a topic for future research to see how well it works in practice.

Another reason, why we believe that an interior point method may be a reasonable approach in solving MOLP-problems is that no premature stopping is expected, because the DM is unlikely to be willing to end up with a dominated solution. Premature stopping is further discussed in Korhonen, Moskowitz, and Wallenius [1990].

Avoiding Only Weakly Efficient Solutions

To generate nondominated and only nondominated solutions is not very simple, because the set of feasible parameter values guaranteeing that all nondominated solutions can be generated (and all the generated solutions are nondominated) is open ($\rho > 0$) for the achievement scalarizing functions (Theorem 1). If the closure of the set of feasible parameter values is used instead, then weakly nondominated solutions may be generated as well (Theorem 2). Moreover, if a closed subset of the set of feasible parameter values is used, some nondominated solutions may be excluded from consideration.

It is difficult to specify precisely, when weakly nondominated solutions cause problems. However, in real applications the presence of weakly nondominated solutions seems to be rather a rule than an exception. Obviously, features such as the sparsity of a coefficient matrix, structural constraints, and the lower and upper bounds of decision variables which are typical to real applications, increase the likelihood to generate solutions which are only weakly nondominated. It is important that a support system is able to avoid those solutions. A DM does not like to evaluate dominated solutions, when he/she assumes that they all are nondominated. In the worst case, a DM may even end up with a dominated solution, if he/she does not recognize that a solution is dominated.

Korhonen and Halme [1995] have developed an approach and the requisite theory to search nondominated and only nondominated solutions in multiple objective linear programming (MOLP), when the reference direction approach is used as a solution procedure. The idea is based on the use of a lexicographic formulation (2.7). The use of the reference direction approach leads to a lexicographic parametric programming formulation:

$$\begin{aligned} &lex\,min \quad \{\varepsilon\ ,\ -1^T z\} \\ &\text{s.t.} \\ &\qquad Cx + \varepsilon w - z = g + tr \\ &\qquad x \in X \\ &\qquad x, z \geq 0. \end{aligned} \tag{4.2}$$

Based on the theory developed in Korhonen and Halme [1995], we may choose a new basis after each pivoting in such a way that the new facet corresponding to the new basis is nondominated. Note that the second objective is only needed, when the optimal solution of the first objective function is not unique. When this is the case, then the new basis is chosen in such a way that the second objective function remains lexicographically optimal after the basis change. When the second objective function is needed, required extra work is very marginal.

Multiple Objective Integer Programming

In Karaivanova et al. [1995], we proposed two methods for solving multiple objective integer linear programming (MOILP) problems. One of the methods is based on using Pareto Race (Korhonen and Wallenius [1988]) to find continuous solutions. Then an integer solution "closest" to the current continuous solution in terms of the achievement scalarizing function (2.2) is identified. The efficient integer solution is produced for the DM's evaluation whenever he/she likes to consider it. The process is repeated until the DM has identified a satisfactory compromise (integer) solution. Since we consider continuous and integer solutions at each iteration, we call this method the continuous/integer method.

The biggest problem in our method is the generation of integer solutions. We can speed up the computations by incorporating heuristics in the method. By not requiring that each integer solution during the process be optimal or even feasible in terms of the achievement scalarizing function, we can improve the computational efficiency of the methods dramatically. However, the method is unfortunately quite time consuming. Therefore, our interactive approach is suitable to solve rather small problems. Of course, we can divide the method into an interactive and a non-interactive part. Pareto Race is used interactively, and the closest integer solution is produced in a batch mode. Thus we can increase the usability of the method.

Nadir Solutions

The problem of finding nadir criterion values (minimum criterion values over the nondominated set) in a multiple objective linear program (2.1) has received increased attention in recent years, because of the need to know such information by analysts and decision makers for mathematical, operational, and behavioral reasons.

Although easy to obtain, the minimum values presented in a payoff table are unreliable and should only be used with caution, especially in problems that have more than a small number of extreme points. To obtain better estimates of the nadir criterion values without adding great complexity to the task, Korhonen, Salo and Steuer [1991] developed an approach based upon the use of reference directions. At each iteration of this approach, a reference direction is chosen that maximally minimizes the objective under consideration. We proceed with reference directions that accomplish this until the objective under consideration reaches a local minimum over the nondominated set. Then a cutting plane is inserted into the problem and another direction, if one can be found, that maximally minimizes the objective under consideration is employed. Although the method is heuristic, computational experience shows that much better estimates of the nadir criterion values can be obtained than from payoff tables.

Reference Direction Approach in Solving Evaluation Problems

In an evaluation problem, we assume that a single DM has a set of n deterministic (existing) decision alternatives, and k criteria ($k > 1$), which define an $n \times m$ decision matrix D. Thus d_i refers to the decision alternative in row i. Thus each decision alternative is a point in the criterion space $\Re^k$.

Assuming that the DM wishes to maximize each of the m criteria, the problem is written:

$$\text{“max”} \; d_i \,, i = 1,2, \ldots, n.$$

Applying the reference direction approach to solve an evaluation problem, we may present it as a multiple objective integer linear programming problem:

$$\begin{aligned} & \min \quad \varepsilon \\ & \text{s.t.} \\ & \quad D^T x + \varepsilon w - z = g + tr, \\ & \quad 1'x = 1 \\ & \quad x_i = 0 \text{ or } 1, \text{ for } i = 1,2, \ldots, n \\ & \quad x, z \geq 0, \end{aligned} \tag{4.3}$$

where $1 = [1, 1, \ldots, 1]^T \in \Re^n$. However, problem (4.3) has a special structure making it possible to solve the problem by using algorithms, which are much more effective than general integer programming algorithms (Korhonen [1988]).

Actually, the procedure used to solve evaluation problems resembles very much the original reference direction approach. For instance, the DM controls the search by means of aspiration levels and the interface is like in the original version.

5. Concluding Remarks

In this paper, we have described the most important technical development phases of the reference direction approach. We have not discussed the several real applications in which the reference direction approach has been used. Descriptions of some of these applications can be found. e.g., in Korhonen [1991] and Korhonen, Moskowitz, and Wallenius [1992].

We have neither discussed the implementation of the reference direction approach. At this moment, there exist two commercially available multiple criteria decision support systems which are based on the reference direction approach: VIG and VIMDA. VIG (A Visual Interactive Goal Programming) is developed to solve multiple objective linear programming problems in the spirit of an evolutionary approach. Pareto Race, where a reference direction is specified in a dynamic way, is an essential part of the system. VIG is user-friendly and is based on visual interaction. The user communicates with the respective systems using spreadsheets, menus, and graphics. The system also provides a DM with the possibility to consider multiple objective problems in which the role of objectives and constraints can be interchanged during an interactive session. For more information about the system, see Korhonen [1987a], Korhonen and

Wallenius [1988], or Korhonen, Siljamäki, and Wallenius [1989]. Another system is named VIMDA (A Visual Interactive Method for Discrete Alternative) (Korhonen [1988]). It was developed to solve evaluation problems (discrete alternative problems) using the modification of the original reference direction approach. Both systems are implemented on microcomputers. In addition to these two commercially available systems, there exist prototypical versions of some other decision support systems..

Over the years, the reference direction approach has provided many interesting and challenging research topics, and it seems that many are still unexplored. Some new interesting ongoing research topics which we would like to mention in conclusion are *Multiple Objective Quadratic-Linear Programming* (Korhonen and Yu [1995]) and a *Structural Comparison of Data Envelopment Analysis (DEA) and Multiple Objective Linear Programming* (Joro, Korhonen, and Wallenius [1995]).

References

Arbel, A. and Korhonen, P. (1996). "Using Aspiration Levels in Interactive Interior Multiobjective Linear Programming Algorithm", *European Journal of Operational Research* 89, pp. 193-201.

Geoffrion, A., Dyer, J., and Feinberg, A. (1972). "An Interactive Approach for Multi-Criterion Optimization, with an Application to the Operation of an Academic Department", *Management Science* 19, pp. 357-368.

Halme, M. (1992). *Local Characterization of Efficient Solutions in Interactive Multiple Objective Linear Programing*, Doctoral Dissertation, Helsinki School of Economics, Series A:84.

Hemming, T. (1981). "Some Modifications of a Large Step Gradient Method for Interactive Multicriterion Optimization", in Morse, J. (Ed.): *Organizations: Multiple Agents with Multiple Criteria*, Springer-Verlag.

Ignizio, J. P. (1983). "Generalized Goal Programming", *Computers and Operations Research* 10, pp. 277-289.

Joro, T., Korhonen, P. and Wallenius, J. (1995). "Structural Comparison of Data Envelopment Analysis and Multiple Objective Linear Programming", Working Papers W-144, Helsinki School of Economics.

Karaivanova, J., Korhonen, P., Wallenius, J. and Vassilev, V. (1995). "A Reference Direction Approach to Multiple Objective Integer Linear Programming", *European Journal of Operational Research* 81, pp. 176-187.

Karmarkar, N.K. (1984). "A new polynomial time algorithm for linear programming," *Combinatorica* 4, pp. 373-395.

Korhonen, P. (1987a). "VIG- A Visual Interactive Support System for Multiple Criteria Decision Making", *Belgian Journal of Operations Research, Statistics and Computer Science* 27, pp. 3-15.

Korhonen, P. (1987b). "The Specification of a Reference Direction Using the AHP", *Mathematical Modelling* 9, pp. 361-368.

Korhonen, P. (1988). "A Visual Reference Direction Approach to Solving Discrete Multiple Criteria Problems", *European Journal of Operational Research* 34, pp. 152-159.

Korhonen, P. (1991). "Two Decision Support Systems for Continuous and Discrete Multiple Criteria Decision Making: VIG and VIMDA", in Lewandowski, A., Serafini, P., and Speranza, M. G. (Eds.): *Methodology, Implementation and Applications of Decision Support Systems*, CISM Courses and Lectures - No. 320, International Centre for Mechanical Sciences, Springer Verlag.

Korhonen, P. and Halme, M. (1995). "Using Lexicographic Parametric Programming for Searching a Nondominated Set in Multiple Objective Linear Programming", (Forthcoming in *Journal of Multi-Criteria Decision Analysis*).

Korhonen, P., and Laakso, J. (1986a). "A Visual Interactive Method for Solving the Multiple Criteria Problem", *European Journal of Operational Research* 24, pp. 277-287.

Korhonen, P. and Laakso, J. (1986b). "Solving Generalized Goal Programming Problems Using a Visual Interactive Approach", *European Journal of Operational Research* 26, pp. 355-363.

Korhonen, P. and Lantto, O. (1986). "An Experimental Comparison of Some Reference Direction Techniques for MCDM-Problems", Working Paper, F-142, Helsinki School of Economics.

Korhonen, P., Moskowitz, H. and Wallenius, J. (1990). "Choice Behavior in Interactive Multiple Criteria Decision-Making", *Annals of Operations Research* 23, pp. 161-179.

Korhonen, P., Moskowitz, H., and Wallenius, J. (1992). "Multiple Criteria Decision Support-A Review", *European Journal of Operational Research* 63, pp. 361-375.

Korhonen, P. and Narula, S. (1993). "An Evolutionary Approach to Support Decision-Making with Linear Decision Models", *Journal of Multi-Criteria Decision Analysis* 2, pp. 111-119.

Korhonen, P., Narula, S. and Wallenius, J. (1989). "An Evolutionary Approach to Decision-Making, with an Application to Media Selection", *Mathematical and Computer Modelling* 12, pp. 1239-1244.

Korhonen, P., Salo, S., and Steuer, R. (1991). "A Heuristic for Estimating Nadir Criterion Values in Multiple Objective Linear Programming", Unpublished Working Paper, Helsinki School of Economics..

Korhonen, P., Siljamäki, A. and Wallenius, J. (1989). *A Multiple Criteria Decision Support System VIG in Corporate Planning*, NumPlan Oy.

Korhonen, P., and Wallenius, J. (1988). "A Pareto Race", *Naval Research Logistics* 35, pp. 615-623.

Korhonen, P., Wallenius, J. and Zionts, S. (1992). "A Computer Graphics-Based Decision Support System for Multiple Objective Linear Programming", *European Journal of Operational Research* 60, pp. 280-286.

Korhonen, P. and Yu G.-Y. (1995). "A Reference Direction Approach to Multiple Objective Quadratic-Linear Programming", Unpublished Working Paper, Helsinki School of Economics.

Sawaragi, Y., Nakayama, H. and Tanino, T. (1985). *Theory of Multiobjective Optimization*, Academic Press, New York.

Steuer, R. (1986). *Multiple Criteria Optimization: Theory, Computation, and Applications*, Wiley.

Steuer, R. E. and Choo, E.-U. (1983). "An Interactive Weighted Tchebycheff Procedure for Multiple Objective Programming", *Mathematical Programming* 26, pp. 326-344.

Wierzbicki, A. (1980). "The Use of Reference Objectives in Multiobjective Optimization", in G. Fandel and T. Gal (eds.), *Multiple Criteria Decision Making Theory and Applications*, Springer, New York.

Wierzbicki, A. (1986). "On the Completeness and Constructiveness of Parametric Characterizations to Vector Optimization Problems", *OR Spektrum* 8, pp. 73-87

Zionts, S. and Wallenius, J. (1976). "An Interactive Programming Method for Solving the Multiple Criteria Problem", *Management Science* 22, pp. 652-663.

Zionts, S. and Wallenius, J. (1983). "An Interactive Multiple Objective Linear Programming Method for a Class of Underlying Nonlinear Utility Functions", *Management Science* 29, pp. 519-529.

Implementing the Tchebycheff Method in a Spreadsheet

Ralph E. Steuer

Management Science, Brooks Hall, University of Georgia, Athens, Georgia 30602-6255 USA

1 Introduction

The Tchebycheff Method [29] is one of a group of interactive procedures [e.g., 3-7, 10, 13-14, 16-19, 21, 23, 25-27, 30-31] for solving multiple objective programming problems. While the procedures in this group represent a diversity of solution strategies, the implementations of most of these procedures posses remarkable similarities [8, 9], with most of these procedures being essentially supervisory routines designed to call commercial-grade linear, integer and nonlinear programming codes as their workhorse software. That is, to probe the solution set at each iteration, one or more (single objective) mathematical programs must be solved, at which point in earlier computing environments, an appropriate solver such as MINOS (Murtagh and Saunders [1977]) for linear problems, MPSX (IBM [1979]) for integer problems, or GRG2 (Lasdon and Waren [1986]) for nonlinear problems would be called.

With such collections of solvers having been available on most every university's central computer, this offered a certain convenience. Without cost or maintenance headaches, a programmer could then concentrate on just the supervisory part of a package that deployed these solvers, rather than get bogged down in solver development best left to others.

A problem, however, with being dependent upon the scenario of centrally-provided solver software is encountered when attempting to transport a package to be run on one's home PC, or at another location. While the supervisory part of a package can be moved anywhere by the person who wrote it, the solvers upon which the package is dependent generally can't because of license restrictions. This goes a long way to explain why, despite considerable research devoted to the development of interactive procedures, little transportable interactive multiple objective programming software is available.

However, with modern spreadsheet software on almost everyone's PC, we have a solution that offers many exciting possibilities. For instance, in the Tools menu of Excel is Solver, a quality capability for solving, among other things, linear, integer and nonlinear programming problems. Thus, by writing the supervisory routine of an interactive procedure in Visual Basic [11, 20] along the lines of [1, 2] and attaching it to Excel as a Add-in, we have a transportability as never before. Now there is an incentive to write an interactive multiple objective programming package in the form of an Add-in because it can then be transported to any place in the world possessing Excel, and nowadays, that means just about anyplace. In this way, we can get the tools of interactive multiple objective programming into the hands of many more managers because, without license restrictions, they would be written for use on the software that they use most.

In this paper, we discuss design possibilities for delivering the interactive Tchebycheff Method for solving the multiple objective program

$$\begin{aligned} &max\{f_1(x) = z_1\} \\ &\quad\vdots \\ &max\{f_k(x) = z_k\} \\ &\quad x \in S, \end{aligned}$$

in an Excel spreadsheet. Because of the generalizability of the approach suggested, it is hoped that features of it will be applicable to other interactive multiple objective programming procedures. In Section 2 is discussed a tutorial to picture the overall nature of the interactive Tchebycheff Method written as an Add-in to a user. Section 3 discusses operation of the Tchebycheff Method Add-in concept, Section 4 provides an illustrative example, and Section 4 comments on future directions.

2 Tutorial Description

To review basic concepts and show their organization to form the interactive Tchebycheff philosophy, the interactive Tchebycheff Method Add-in will posses a tutorial to portray the procedure. Concepts to be covered include decision space vs. criterion space; nondominated criterion vectors; graphical detection of nondominated criterion vectors; z^{**} reference criterion vectors; weighting vector space, λ-vectors, and probing directions; and the augmented Tchebycheff program.

Decision Space vs. Criterion Space. Whereas single objective programming is studied in *decision space*, multiple objective programming is mostly studied in *criterion space*. With $S \subset R^n$ the feasible region in decision space, let $Z \subset R^k$ be the feasible region in criterion space where $z \in Z$ if and only if there exists an $x \in S$ such that $z = (f_1(x), \ldots, f_k(x))$. Note that regardless of the dimension of S, the dimension of Z is k, usually a much smaller number than n.

Nondominated Criterion Vectors. Let $K = \{1, \ldots, k\}$. Criterion vector $\bar{z} \in Z$ is *nondominated* iff there does not exist another $z \in Z$ such that $z_i \geq \bar{z}_i$ for all $i \in K$ and $z_i > \bar{z}_i$ for at least one $i \in K$. We are interested in the set of all nondominated criterion vectors, called the *nondominated set*, because under the assumption that "more is always better than less of each criterion," a decision maker's optimal point is nondominated.

Graphical Detection of Nondominated Criterion Vectors. Let $\bar{z} \in Z$ be a test point and consider the nonnegative orthant translated to the test point. Note that a point is in the translated nonnegative orthant iff the point dominates $\bar{z}$. Therefore, $\bar{z}$ is nondominated iff there are no points of intersection between the translated nonnegative orthant and Z, other than for the test point. In Figure 1, because of the points of intersection between the translated nonnegative orthant and $\bar{z}$, z^a is dominated while z^b is nondominated.

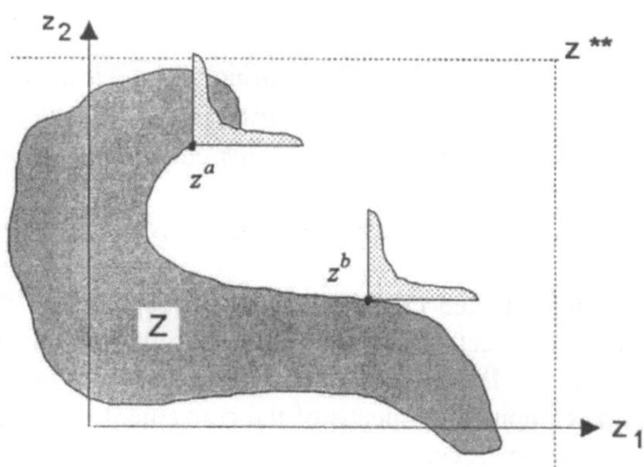

Figure 1

z Reference Criterion Vectors.** The components of a $z^{**} \in R^k$ reference criterion vector are given by

$$z_i^{**} = max\{f_i(x) \mid x \in S\} + \varepsilon_i$$

where the ε_i are small positive values. An ε_i value that raises z_i^{**} to the smallest integer greater than $max\{f_i(x) \mid x \in S\}$ is often convenient [28, Section 14.2]. After establishing a z^{**} reference criterion vector, z^{**} is used as a point from which *downward probes* of the nondominated set are made.

Weighting Vector Space, λ-Vectors, and Probing Directions. Let *weighting vector space* be defined as

$$\Lambda = \{\lambda \in R^k \mid \lambda_i \in (0, 1), \sum_{i \in K} \lambda_i = 1\}$$

In this way, we say that any vector from Λ is a *λ-vector*, and for each λ-vector from Λ, there is a downward probe emanating from z^{**} whose direction is given by

$$-\left(\frac{1}{\lambda_1},\ldots,\frac{1}{\lambda_k}\right)$$

and vice versa.

The nondominated criterion vector obtained from a downward probe is determined by imagining a translated nonnegative orthant, whose origin is always on the downward probe, starting at z^{**} and sliding down the downward probe until it first intersects feasible region Z. If there is only one point of *first intersection*, the point is nondominated. If there is more than one point, among them is at least one nondominated criterion vector.

Supported and Unsupported Nondominated Criterion Vectors. A nondominated criterion vector $\bar{z} \in Z$ is *unsupported* iff there does not exist a nonnegative $\lambda \in \Lambda$ such that $\bar{z}$ solves the *weighted-sums program*

$$max\{\lambda^T z \mid f_i(x) = z_i,\ x \in S\}$$

Otherwise, the nondominated criterion vector is *supported.* While unsupported nondominated criterion vectors can only occur in integer and nonlinear multiple objective programs, the lesson here is that unsupported nondominated criterion vectors can not be computed by means of the convenient weighted-sums program. Other methods utilizing downward probes and sliding nonnegative orthants are better.

Augmented Tchebycheff Program. To carry out the effect of sliding a nonnegative orthant down a downward probe until it first intersects Z, we solve the *augmented Tchebycheff program*

$$\begin{aligned}
&min\{\alpha - \rho \sum_{i \in K} z_i\} && \\
\text{s.t.}\ \ &\alpha \geq \lambda_i\,(z_i^{**} - z_i) && i \in K \\
&f_i(x) = z_i && i \in K \\
&x \in S && \\
&\alpha \geq 0 \text{ and } z \in R^k \text{ unrestricted} &&
\end{aligned}$$

where the λ_i are from a $\lambda \in \Lambda$ that specifies the direction of the downward probe, and the "augmentation" term, in which ρ is a sufficiently small positive scalar, is present to ensure that the z-vector returned by the program is nondominated. Note that the smaller the value of the objective function, the closer the sliding nonnegative orthant is to z^{**}.

In Figure 2, we have the downward probe corresponding to $\lambda = (.4, .6)$ which would cause the augmented Tchebycheff program to return nondominated criterion vector z^c as shown. Note that, because z^c is under the dashed targent line, z^c is unsupported and is hence not computable by the weighted-sums program for any $\lambda \in \Lambda$.

A tutorial description of topics such as the above, but with many more graphical illustrations, forms an important part of the Tchebycheff Method Add-in in order to be always available to refresh for the user the ideas and mechanics of the procedure implemented.

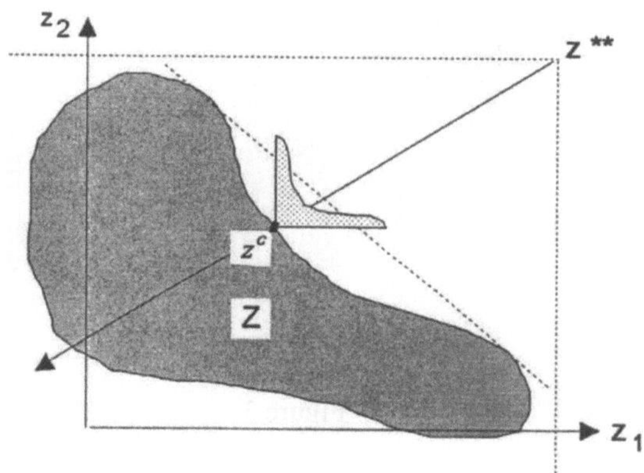

Figure 2

3 Tchebycheff Method Add-in

In interactive multiple objective programming, a solution that is either optimal, or close enough to being optimal to terminate the decision process, is called a *final solution*. To locate to a final solution by means of the Tchebycheff Method, we begin with a group of dispersed λ-vectors from Λ that create a group of radially dispersed probing directions emanating from z** and solve the augmented Tchebycheff program once for each of them. From the nondominated solutions generated, the most preferred is selected. From a neighborhood about the λ-vector that generated the most preferred solution, a group of dispersed λ-vectors is obtained [28, Section 11.9]. Solving the augmented Tchebycheff program once for each of them, we have a new, but more concentrated group of nondominated solution candidates. Selecting the most preferred from the new group, we obtain a group of dispersed λ-vectors from a neighborhood about the λ-vector that produced it, and so forth. As long as the neighbordhoods keep getting smaller, the algorithm will converge to a final solution.

The first iteration of the Tchebycheff Method is signified in Figure 3. After obtaining three dispersed λ-vectors from Λ (e.g., (.25, .75) (.50, .50) (.75, .25)), they are applied via the augmented Tchebycheff program to generate the firstitatation group of nondominated criterion vectors. Selecting the upper leftmost of the generated solutions in Figure 3 as the most preferred, we note that the λ-vector

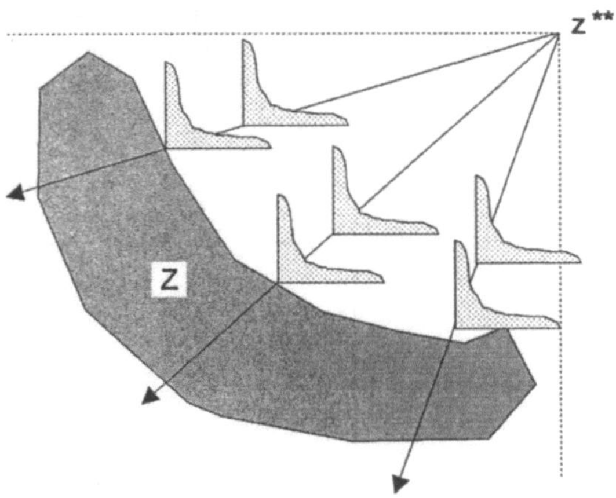

Figure 3

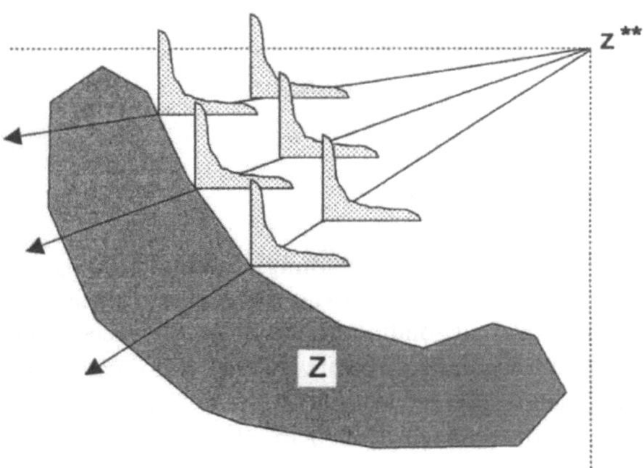

Figure 4

that generated this solution was (.25, .75). About a neighborhood of (.25, .75), three dispersed λ-vectors (e.g., (.16, .85) (.23, .78) (.35, .65)) are obtained. Applying these λ-vectors via the augmented Tchebycheff program we have the results of Figure 4, and so forth.

While the above is for the Tchebycheff Method, descriptions are similar for other procedures in the class of reference point methods of interactive multiple objective programming [3, 6, 13-14, 16, 19, 23].

4 Illustrative Example

Consider the multiple objective program

$$
\begin{aligned}
& max\{2x_1 - x_2 = z_1\} \\
& max\{-x_1 + 3x_2 = z_2\} \\
& \text{s.t.} \quad 3x_1 + x_2 \le 9 \\
& \qquad\quad x_1 + 4x_2 \le 16 \\
& \qquad\quad x_1, x_2 \ge 0
\end{aligned}
$$

Using for convenience the augmented Tchebycheff program variant with w-variables

$$
\begin{aligned}
& min\{\alpha + \rho \sum_{i \in K} w_i \} \\
& \text{s.t.} \quad \alpha \ge \lambda_i w_i && i \in K \\
& \qquad w_i = z_i^{**} - z_i && i \in K \\
& \qquad f_i(x) = z_i && i \in K \\
& \qquad x \in S \\
& \alpha,\ w_i \ge 0 \text{ and } z \in R^k \text{ unrestricted}
\end{aligned}
$$

as recommended in [28, Section 15.1], with $z^{**} = (7, 13)$ and $\rho = .001$, we have the formulation of Figure 5. Employing Solver from the Tools menu of Excel, the thin horizontal rectangle above the variables is reserved for the *changing cells* (i.e., variable values). Reserved at the top of the column to the left is the *target cell* (i.e., objective function value), and below it in the same column are the *constraint cells*, reserved for the expressions of the left-hand sides of the constraints (see [24, pp. 46-56]).

Getting ready to do an iteration of the Tchebycheff Method, in Figure 6 we have the Solver dialog box alongside the eight λ-vectors to be utilized. With the first of the λ-vectors already loaded into the formulation of Figure 5, we have the nondominated solution generated by the augmented Tchebycheff program in the first four of the changing cells as in Figure 7. The nondominated solutions of the other seven λ-vectors are obtained similarly. In this way, we have an outline of an implementation of the Tchebycheff Method.

5 Future Directions

There appear to be many unexploited possibilities for interactive multiple objective programming. That is, anyone who has Excel, for instance, can build an Add-in similar to that outlined in this paper given the collection of operations research software contained in the spreadsheet's Solver capability.

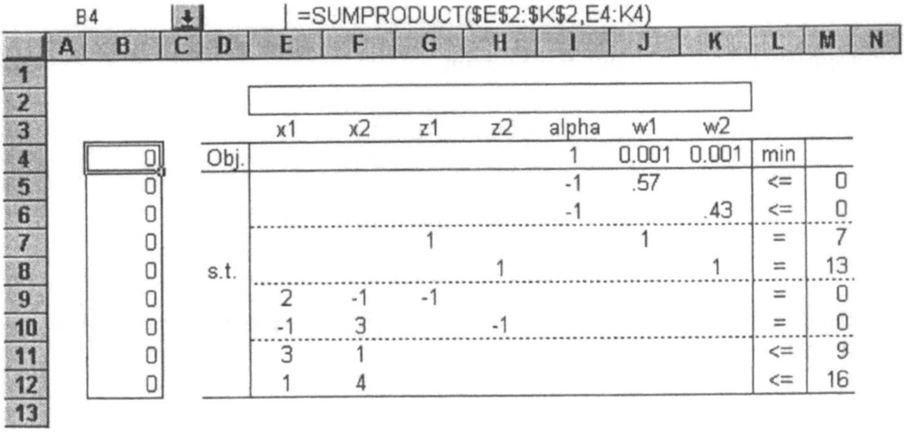

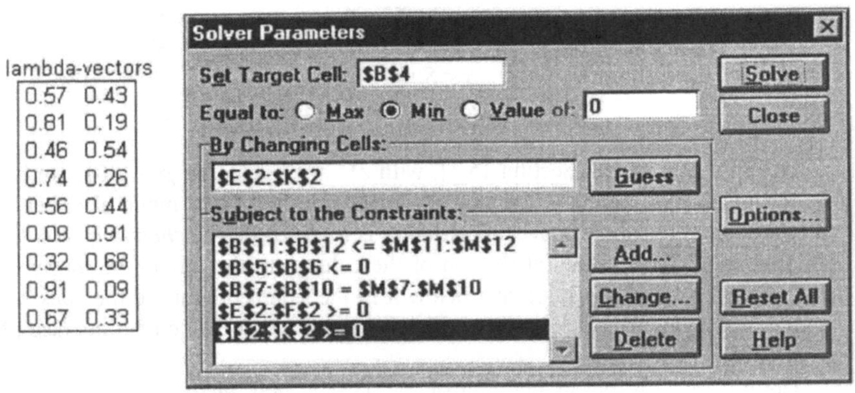

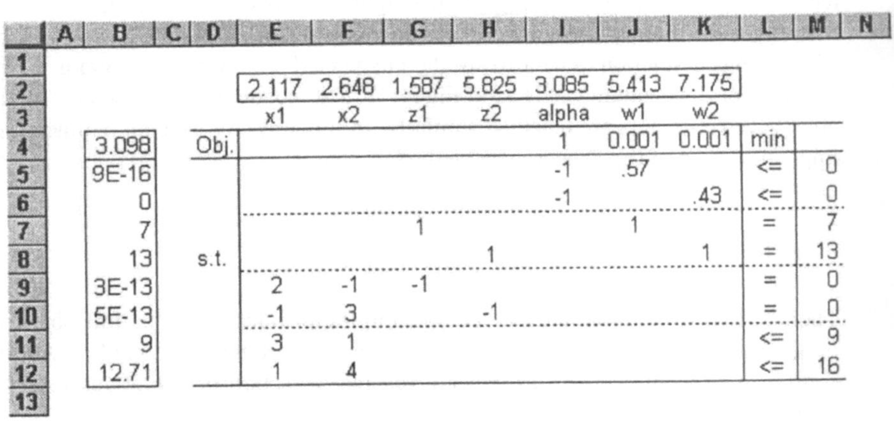

Figure 5 (top), Figure 6 (middle), Figure 7 (bottom)

References

[1] Agrell, P. J. (1995). "A Multicriteria Framework for Inventory Control," *International Journal of Production Economics* 41, 59-70.

[2] Agrell, P. J. and J. Wikner (1997). "An MCDM Framework for Dynamic Systems," *International Journal of Production Economics*, forthcoming.

[3] Benayoun, R., J. de Montgolfier, J. Tergny and O. Larichev (1971). "Linear Programming with Multiple Objective Functions: Step Method (STEM)," *Mathematical Programming* 1/3, 366-375.

[4] Chankong, V. and Y. Y. Haimes (1978). "The Interactive Surrogate Worth Trade-Off (ISWT) Method for Multiobjective Decision-Making." In: *Multiple Criteria Problem Solving* (S. Zionts, ed.), Lecture Notes in Economics and Mathematical Systems, Vol. 155, 42-67.

[5] Climaco, J. C. N. and C. H. Antunes (1987). "TRIMAP - An Interactive Tricriteria Linear Programming Package," *Foundations of Control Engineering* 12/3, 101-120.

[6] Franz, L. S. and S. M. Lee (1981). "A Goal Programming Based Interactive Decision Support System." In: *Organizations: Multiple Agents with Multiple Criteria* (J. N. Morse, ed.), Lecture Notes in Economics and Mathematical Systems, Vol. 190, 110-115.

[7] Gabbani, D. and M. Magazine (1986). "An Interactive Heuristic Approach for Multi-Objective Integer-Programming Problems," *Journal of the Operational Research Society* 37, 285-291.

[8] Gardiner, L. R. and R. E. Steuer (1994). "Unified Interactive Multiple Objective Programming," *European Journal of Operational Research* 74/3, 391-406.

[9] Gardiner, L. R. and R. E. Steuer (1994). "Unified Interactive Multiple Objective Programming : An Open Architecture for Accommodating New Procedures," *Journal of the Operational Research Society*, 45/12, 1456-1466.

[10] Geoffrion, A. M., J. S. Dyer and A. Feinberg (1972). "An Interactive Approach for Multicriterion Optimization, with an Application to the Operation of an Academic Department," *Management Science* 19/4, 357-368.

[11] Heyman, M. S. (1995). *Essential Visual Basic 4*, SAMS Publishing, Indianapolis, Indiana.

[12] IBM Document No. GH19-1091-1 (1979). "IBM Mathematical Programming System Extended/370: Primer," IBM Corporation, Data Processing Division, White Plains, New York.

[13] Karaivanova, J. N., S. C. Narula and V. Vassilev, (1993). "An Interactive Procedure for Multiple Objective Integer Linear Programming Problems," *European Journal of Operational Research* 68/3, 344-351.

[14] Korhonen, P. J. and J. Wallenius (1988). "A Pareto Race," *Naval Research Logistics* 35/6, 615-623.

[15] Lasdon, L. S. and A. D. Waren (1986). "GRG2 User's Guide," University of Texas, Austin, Texas.

[16] Lewandowski, A., T. Kreglewski, T. Rogowski and A. P. Wierzbicki (1987). "Decision Support Systems of DIDAS Family (Dynamic Interactive Decision Analysis and Support)," *Archiwum Automatyki i Telemechaniki* 32/4, 221-246.

[17] Lotov, A. V. (1989). "Generalized Reachable Sets Method in Multiple Criteria Problems." In: *Methodology and Software for Interactive Decision Support* (A. Lewandowski and I. Stanchev, eds.), Lecture Notes in Economics and Mathematical Systems, Vol. 337, 65-73.

[18] Marcotte, O. and R. M. Soland (1986). "An Interactive Branch-and-Bound Algorithm for Multiple Criteria Optimization," *Management Science* 32/1, 61-75.

[19] Michalowski, W. and T. Szapiro (1992). "A Bi-Reference Procedure for Interactive Multiple Criteria Programming," *Operations Research* 40/2, 247-258.

[20] Microsoft Document No. XL57927-0694 (1994). *Visual Basic User's Guide: Automating, Customizing, and Programming in Microsoft Excel*, Microsoft Corporation.

[21] Mikhalevich, V. S. and V. L. Volkovich (1987). "Methods for Constructing Interactive Procedures in Multiobjective Optimization Problems." In: *Toward Interactive and Intelligent Decision Support Systems, Volume 1* (Y. Sawaragi, K. Inoue and H. Nakayama, eds.), Lecture Notes in Economics and Mathematical Systems, Vol. 285, 105-113.

[22] Murtagh, B. A. and M. A. Saunders (1987). "MINOS 5.1 User's Guide," Report SOL 83-20R, Department of Operations Research, Stanford University, Stanford, California.

[23] Nakayama, H. (1991). "Interactive Multi-Objective Programming and Its Applications." In: *Methodology, Implementation and Applications of Decision Support Systems* (A. Lewandowski, P. Serafini, and M. G. Speranza, eds.), Springer-Verlag, Vienna, 75-197.

[24] Ragsdale, C. T. (1995). *Spreadsheet Modeling and Decision Analysis*, Course Technology, Inc., Cambridge, Massachusetts.

[25] Reeves, G. R. and L. S. Franz (1985). "A Simplified Interactive Multiple Objective Linear Programming Procedure," *Computers & Operations Research* 12/6, 589-601.

[26] Sakawa, M. and H. Yano (1990). "An Interactive Fuzzy Satisfying Method for Generalized Multiobjective Programming Problems with Fuzzy Parameters," *Fuzzy Sets and Systems* 35/2, 125-142.

[27] Spronk, J. (1981). *Interactive Multiple Goal Programming*, Martinus Nijhoff, Boston.

[28] Steuer, R. E. (1986). *Multiple Criteria Optimization: Theory, Computation, and Application*, John Wiley & Sons, New York.

[29] Steuer, R. E. and E.-U. Choo (1983). "An Interactive Weighted Tchebycheff Procedure for Multiple Objective Programming," *Mathematical Programming* 26/1, 326-344.

[30] Zionts, S. and J. Wallenius (1976). "An Interactive Programming Method for Solving the Multiple Criteria Problem," *Management Science* 22/6, 652-663.

[31] Zionts, S. and J. Wallenius (1983). "An Interactive Multiple Objective Linear Programming Method for Class of Underlying Nonlinear Utility Functions," *Management Science*, 29/5, 519-529.

PART 2:

PREFERENCES AND LEARNING

From Maximization to Optimization: MCDM and the Eight Models of Optimality

Milan Zeleny, Graduate School of Business Administration, GBA 626E, Fordham University at Lincoln Center, New York, New York 10023, U.S.A.

Abstract. Since its inception, Multiple Criteria Decision Making (MCDM) was destined to evolve alternative notions of optimality. Its challenges to the traditional OR/MS single-objective maximization have crossed the generations, continuing for some 25 years. Yet, its alternative notions of optimality still remain mostly undeveloped.

It is the multiple rather than single criteria that characterize "the best" (optimal) in the areas of economics, engineering, management and business. In this paper we develop the notion of optimum conceived as a balance among multiple criteria. We propose a classificational scheme of eight different, separate and mutually irreducible optimality concepts, with the traditional single-objective "optimality" representing a one special case. The eight optimality concepts provide a useful foundation and framework for the future of MCDM in both research and applications.

1. Introduction

Multiple Criteria Decision Making (MCDM) remains intimately related to the traditional single-objective optimality concept of OR/MS, based on a single scalar function maximization with respect to a priori given constraints. In spite of sometimes admitting multiple criteria via proper vector function, MCDM has often resorted to its subsequent aggregation and scalarization. Multiple criteria are often replaced by a single composite or aggregate, and thus reduced to the traditional single-criterion maximization. Individual criteria are seldom pursued in their true vector sense, i. e., non-scalarized, in their full dimensionality and autonomy - as would be required by most practical considerations.

MCDM problem is not a problem of scalar maximization subject to constraints. It is a vector optimization problem in the sense of Von Neumann and Morgenstern [1]:

> *This is certainly no maximum problem, but a peculiar and disconcerting mixture of several conflicting maximum problems ... This kind of problem is nowhere dealt with in classical mathematics. We emphasize at the risk of being pedantic that this is no conditional maximum problem, no problem of the calculus of variation, of functional analysis, etc. It arises in full clarity, even in the most "elementary" situations, e. g., when all variables can assume only a finite number of values.*

The vector optimization problem [2] arises when two or more (K ≥ 2) scalar-valued objective functions (or criteria) are to be maximized over a set of feasible solutions:

$$\text{opt } \{f(x) \mid x \in X\}, \qquad (1)$$

where $X \in R^n$ and $f(x) = [f_1(x), ..., f_K(x)]$.

A feasible solution x" is *nondominated* with respect to X and f(x) if and only if there exists no $x' \in X$ such that $f(x') \geq f(x")$ and $f(x') \neq f(x")$. Solving a vector optimum problem means identifying all such x", i. e., the set N(X) of solutions nondominated with respect to X and f(x).

There are K associated scalar-optimum problems,

$$\text{opt}^+ \{f_k(x) \mid x \in X\}, k = 1, ..., K \qquad (2)$$

with an optimal solution x_{k+} and $f_{k+} = f_k(x_{k+})$. Vector $f^+ = (f_{1+}, ..., f_{K+})$ represents an ideal value of f(x) *with respect* to X. If $x_{k+} = x^+$, k = 1, ..., K, the vector optimum problem is said to have a *perfect (or ideal) solution* $f^+ = f(x^+)$. This case excepted, no feasible x yields vector f^+.

The perfect solution f^+ is mirrored by its opposite: lowest but separately acceptable values achieved by "anti-optimizing" over the nondominated set N(X) of X. Each such scalar anti-optimum problem,

$$\text{opt}^- \{f_k(x) \mid x \in N(X)\}, k = 1, ..., K \qquad (3)$$

has anti-optimal solution x^{k-} with $f_{k-} = f_k(x_{k-})$. The vector $f^- = (f_{1-}, ..., f_{K-})$ represents anti-ideal value of f(x) with respect to N(X). If $x_{k-} = x^-$, k = 1, ..., K, the vector optimum problem has the *imperfect solution* $f^- = f(x^-)$.

Imperfect solution is not acceptable as a vector (because it is necessarily dominated), but its individual component values could be acceptable separately (as they could be a part of N(X)).

Performance values of perfect and imperfect solutions identify the ranges of achievable values $[f_{k-}, f_{k+}]$, k = 1, ..., K, identified over N(X).

This summary covers most of the advances made by MCDM in the area of vector optimization. The vector optimization problem has been often transformed into an aggregate *scalar* optimization problem and most research concentrated on determining the multiplier-like "weights of importance" for component criteria functions entering into the aggregate scalar function.

Vector problem (1) was often conveniently replaced by a scalar problem:

$$\max \{U[f(x)] \mid x \in X\}, \qquad (4)$$

or by

$$\max \{U[w, f(x)] = w_1 f_1(x) + ... + w_K f_K(x) \mid x \in X\}, \qquad (5)$$

or by any other utility-function variation of U[w, f(x)].

Such scalarization of MCDM has had clear effects on the treatment of criteria weights of importance. The U[w, f(x)] is a single-criterion aggregation structure which supersedes individual decision criteria and interprets their weights of importance, w_k, k=1, ..., K, as normalizing multipliers or aggregation "discount factors". The larger the weight w_k, the more valued is criterion's *contribution* to the overall aggregate performance of U[w, f(x)].

2. Eight Concepts of Optimality

What is a priori given, fixed or determined cannot be subsequently optimized - because it is given. Only what is not given still remains to be selected, chosen or otherwise identified and it is therefore, by definition, still subject to optimization. Consequently, different optimality concepts reflect the distinctions between what is given and what is yet to be (optimally) determined.

We have chosen to present each of the eight optimality concepts, summarized in Table 1, through a small real-life example and through a graphical representation of two-dimensional linear programming (LP) problem.

2.1. Single-objective "optimality"

This refers to conventional maximization (or "optimization"). This is not strictly a multiple criteria or tradeoff balancing problem, but it should be included for the sake of completeness.

To maximize any single criterion, it is entirely sufficient to perform technical measurement and search processes only. Once both X and f are formulated or given, the "optimum" (e. g., maximum) is explicated by mere computation, no optimization, balancing, decision or conflict dissolution processes are needed. The search for optimality is replaced by *scalarization*, assigning each alternative a number (scalar) and then identifying the highest-numbered alternative.

Example: From a given list of five places (X), find the one that is the cheapest (Min f) for vacations.

LP problem:

$$\text{Max } f = cx$$
$$\text{s. t. } Ax \leq b,\ x \geq 0,$$

where $c \in R^n$ and $A \in R^{m \times n}$ are coefficient vector and matrix of dimensions 1xn and mxn respectively, $b \in R^m$ is mx1 dimensional vector of *given* resources (constraint levels) and $x \in R^n$ is nx1 vector of decision variables (solutions). This problem, for n = 2, is graphically represented in Figure 1.

2.2. Multiobjective optimality

Optimization is not just a single-objective measurement-search maximizing but must involve balancing *multiple criteria* and *their tradeoffs*. In reality, humans continually resolve conflicts among multiple criteria which are "competing" for attention and assignation of importance.

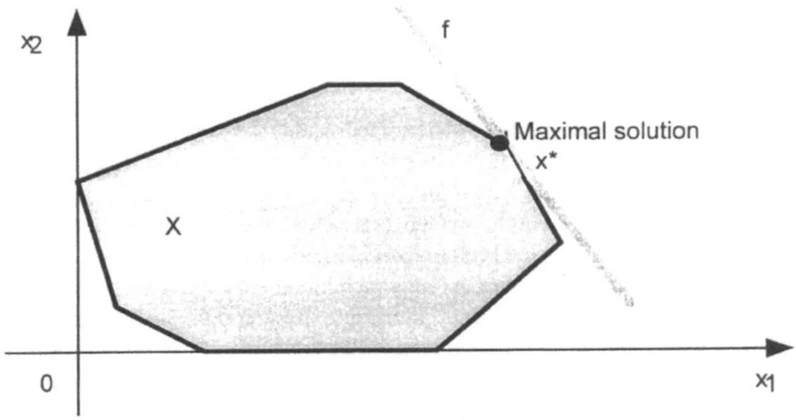

Figure 1. Single-objective "optimality"

Parallel maximization of individual criteria functions should remain non-scalarized, separate and independent, i. e., not subject to aggregation. Forming and maximizing a superfunction $U[f_1(x), f_2(x), ..., f_K(x)]$ effectively reduces multiobjective optimization to the previous case of single-objective maximization.

Example: From a given list of five places (X), select the one that is the cheapest (Min f_1) and also safest (Max f_2) for vacations.

LP problem:

$$\text{Max } F = Cx$$

$$\text{s. t. } Ax \leq b, x \geq 0,$$

where $C \in R^{Kxn}$ is coefficient matrix of dimensions Kxn. The solution set X* is a set of nondominated solutions. Each nondominated solution is potentially optimal, representing an equilibrium among multiple criteria. The situation is graphically portrayed in Figure 2.

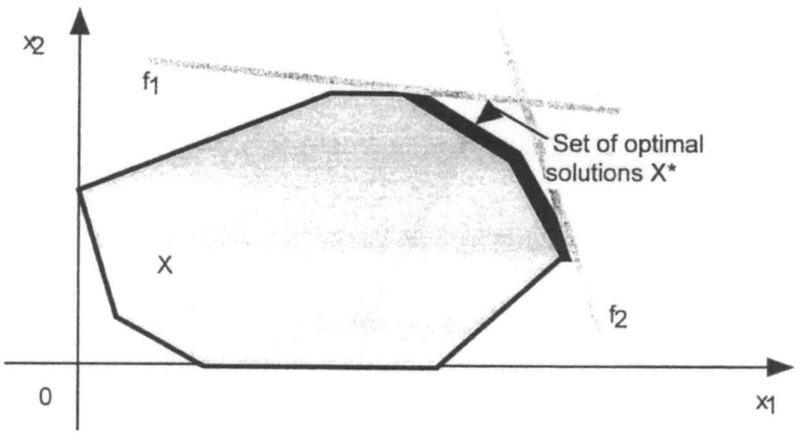

Figure 2. Multiobjective optimality

2.3. Optimal system design: single criterion

Instead of optimizing a priori given system X with respect to selected criteria, humans often seek to form or construct an optimal system X of decision alternatives, designed with respect to chosen criteria [3, 5]. Single-criterion design is analogous to single-criterion "optimization," producing the best (optimal) set of alternatives X at which a given, single objective function f(x) is maximized subject to the budgeted cost of such design (affordability).

Example: Design a list of affordable places (X) which would assure the cheapest (Min f) vacations.

LP problem:

$$\text{Max } f = cx$$
$$\text{s. t. } Ax \le b,\ x \ge 0,$$
$$pb \le B$$

where $p \in R^m$ is $1\text{x}m$ vector of unit prices of m resources and B is total available budget (or cost).

Solving the above problem requires finding the optimal allocation of budget B so that the purchased portfolio of resources b assures the feasibility of x* while f(x*) = Max f(x). The resulting X* (the intersection of active constraints) is reduced to a single point: maximal solution x*. This situation is represented in Figure 3.

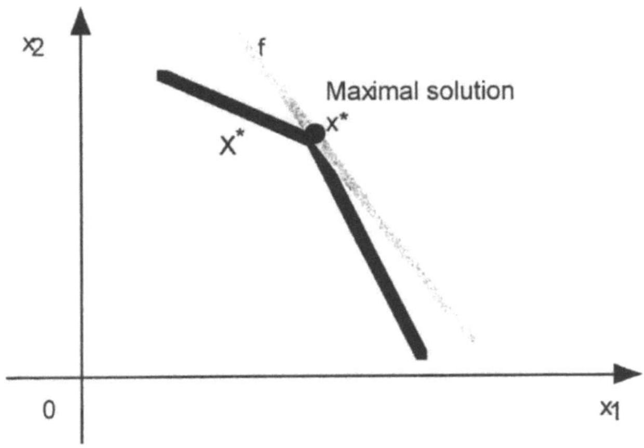

Figure 3. Optimal system design: single criterion

2.4. Optimal system design: multiple criteria

This optimality concept refers to the best system design, defined as above, but with respect to multiple criteria.

Example: Design a list of affordable places (X), which assures the cheapest (Min f_1) and also the safest (Max f_2) for vacations.

LP problem:

$$\text{Max } F = Cx$$
$$\text{s. t. } Ax \leq b, x \geq 0,$$
$$pb \leq B$$

Solving the above problem implies finding the optimal allocation of budget B so that the purchased portfolio b of resources assures the feasibility of x* and F(x*) at cost B and remains "as close as possible" to the metaoptimum design F* at cost B*, B* ≥ B.

Linear inequalities become linear equations and the resulting X* (their intersection) is reduced to a single point: maximal solution x* at cost B. This situation is portrayed in Figure 4.

Instead of the set of nondominated solutions, we now face a set (or a family) of optimal system designs, characterized by the equivalent cost B and the differential importance of objective functions f_1 and f_2.

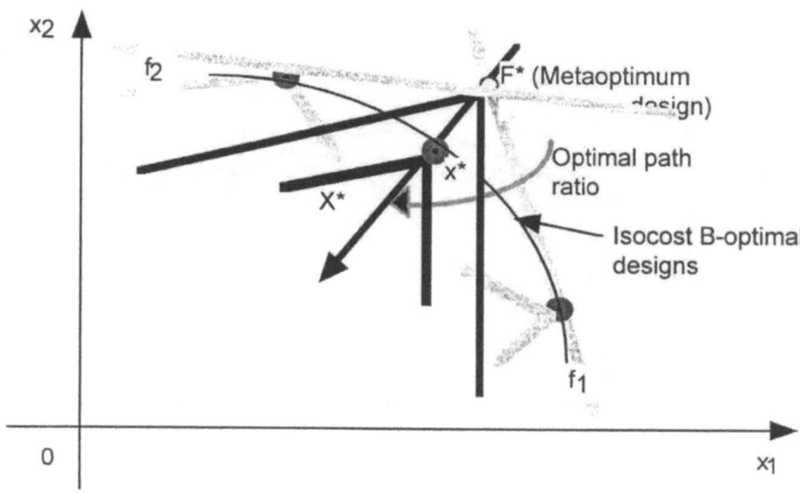

Figure 4. Optimal system design: multiple criteria

2.5. Optimal valuation: single criterion

The so far discussed optimization concepts assumed the criteria to be given a priori. However, in human decision-making processes, different criteria are continually being tried and reapplied, some are discarded, new ones added, until a properly balanced mix (or portfolio) of both quantitative and qualitative criteria is achieved. There is nothing more wasteful than engaging perfectly good means X towards unworthy, ineffective or only habitually determined criteria (goals and objectives).

If the set of alternatives X is given and fixed a priori, but the criteria are not - we can speak of the problem of valuation: How should the alternatives be ordered? According to criterion f_1 or f_2 or f_3? Which criterion best captures decision maker's values and purposes?

Example: Select one single criterion (f), perhaps from a given list (f_1, f_2, f_3, ...), like entertainment, education, privacy, cost, etc., which would best evaluate a given list of affordable places (X) in order to attain most satisfactory valued or fulfilling (through Max or Min f) vacations.

LP problem:

Find f such that

Max $f = cx$

s. t. $Ax \leq b$, $x \geq 0$,

provides the best valuation of X with respect to a given value complex.

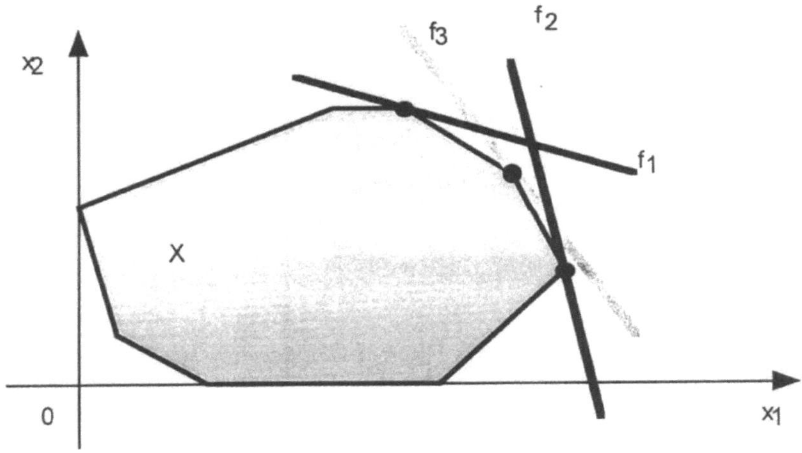

Figure 5. Single-objective valuation

2.6. Optimal valuation: multiple criteria

If the set of alternatives X is given and fixed a priori, but the set of multiple criteria is still to be selected for the valuation and ordering of X, we can speak of multiple-criteria valuation: Which set (or subset) of criteria best captures decision maker's complex of values and purposes? Is it (f_1 and f_2)? Or (f_3 and f_4)? Or perhaps (f_1 and f_2 and f_3)?

Example: Select a set of criteria, perhaps a combination from a list (f_1, f_2, f_3, f_4, ...), like entertainment, education, privacy, cost, etc., which would best valuate a given list of affordable places (X) in order to attain most satisfactory or fulfilling (through Max or Min f) vacations.

LP problem:

Find F such that

Max $F = Cx$

s. t. $Ax \leq b$, $x \leq 0$,

provides the best valuation of X with respect to a given value complex.

2.7. Optimal pattern matching: single criterion

Here we optimize both X and f simultaneously. There is a problem formulation representing an "optimal pattern" of interaction between alternatives and criteria. It is this optimal, ideal or balanced problem formulation or pattern that is to be approximated or matched by decision makers.

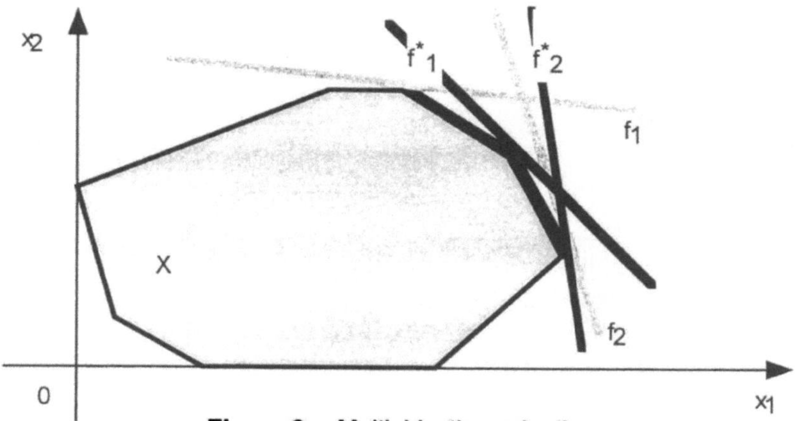

Figure 6. Multiobjective valuation

Example: Select a criterion (f), like entertainment, education, privacy, cost, etc., and design a list of affordable places (X), which assures the most satisfactory or fulfilling (through Max or Min f) vacations.

LP problem:

Find f and b such that

Max $f = cx$

s. t. $Ax \leq b$, $x \leq 0$,

matches as closely as possible the optimal but infeasible (or unaffordable) pattern f^* and b^*, at $pb \leq B$.

Solving the above problem means establishing the optimal pattern (f^*, b^*, x^*) and its budgetary level B^*, then finding the optimal allocation of actual budget B so that the purchased portfolio of resources b, chosen objective function f and the implied solution x assures that pattern (f, b, x) matches the optimal pattern (f^*, b^*, x^*) as closely as possible. This situation is graphically represented in Figure 7.

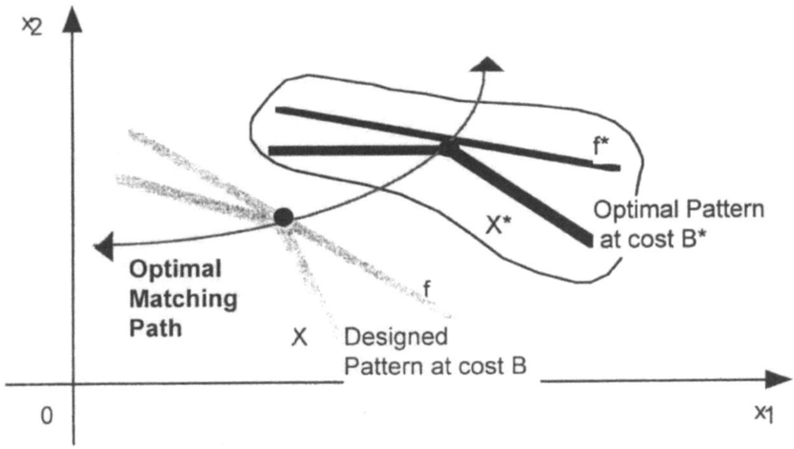

Figure 7. Optimal pattern matching: single criterion

2.8. Optimal pattern matching: multiple criteria

Pattern matching with multiple criteria is more involved and the most complex optimality concept examined so far. In all "matching" optimality concepts there is a need to evaluate the closeness (resemblance or match) of a proposed problem formulation (single- or multi-criterion) to the optimal problem formulation (or pattern).

Example: Choose the necessary criteria (f_1, f_2, ..., f_K), like entertainment, education, privacy, cost, etc., and design a list of affordable places (X), which assure the most satisfying or fulfilling (through Max f_1 and Min f_2) vacations.

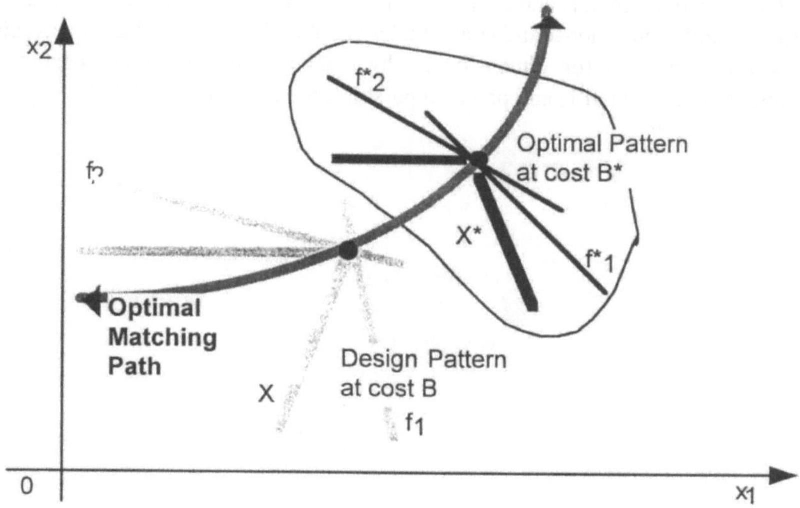

Figure 8. Optimal pattern matching: multiple criteria

LP problem:

Find F and b such that

Max F = Cx

s. t. $Ax \leq b$, $x \geq 0$,

matches as closely as possible optimal but infeasible (unaffordable) pattern F* and b*, at $pb \leq B$.

Solving the above problem means establishing the optimal pattern (F*, b*, x*) at its budgetary level B*, then finding the optimal allocation of actual budget B so that the purchased portfolio of resources b and the chosen set of objective functions F imply solution x, assuring that pattern (F, b, x) matches the optimal pattern (F*, b*, x*) as closely as possible. The situation is graphically represented in Figure 8.

3. Conclusion

The eight optimality concepts of Table 1 are all distinct and essentially irreducible to each other. The field of MCDM should derive its contributory charge and challenge from pursuing and implementing all eight optimality concepts in all their practical combinations. MCDM should not limit itself to the top left-hand corner of Traditional "Optimality" and thus cut itself off the significant part of management and business reality. Otherwise, MCDM shall remain a philosophical copy of the traditional OR/MS, sharing in its fate, reputation and practical relevance even in the modern space of world-class management and business.

Optimization remains and shall remain at the core of MCDM. But optimization itself is multidimensional and contextually variable. In reality, it has at least eight key manifestations (or dimensions) which should neither be aggregated nor reduced. Their scientific and practical pursuit is the challenge to MCDM.

Number of Criteria / Given	Single	Multiple
Criteria & Alternatives	Traditional "Optimality"	MCDM
Criteria Only	Optimal Design (De Novo Programming)	Optimal Design (De Novo Programming)
Alternatives Only	Optimal Valuation (Limited Equilibrium)	Optimal Valuation (Limited Equilibrium)
"Value Complex" Only	Cognitive Equilibrium (Matching)	Cognitive Equilibrium (Matching)

Table 1. Eight Concepts of Optimality

The difference between constraints, goals and objectives is merely technical. Imposing a limit, bound or value on an objective will turn it into a goal if approachable by degrees, or into a constraint if it is to be strictly satisfied or adhered to. Constraints are always handled individually, only rarely (for purely technical reasons) summed up into meaningless aggregates or "superconstraints". Yet, the same economic variables, the same "constraints" are freely combined, scalarized, added up, weighted and aggregated as soon as their constraining values were relaxed or made approachable. This represents a profound dilemma: how can criteria become additive and constraints not - even though they may have identical economic interpretation and contents - simply by the virtue of arbitrarily chosen values?

We conclude with a working conjecture, intended to predict and shape the future of MCDM:

Multiple criteria - as is with multiple constraints - are not to be aggregated.

Bibliography

1. **Von Neumann, J. and O. Morgenstern,** Theory of Games and Economic Behavior, 3rd Edition, Princeton University Press, Princeton, N. J., 1953, pp. 10-11.

2. **Kuhn, H. W. and A. W. Tucker**, "Nonlinear Programming," in: J. Neyman (ed.), Proceedings of the Second Berkeley Symposium on Mathematical Statistics and Probability, University of California Press, Berkeley, CA, 1951, p. 488.

3. **Zeleny, M.**, "Optimizing Given Systems vs. Designing Optimal Systems: The De Novo Programming Approach," General Systems, 17(1990)4, pp. 295-307.

4. **Zeleny, M.**, Trade-Offs-Free Management via De Novo Programming," International Journal of Operations and Quantitative Management, 1(1995)1, pp. 3-13.

5. **Zeleny, M.**, "Optimal System Design with Multiple Criteria: De Novo Programming Approach," Engineering Costs and Production Economics, 10, 1986, pp. 89-94.

A General Model of Preference Aggregation

D. Bouyssou[1], M. Pirlot[2] and Ph. Vincke[3]

[1] Groupe ESSEC, F-95021 Cergy, France
[2] Faculté Polytechnique de Mons, B-7000 Mons, Belgium
[3] Université Libre de Bruxelles, CP 210/01, B-1050 Bruxelles, Belgium

Abstract. A general model is presented which encompasses many procedures used for aggregating preferences in multicriteria decision making (or decision aid) methods. Are covered in particular: MAUT, ELECTRE and several other outranking methods. The main interest of the model is to provide a key for understanding the differences between methods. Methods are analyzed in terms of their way of dealing with "preference differences" on each criterion/attribute. The more or less large number of equivalence classes of preference differences that can be distinguished in a method helps to situate it in a continuum going from compensatory to noncompensatory procedures, from cardinal to ordinal methods.

Keywords. Multicriteria decision analysis, aggregation of preferences, compensation, ordinality.

1 Introduction

The classical way of modelling the preferences of a Decision Maker, consists in assuming the existence of a value function u such that an alternative a is at least as good as an alternative b ($a \succsim b$) if and only if $u(a) \geq u(b)$. This leads to a model of preference in which $\succsim$ is complete and transitive. Using such a preference model to establish a recommendation in a decision-aid study is straightforward and the main task of the Analyst is to assess u. In a multicriteria/multiattribute (we will use these terms interchangeably in this paper) context (for a review, see Zionts 1992), each alternative a is usually seen as a vector $\overline{g}(a) = (g_1(a), \ldots, g_n(a))$ of evaluations of a w.r.t n points of view. Under some well-known conditions (see e.g., Krantz et al. 1978 or Wakker 1989), u can be obtained in an additive manner, i.e. there are functions u_i such that

$$u(a) = \sum_{i=1}^{n} u_i\left(g_i(a)\right).$$

Modelling preferences therefore amounts to assess the partial value functions u_i. Several techniques have been proposed to do so (see Keeney and

Raiffa 1976 or von Winterfeldt and Edwards 1986). It should be noticed that the additive model implies independence of each attribute, i.e. that the preference betwen alternatives which only differ on an attribute does not depend on their evaluations on the other attributes, and that individual (also called partial) preferences $\succeq_i$ deduced from $\succeq$ through independence are complete and transitive. In some situations, such a model might not appear to be appropriate, for instance because:

- indifference (seen as the symmetric part of $\succeq$) may not be transitive;
- $\succeq$ may not be a complete relation, i.e. some alternatives may be incomparable;
- compensation effects between criteria are more complex than with an additive model;
- criteria interact (there is no preference independence).

This calls for an extension of the additive utility framework allowing to better deal with some of these cases. Such an extension is also called for by a number of approaches developed since the early seventies. In those methods, the overall preference of a over b is usually determined by looking at the evaluation vectors $\overline{g}(a)$ and $\overline{g}(b)$ independently of the rest of the alternatives and treating the difference $g_i(a) - g_i(b)$ in rather an ordinal way by comparing the difference to a limited number of thresholds. This simple option usually leads to a global preference relation that is not a complete preorder (this being not unrelated to Arrow's theorem). This implies that the aggregation procedure results in structures from which it might not be easy to derive a recommendation (choice of an alternative, ranking of all alternatives). Elaborating a recommendation usually calls for the application of specific "exploitation techniques" (see Roy 1993). The perspective in which such methods were conceived is neither normative (what should the Decision Maker decide in order to be rational) nor descriptive (what are possibly the mechanisms at work in a Decision Maker's mind when he makes a decision); they claim to be constructive in the sense that, the resulting global preference is built or learnt through a dialog between the Decision Maker and the Analyst based on supposedly intuitive concepts. For more information on normative vs descriptive approaches, the reader is referred to Bell et al. 1988; for the constructive approach, see Roy 1993.

Among the methods alluded to, are the so-called *outranking* methods, where $\succeq$ is the outranking relation; the semantic content of a statement like "a outranks b" has been expressed by B. Roy in Roy 1985 :

> "An outranking relation is a binary relation S defined in A such that $a\ S\ b$ if, given what is known about the decision maker's preferences and given the quality of the valuations of the alternatives and the nature of the problem, there are enough arguments

to decide that a is at least as good as b, while there is no essential reason to refute that statement."

There has been relatively little interest in these methods outside Europe. There are several reasons to that. Two of them might be that

- they are not well founded from a formal point of view (no axiomatization);
- they may lead to preference structures from which it is not easy to derive a recommendation.

What we aim at doing in this paper is to show a sort of continuity between the dominant "value function" model and a number of pairwise comparisons approaches. This is done through exhibiting a very general model of preference aggregation and showing that a variety of methods fits into the model. Finally we are able to situate the different aggregation procedures as more or less compensatory, the utility approach being compensatory whereas outranking methods tend to be less compensatory.

2 Models

The models presented below are built on a product space $X = \prod_{i=1}^{n} X_i$, where X_i can be viewed as the evaluations of a set $\mathcal{A}$ of alternatives with respect to criterion i. In this paper we only consider the case where $\mathcal{A}$ is finite (the analysis may easily be extended to denumerable sets of alternatives, see Bouyssou and Pirlot 1996). Working on the product space X usually amounts to extend the set of alternatives since it is implicitly considered that any combination $x = (x_1, \ldots, x_n) \in X$ of evaluations corresponds to some alternative.

Denoting by $x = (x_1, \ldots, x_n)$, $y = (y_1, \ldots, y_n)$, elements of X, the classical conjoint measurement model, alluded to above, reads;

$$(M_1) \qquad x \succeq y \text{ iff } \sum_{i=1}^{n} u_i(x_i) \geq \sum_{i=1}^{n} u_i(y_i),$$

where u_i is a real valued function defined on X_i, for all $i = 1, \ldots, n$.

Combining the u_i functions in a non necessarily additive manner yields the transitive decomposable model:

$$(M_2) \qquad x \succeq y \text{ iff } F\left((u_i(x_i))_{i=1,\ldots,n}\right) \geq F\left((u_i(y_i))_{i=1,\ldots,n}\right),$$

with u_i's as in M_1 and $F : \mathcal{R}^n \longrightarrow \mathcal{R}$, strictly increasing in each argument. As shown by Krantz et al. (1978), replacing the additivity requirement of M_1 by the more general decomposability requirement used in M_2 allows to drastically simplify the axiomatic analysis of the model while severely weakening the unicity properties of the numerical representation; note in particular, that the model does not imply that the u_i's define an interval scale whereas M_1 does imply it when the structure of X is "continuous".

Another generalization of M_1, the so called *additive difference model* (Tversky 1969), is defined as

$$(M_3) \qquad x \succeq y \text{ iff } \sum_{i=1}^{n} \phi_i(u_i(x_i) - u_i(y_i)) \geq 0,$$

with u_i as above and ϕ_i, a strictly increasing odd function $\mathcal{R} \longrightarrow \mathcal{R}$, for $i = 1, \ldots, n$. In M_3, the preference differences on each axis, $u_i(x_i) - u_i(y_i)$ are recoded and additively combined. An interesting feature is that such a model encompasses *nontransitive* global preferences. The need for nontransitive models for rational decision has been stressed by several authors (see Fishburn 1991b).

A far-reaching generalization of M_3 dropping at the same time additivity and subtractivity is

$$(M_4) \qquad x \succeq y \text{ iff } F\left(\psi_i(u_i(x_i), u_i(y_i)), i = 1, \ldots, n\right) \geq 0,$$

with u_i's as above, $F : \mathcal{R}^n \longrightarrow \mathcal{R}$, a strictly increasing function and ψ_i: $\mathcal{R}^2 \longrightarrow \mathcal{R}$, nondecreasing in its first argument and nonincreasing in the second, for $i = 1, \ldots, n$.

Model M_4, though very general, shows fundamental features. A key concept emerging from M_4 is the quaternary relation $\succeq_i^*$ defined below. Relations $\succeq_i^*$ encode the comparison of pairs of levels on each criterion; we will refer to the comparison of pairs of them as the comparison of *differences of preference*. For all $i = 1, \ldots, n$, the relation $\succeq_i^*$ is defined as follows: for all $x_i, y_i, x_i', y_i' \in X_i$,

$$(x_i, y_i) \succeq_i^* (x_i', y_i')$$

iff for all $z, w \in X$,

$$(x_i', z_{i^-}) \succeq (y_i', w_{i^-}) \text{ implies } (x_i, z_{i^-}) \succeq (y_i, w_{i^-}),$$

where, for instance, (x_i, z_{i^-}) denotes an element of X equal to z except for its ith coordinate which is equal to x_i. The relation $(x_i, y_i) \succeq_i^* (x_i', y_i')$ reads "the difference of preference between x_i and y_i is at least as large as that between x_i' and y_i'".

It is easy to show that in model M_4, $\succeq_i^*$ is a complete preorder even if $\succeq$ is noncomplete and/or nontransitive. The number of equivalence classes of this relation may be considered as reflecting discrimination power in the perception of degrees of difference of preference. This point will be abundantly illustrated in the sequel.

Another important characteristic of model M_4 is that it implies that individual preference relations $\succeq_i$ on each criterion defined by, for all $i = 1, \ldots, n$, for all $x_i, y_i \in X_i$,

$$x_i \succsim_i y_i \text{ iff } \forall z \in X, (x_i, z_{i^-}) \succsim (y_i, z_{i^-}),$$

are welll behaved. Though model M_4 does not necessarily imply independence of each attribute, it is not difficult to prove that (as soon as $\succsim$ is reflexive) the relations $\succsim_i$ are semiorders, i.e. complete semi-transitive and Ferrers relations (see Luce 1956, Roubens and Vincke 1985). Such an ordered structure appears a particularly desirable generalization of the usual complete preorder for at least two reasons:

- it encompasses the idea that there is a threshold under which differences of performance on a point of view are not perceived as implying definite preference; it thus allows to model preferences in which indifference is not transitive;
- it actually appears in one of the oldest and most famous family of methods based on pairwise comparisons and majority, the ELECTRE family (Roy 1968, Vincke 1992, Roy and Bouyssou 1993).

3 A characterization of M_4

The axioms for model M_1 are well-known (see Krantz et al. 1978 or Wakker 1989). Model M_2 has been proposed and axiomatized in Krantz et al. (1978), Chap. 7. Axioms for model M_3 may be found in Fishburn (1992). Model M_4 is closely related to the nontransitive additive conjoint measurement model proposed in Bouyssou (1986), Fishburn (1990), Fishburn (1991a) and Vind (1991) and is fully discussed in Bouyssou and Pirlot (1996).

Although very general, model M_4 places nontrivial restrictions on $\succsim$ without imposing its completeness and/or the transitivity of $\succ$ or $\sim$. Central to many aggregation procedures is the way a "difference" on one attribute can be compensated by a "difference" of opposite sign on another attribute. Though the way of modelling these "differences"may vary, in most aggregation procedures they are computed with reference to an underlying ordering of each attribute. Model M_4 allows to capture in a simple way this idea of "differences" computed on the basis of an underlying ranking via the functions ψ_i and u_i.

A few elementary properties of preference relations in model M_4 may easily be derived. We state a few of them, which we name "weak cancellation". In the sequel we denote by K $(K', L, L', \ldots)$, elements of $X_{-i} = \prod_{j \neq i} X_j$; for all $x_i \in X_i$, $K \in X_{-i}$, we have $(x_i, K) \in X$.

(WC_i) For all $x_i, y_i, z_i, w_i \in X_i$ and for all $K, K', L, L' \in X_{-i}$,

$$\left.\begin{array}{ccc} (x_i, K) & \succsim & (y_i, L) \\ & \text{and} & \\ (z_i, K') & \succsim & (w_i, L') \end{array}\right\} \Longrightarrow \left\{\begin{array}{ccc} (x_i, K') & \succsim & (y_i, L') \\ & \text{or} & \\ (z_i, K) & \succsim & (w_i, L). \end{array}\right.$$

We say that $\succsim$ satisfies (WC) iff it satisfies (WC_i) for all $i = 1, \ldots, n$. The (WC_i) property is linked to the fact that $\succsim_i^*$ is an ordering on the differences of preference on attribute i, i.e. that (x_i, y_i) is at least as "large" as (z_i, w_i) or the contrary.

A second kind of weak cancellation properties (WC') are in connection with the fact that the relations $\succsim_i$ are semiorders. The (WC') cancellation property splits into three conditions for each criterion i, $(WC'1)_i$, $(WC'2)_i$ and $(WC'3)_i$.

$(WC'_1)_i$ For all $x_i, y_i, z_i, w_i \in X_i$ and for all $K, K', L, L' \in X_{-i}$,

$$\left.\begin{array}{ccc} (x_i, K) & \succsim & (y_i, L) \\ & \text{and} & \\ (z_i, K') & \succsim & (w_i, L') \end{array}\right\} \Longrightarrow \left\{\begin{array}{ccc} (z_i, K) & \succsim & (y_i, L) \\ & \text{or} & \\ (x_i, K') & \succsim & (w_i, L'). \end{array}\right.$$

$(WC'_2)_i$ For all $x_i, y_i, z_i, w_i \in X_i$ and for all $K, K', L, L' \in X_{-i}$,

$$\left.\begin{array}{ccc} (x_i, K) & \succsim & (y_i, L) \\ & \text{and} & \\ (z_i, K') & \succsim & (w_i, L') \end{array}\right\} \Longrightarrow \left\{\begin{array}{ccc} (x_i, K) & \succsim & (w_i, L) \\ & \text{or} & \\ (z_i, K') & \succsim & (y_i, L'). \end{array}\right.$$

$(WC'_3)_i$ For all $x_i, y_i, z_i, w_i \in X_i$ and for all $K, K', L, L' \in X_{-i}$,

$$\left.\begin{array}{ccc} (x_i, K) & \succsim & (y_i, L) \\ & \text{and} & \\ (z_i, K') & \succsim & (x_i, L') \end{array}\right\} \Longrightarrow \left\{\begin{array}{ccc} (w_i, K) & \succsim & (y_i, L) \\ & \text{or} & \\ (z_i, K') & \succsim & (w_i, L'). \end{array}\right.$$

We say that $\succsim$ satisfies (WC') if it satisfies $(WC'_1)_i$, $(WC'_2)_i$, $(WC'_3)_i)$ for all $i = 1, \ldots, n$. Property $(WC'_1)_i$ can be interpreted as telling that there is an ordering on the values taken by the alternatives on criterion i: x_i, is either "larger" than z_i or the converse; $(WC'_2)_i$ tells a similar thing. Considering $(WC'_1)_i$ to $(WC'_2)_i$ suggests that the orderings on X_i may differ depending on whether the ith coordinate belongs to the description of an alternative which dominates another or is dominated by another. Taken together with $(WC'_1)_i$ or $(WC'_2)_i$, $(WC'_3)_i$ imply that both orderings are compatible.

The main result of this paper is a characterization of the global preferences of model M_4. The interested reader is referred to Bouyssou and Pirlot (1996) for the proof.

Theorem

A reflexive preference relation $\succeq$ on $X = \prod_{i=1}^{n} X_i$ is representable as in model M_4 iff $\succeq$ satisfies the weak cancellation properties WC and (WC').

This result can easily be extended to denumerable and nondenumerable infinite sets X. It should be noted replacing additivity by a mere decomposability requirement allows a simple necessary and sufficient axiomatization even in the finite case.

4 Methods

In order to illustrate how the framework of model M_4 helps to contrast aggregation procedures,

- we recall the aggregation mechanisms used in a few popular MCDA methods (for more detail, the reader is referred to Vincke (1992));
- we show how they fit into M_4;
- we interpret their differences in terms of the structure of equivalence classes of $\succeq_i^*$.

4.1 Conjoint measurement (model M_1)

We have $F(\psi_i\,(u_i(x_i), u_i(y_i)), i = 1, \ldots, n) = \sum_{i=1}^{n} (u_i(x_i) - u_i(y_i))$.

4.2 ELECTRE I (Roy 1968)

$x \succeq y$ if and only if (x, y) belongs to the *concordance* relation, i.e. there is a majority of viewpoints on which x is at least as good as y, and there is no *veto* against declaring x at least as good as y. More precisely,

- **veto** against $x \succeq y$ occurs if for at least one i, $u_i(y_i) - u_i(x_i)$ is too large, i.e. is at least equal to some *veto threshold* Q_i; then one may not have $x \succeq y$;
- (x, y) belongs to the **concordance** relation if

$$\frac{1}{\sum_{i=1}^{n} w_i} \sum_{i:x_i \succeq y_i} w_i \geq s,$$

  where w_i denotes a nonnegative weight associated to criterion i, s is the so-called *concordance threshold* ($\frac{1}{2} \leq s \leq 1$) and $x_i \succeq_i y_i$ if $u_i(x_i) - u_i(y_i) \geq -q_i$, q_i, a non-negative threshold ($q_i << Q_i$).

From the definition of $\succeq_i$, by means of a numerical representation u_i with constant threshold q_i, it is clear that $\succeq_i$ is a semi-order.

The procedure just described is covered by model M_4 ; with M denoting an arbitrarily large positive number, take

$$\psi_i(x_i, y_i) = \begin{cases} (1-s)w_i & \text{if } \; u_i(x_i) - u_i(y_i) \geq -q_i, \\ -sw_i & \text{if } \; -Q_i < u_i(x_i) - u_i(y_i) < -q_i, \\ -M & \text{if } \; u_i(x_i) - u_i(y_i) \leq -Q_i \end{cases}$$

and for F, the summation operator.

4.3 TACTIC (Vansnick 1986)

We present here a variant of TACTIC, itself a variant of ELECTRE I. Using the formalism of ELECTRE I, we have $x \succeq y$ according to TACTIC if there is no veto against this assertion (as in ELECTRE) and (x, y) belongs to the TACTIC concordance relation defined by

$$\sum_{i:x_i \succ_i y_i} w_i \geq \frac{1}{s} \sum_{j:y_j \succ_j x_j} w_j,$$

with $s \geq 1$, the concordance threshold. Of course, one has $x_i \succ_i y_i$ if not $(y_i \succeq_i x_i)$, i.e. iff $u_i(x_i) - u_i(y_i) > q_i$; $\succ_i$ is the asymmetric part of a semiorder.

The TACTIC method enters into M_4 formalism if one considers F as the summation operator and

$$\psi_i(x_i, y_i) = \begin{cases} w_i & \text{if } \; u_i(x_i) - u_i(y_i) > q_i, \\ 0 & \text{if } \; |u_i(x_i) - u_i(y_i)| \leq q_i, \\ -(1/s)w_i & \text{if } \; -Q_i < u_i(x_i) - u_i(y_i) < -q_i, \\ -M & \text{if } \; u_i(x_i) - u_i(y_i) \leq -Q_i. \end{cases}$$

4.4 Valued global preferences

An interesting extension of the M_4 model consists in considering the global preference as a valued relation; in the *valued* M_4 model, the global preference attached to any pair $(x, y) \in X^2$ is computed as

$$p(x, y) = F(\psi_i(u_i(x_i), u_i(y_i)), i = 1, \ldots, n),$$

with F and ψ as in M_4. An example of a method of that type is PROMETHEE (Brans and Vincke 1985); we have, for all $x, y \in X$,

$$p(x, y) = \sum_{i=1}^{n} w_i \phi_i(u_i(x_i) - u_i(y_i))$$

where ϕ_i can take the forms shown in Figure 1.

Obviously, $\psi_i(x_i, y_i) = \phi_i(u_i(x_i) - u_i(y_i))$ and F can be interpreted as a weighted sum operator.

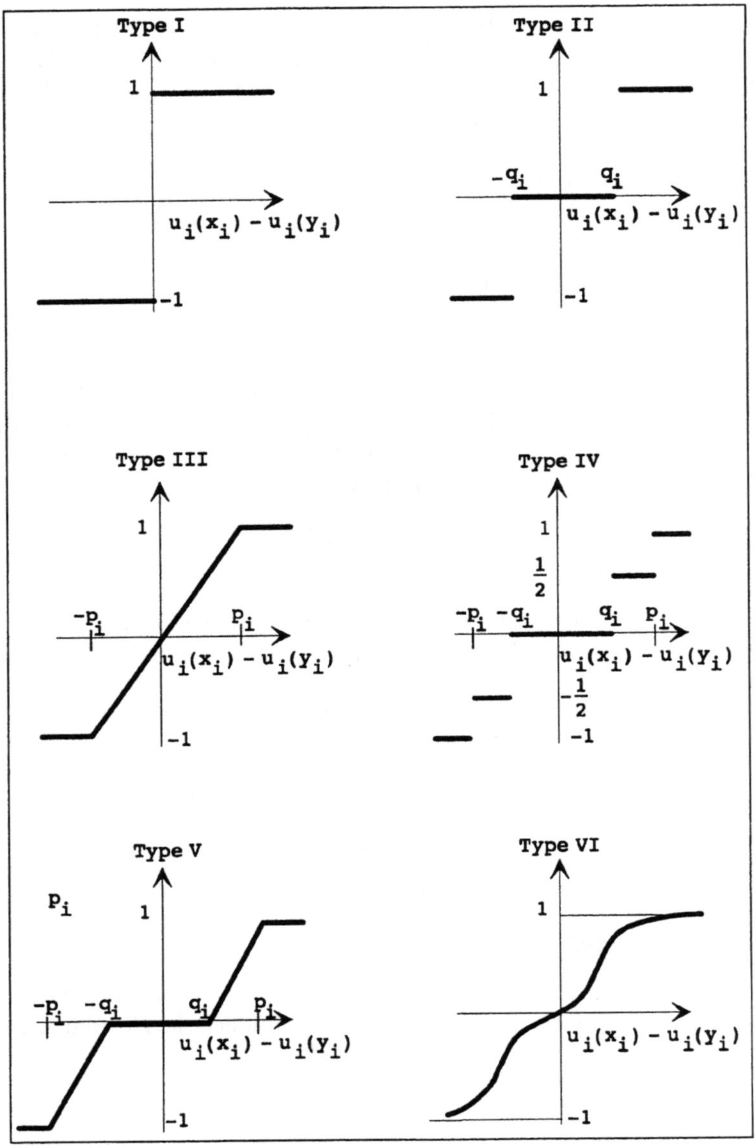

Fig. 1: The six types of recoding of the difference of evaluations used in PROMETHEE

5 Compensation versus non-compensation

Since in all examples above, F is the summation operator, the procedures formally differ only in the manner they code "preference differences", i.e. through the ψ_i functions. In all considered examples $\psi_i(x_i, y_i)$ is a function of the difference $u_i(x_i) - u_i(y_i)$ which is graphed in Figure 2 (and in Figure 1 for the valued preference relation of PROMETHEE). Each particular coding ψ_i induces a complete preorder on the pairs (x_i, y_i); in Figure 3, the hierarchy of equivalence classes of pairs (x_i, y_i) is shown on the same three examples.

ELECTRE I is characterized by a very rough structure on preference differences; in the absence of veto threshold ($Q_i = +\infty$), only two classes can be distinguished. TACTIC relies on essentially the same perception but explicitly distinguishes strict preference from indifference in its aggregation procedure. TACTIC (without veto) is the prototype of noncompensatory aggregation procedures. Intuitively, a method is compensatory when a difference on some attribute may be compensated by a "sufficiently large" difference in the opposite direction on another attribute.

In other terms, in a noncompensatory procedure, the only things that matter in the comparison of x and y are the lists of criteria $P(x, y)$ (resp. $P(y, x)$) on which x (resp. y) is better than y (resp. x). The notion of noncompensation, introduced and studied in Fishburn (1976) was generalized by Bouyssou and Vansnick (Bouyssou and Vansnick 1986; see also Bouyssou 1986) for taking vetoes into account.

The precise definition reads as follows. A preference relation is *non compensatory* if for all $x, y, z, w \in X$,

$$\left.\begin{array}{rcl} P(x,y) & = & P(z,w) \\ P(y,x) & = & P(w,z), \end{array}\right\} \Rightarrow [x \succeq y \text{ iff } z \succeq w],$$

with $P(x, y) = \{i : x_i \succ y_i\}$.

Formally very similar is the following consequence of the weak cancellation axiom WC: for all $x, y, z, w \in X$,

$$[\forall\, i, (x_i, y_i) \sim_i^* (z_i, w_i)] \Rightarrow (x \succeq y \text{ iff } z \succeq w).$$

The latter condition can be interpreted as a generalization of the former: if on each criterion, the difference of preference between x and y belongs to the same class as the difference of preference between z and w, then the two pairs (x, y) and (z, w) must compare in the same manner in the global preference $\succeq$. In case there are only three classes, strict preference, indifference and the opposite of strict preference (as in TACTIC without veto), the noncompensatory property is satisfied. So, the more or less compensatory character of a method can be viewed as the possibility of actually taking into account a larger or a smaller number of differences of preference classes on each criterion in the aggregation procedure.

Note that in our model, preference differences do not need necessarily to be reversible. We may have $(x_i, y_i) \sim_i^* (z_i, w_i)$ without having $(y_i, x_i) \sim_i^*$

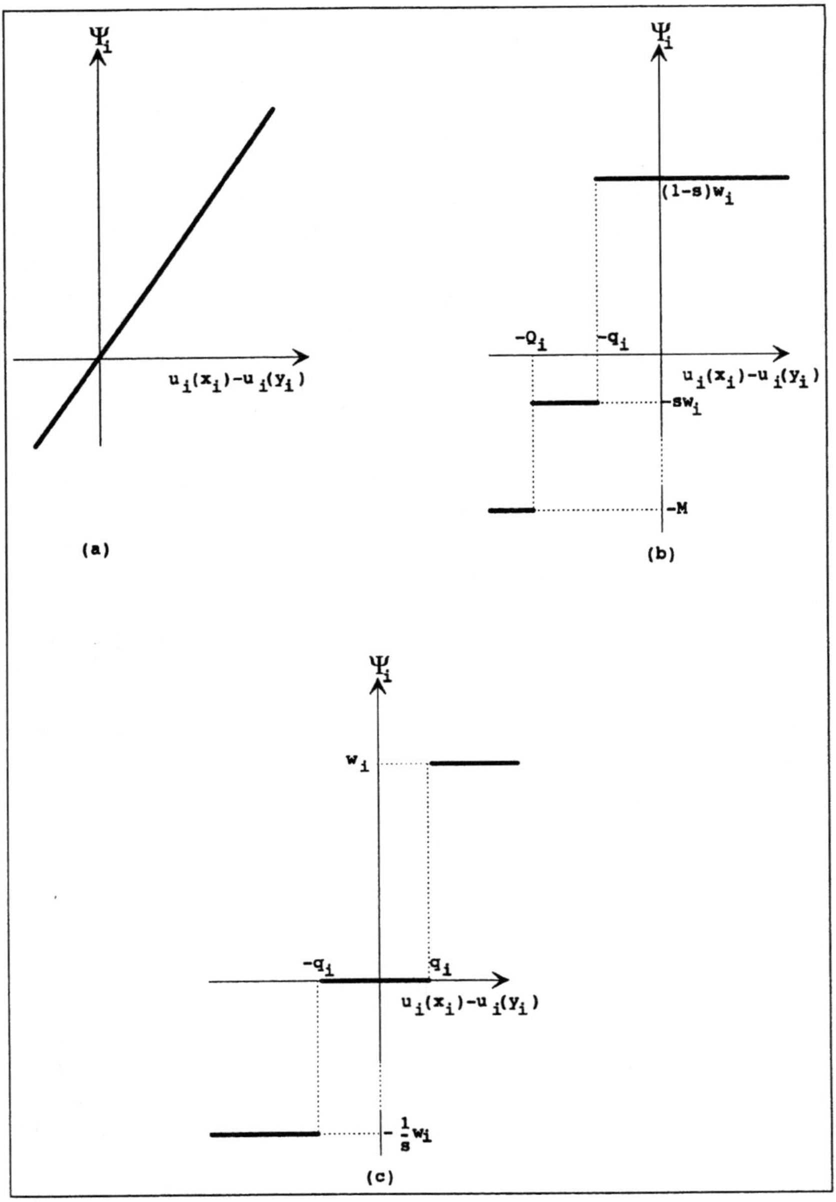

Fig. 2: The function ψ_i in conjoint measurement (a), ELECTRE I (b) and TACTIC without veto (c)

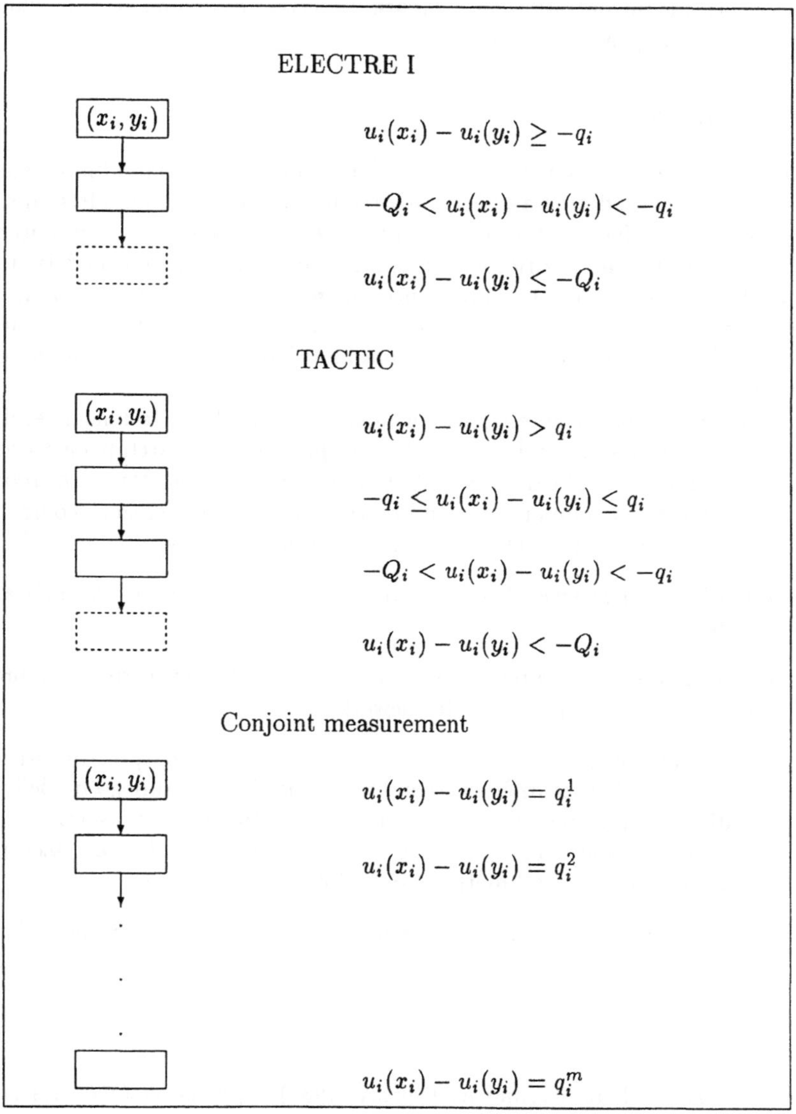

Fig. 3: Equivalence classes of pairs (x_i, y_i) in ELECTRE I, TACTIC and Conjoint measurement

(w_i, z_i); this is in contrast with Fishburn's noncompensation axiom in which "differences" appear to be reversible.

6 Conclusion

With the M_4 model, we have a flexible aggregation scheme that admits a simple axiomatic foundation and encompasses many aggregation models; moreover, we believe that the comparison of preference differences is a key concept for analysing the similarities and dissimilarities of aggregation models (the importance of the concept of preference differences in conjoint measurement has been clearly stressed in Wakker (1989)). A particularly appealing feature of our scheme is that it shows the "continuity" between full compensation and non-compensation.

The present paper emphasizes an interpretation of the technical results obtained in Bouyssou and Pirlot (1996). It opens some research perspectives both on axiomatic and experimental grounds. In the latter, particular models and conditions compatible with observed intuitive preferences could be searched for. On the theoretical side, it would be of interest

- to characterize special models where, e.g., F is additive, the ψ_i's are differences,...;
- to characterize, in a more precise manner, well-known aggregation procedures within our general framework;
- to examine in depth the interconnections of the complete preorder structures on preference differences and the semiorder structure $\succsim_i$ modelling individual preferences on each criterion. Note that the semiordinal character of individual preferences was already stressed as an essential feature of ELECTRE methods in Pirlot (1996);
- to further investigate valued preference relations in the framework of the valued M_4 model.

References

[1] Bell, D.E., H. Raiffa, and A. Tversky (1988). *Decision making: descriptive, normative and prescriptive interactions*. Cambridge University Press.

[2] Bouyssou, D. (1986). Some remarks on the notion of compensation in MCDM. *European Journal of Operational Research 26*, 150–160.

[3] Bouyssou, D. and M. Pirlot (1996). A general framework for the aggregation of semiorders. Technical report, ESSEC, Cergy-Pontoise. Submitted.

[4] Bouyssou, D. and J.-C. Vansnick (1986). Noncompensatory and generalized noncompensatory preference structures. *Theory and Decision 21*, 251–266.

[5] Brans, J. P. and Ph. Vincke (1985). A preference ranking organization method. *Management Science 31*(6), 647–656.

[6] Fishburn, P. C. (1976). Noncompensatory preferences. *Synthese 33*, 393–403.

[7] Fishburn, P. C. (1990). Continuous nontransitive additive conjoint measurement. *Mathematical Social Sciences 20*, 165–193.

[8] Fishburn, P. C. (1991a). Nontransitive preferences in decision theory. *Journal of Risk and Uncertainty 4*, 113–134.

[9] Fishburn, P. C. (1991b). Nontransitive additive conjoint measurement. *Journal of Mathematical Psychology 35*, 1–40.

[10] Fishburn, P. C. (1992). Additive differences and simple preference comparisons. *Journal of Mathematical Psychology 36*, 21–31.

[11] Keeney, R. and H. Raiffa (1976). *Decisions with multiple objectives; preferences and valued trade-offs*. Wiley.

[12] Krantz, D.M., R.D. Luce, P. Suppes, and A. Tversky (1978). *Foundations of Measurement I*. Academic Press.

[13] Luce, R. D. (1956). Semi-orders and a theory of utility discrimination. *Econometrica 24*, 178–191.

[14] Pirlot, M. (1996). A common framework for describing some outranking methods. To appear in *Journal of Multicriteria Decision Analysis*.

[15] Roubens, M. and Ph. Vincke (1985). *Preference modelling*, Volume 250 of *Lecture Notes in Economics and Mathematical Systems*. Springer.

[16] Roy, B. (1968). Classement et choix en présence de points de vue multiples (la méthode Electre). *Revue Française d'Informatique et de Recherche Opérationnelle 8*, 57–75.

[17] Roy, B. (1985). *Méthodologie Multicritère d'Aide à la Décision*. Economica.

[18] Roy, B. (1993). Decision science or decision-aid science ? *European Journal of Operational Research 66*(2), 184–204.

[19] Roy, B. and D. Bouyssou (1993). *Aide Multicritère à la Décision: Méthodes et cas*. Economica.

[20] Tversky, A. (1969). Intransitivity of preferences. *Psychological Review 76*, 31–48.

[21] Vansnick, J.-C. (1986). On the problem of weights in multiple criteria decision making. *European Journal of Operational Research 24*, 288–294.

[22] Vincke, Ph. (1992). *Multicriteria Decision-Aid*. Wiley.

[23] Vind, K. (1991). Independent preferences. *Journal of Mathematical Economics 20*, 119–135.

[24] von Winterfeldt, D. and W. Edwards (1986). *Decision analysis and behavioral research*. Cambridge University Press.

[25] Wakker, P. P. (1989). *Additive representations of preferences. A new foundation of decision analysis*. Kluwer.

[26] Zionts, S. (1992). Some thoughts on research in multiple criteria decision making. *Computers and Operations Research 19*(7), 567–570.

Issues In Supporting Intertemporal Choice

Elizabeth Atherton and Simon French

Department of Computer Studies, University of Leeds, Leeds. LS2 9JT. UK.

Abstract. Intertemporal decision making refers to contexts in which the consequences accumulate in stages over time. Attention is confined to cases in which the stages are discrete. The purpose of this paper is to highlight some of the anomalies between how people 'should' make and how they 'do' make intertemporal decisions: i.e. between findings in the normative and descriptive literatures. The paper indicates some of the framing issues which must be considered when trying to obtain a decision maker's preferences, especially with respect to the way the questions are posed. The intention is to identify some of the bridges which need to be built between descriptive and normative ideas if decision makers are to be supported effectively in making intertemporal decisions.

The concepts and ideas are illustrated in the context of the decisions faced after a nuclear accident to indicate how information should be presented in order to obtain 'unbiased' preferences from decision makers – or, at least, to reduce any bias. These issues are currently being faced in the design of RODOS, a decision support system being built to aid emergency management in the event of a nuclear accident with off-site and, hence, probable long term consequences.

Keywords. Decision analysis and support; descriptive, normative and prescriptive decision analysis; equity; framing; discounting; intertemporal choice; multi-attribute value and utility theories; RODOS.

1. Introduction

Intertemporal decision making refers to contexts in which the consequences accumulate in stages over long periods of time. For this discussion, we confine attention to cases in which the stages are discrete (or may be modelled as being so), thus avoiding technical issues which are introduced when the accumulation is continuous.

Studies of intertemporal decision making draw upon skills in many disciplines including psychology, mathematics and economics. Detailed studies have been carried out in psychology to determine how people make long term decisions and what affects their choices; mathematicians have formulated axioms to try to model how people should make intertemporal decisions and economists have debated whether discounting should be used on long term outcomes. This paper attempts to bridge some of the gaps between normative theories of how people 'should' make decisions and descriptive theories of how people 'do' make decisions. There is a need to provide such bridging in order to develop prescriptive decision support for real decision makers who wish to be guided towards the rationality of normative theories despite being prey to some of the 'biases' discovered in descriptive studies.

We begin by reflecting on normative, descriptive and prescriptive approaches to decision modelling, before surveying some of the complexity of intertemporal decision making that has been discussed in the literature. We then consider how a decision maker's preferences might be elicited to avoid some of the anomalies which can occur due to the framing of the questions; and we discuss related issues which need to be taken into account in establishing elicitation protocols. To focus our discussion, we consider a specific decision context: the actions to be taken after a nuclear accident.

2. Normative, Descriptive and Prescriptive Analysis

Our ultimate aim is to develop decision models and methodologies to help decision makers in evaluating alternatives in intertemporal contexts. There are three general interpretations or uses of decision modelling.

Normative decision theories suggest how people 'should' make decisions if they are to act rationally. They are usually mathematically based and have their origins in philosophy, economics and statistics. An example is Bayesian decision theory which characterises how one should make decisions: *viz.*

- encode beliefs as subjective probabilities
- encode preferences as utilities
- update beliefs in the light of data via Bayes' Theorem
- maximise expected utilities to take decisions
- iterate as more data becomes available

In reality there are anomalies between the behaviour which the Bayesian theory encodes and how people actually 'do' make decisions. These anomalies mean that people often do not act in the way the theory predicts, and thus, it is argued, there may be limited application to guiding real decisions. This discrepancy between the normative theory and actual behaviour is not unique to the Bayesian approach. Other normative theories suffer similarly.

In intertemporal decision making one of the most well known normative models is *discounted utility theory* (DUT), which was first presented by Samuelson (1937). The axiomatic basis of DUT is surveyed in French (1986). The theory assumes that people discount future outcomes at a constant rate. Therefore, the utility of a *timestream* of consumption levels, $\boldsymbol{c} = (c_0,\ldots\ldots,c_T)$ will be:

$$U(c_0,\ldots c_T) = \sum_{t=0}^{T} \delta^t u(c_t) \tag{1}$$

Where $u(c)$ is a concave utility function, δ the discount factor and t the time. It should be noted that DUT is entirely compatible with the Bayesian School in that it provides a particular form of utility function for intertemporal contexts. However, empirical studies have shown many anomalies which people exhibit against this theory and the assumptions it makes (Loewenstein and Prelec, 1992).

More complex utility functions for preferences over timestreams can be built upon multi-linear and other multi-attribute functional forms: see Meyer (1976) and French (1986, 1996a). The multi-linear form is:

$$\begin{aligned} u(\mathbf{c}) = & u_1(c_1) + u_2(c_2) + \ldots + u_T(c_T) \\ & + k_{12}u_1(c_1)u_2(c_2) + k_{13}u_1(c_1)u_3(c_3)\ldots \\ & + k_{23}u_2(c_2)u_3(c_3) + \ldots \\ & + k_{123}u_1(c_1)u_2(c_2)u_3(c_3) + \ldots + \ldots \\ & + k_{12\ldots T}u_1(c_1)u_2(c_2)\ldots u_T(c_T) \end{aligned} \tag{2}$$

Such forms allow the value assigned in one period to depend upon the consumption in other periods and, perhaps more importantly, they allow the risk attitude to consumption in one period to depend upon consumption in other periods.

Empirical studies belong to the realm of *descriptive decision analysis.* Such studies come mainly from psychology and seek to describe how people *do* make decisions. Descriptive models do not seek to aid people in making 'rational' decisions. They do not indicate how people can change the way they view decisions in order to avoid 'inconsistencies' or 'biases' in their choices. Indeed, words such as 'inconsistency', 'bias', and 'anomaly' only take meaning when descriptive theories are compared with normative theories. Loewenstein and Prelec (1992) present a model of intertemporal choice which can accommodate anomalies between DUT and observed decision making:

$$U(x_1,t_1;\ldots\ldots;x_n,t_n) = \sum_i v(x_i)\phi(t_i) \tag{3}$$

Where $v(x)$ is a *value function* and $\phi(t)$ is a *discount function.* The model assumes that intertemporal choices are defined with respect to deviations from a reference point or status quo. However, although this model describes behaviour and possibly explains anomalies, of itself it does not suggest how people can be helped to make more rational choices.

Prescriptive decision analyses seeks to guide decision makers toward consistent, rational choices, while recognising their cognitive limits. They use the descriptive theories of how people 'do' make decisions to understand people's cognitive processes, while using normative theories of decision making as the ideal way to make decisions. Prescriptive theories try to guide people to analyse their decisions in the correct way and make rational choices. Further discussion of how prescriptive analysis brings descriptive and normative theories together can be found in Bell *et al* (1988) and French (1996a).

Prescriptive analyses focus on trying to aid decision makers. They recognise that care is needed to avoid decision makers' choices being biased through poor framing of the questions asked in the elicitation of their beliefs and preferences. Prescriptive decision analysis is invariably an interactive process which guides the evolution of the decision makers' judgements and builds their understanding of their problem. The modelling is cyclic. The decision makers' beliefs and preferences are analysed and modelled; which gives insight into their judgements, often leading to revisions of the model. The process continues until no new insights are found. Thus the decision makers' preferences and beliefs evolve as they understand both the problem and themselves better, thus helping them towards more informed, rational and consistent choices.

Prescriptive analysis can be built on the Bayesian characterisation of decision making, but extends this to include: elicitation procedures, sensitivity analysis and remodelling cycles to enhance understanding until the modelling is *requisite*. Phillips (1984) describes a theory of requisite modelling: "A requisite decision model is defined as a model whose form and content are sufficient to solve a particular problem. The model is constructed through an interactive and consultative process between problem owners and specialists."

3. The Complexity of Intertemporal Decisions

3.1 Spanning Generations

Complexity is introduced when decisions are made on behalf of other people, especially if they are not yet living. An individual's tastes and needs change over time, and these in turn change their preferences. It is, therefore, even more difficult for people to make decisions which affect others yet to be born, and whose future needs are extremely difficult to estimate. It is difficult to define 'equity' between generations; and, if it can be defined, it is difficult to treat future generations equitably because their circumstances cannot be fully predicted. We are concerned with precisely such decisions with long reaching effects. Our interest arises in the nuclear safety field, but there are many other similar contexts, particularly related to environmental issues.

Even if the consequences of decisions do not span generations of people, they can span generations in companies. Many decisions have long life cycles, for example investments in equipment or machinery. Therefore there is a need to take into account the changes in the company which may occur after the decision is made. Forecasting techniques are often very poor: they can predict situations for the next few time periods, but are less useful after that. This adds another element of difficulty, as the future circumstances cannot be predicted.

3.2 Status Quo Bias

People have a tendency to anchor their decisions on the way things are now and their perception of the present status quo. Yet the way things are now may not be the way that they are in a few years or perhaps a few months. Changes in management or government policy can radically affect how companies view situations: see Phillips (1982). Yet, although they know this, decision makers find it very difficult to anticipate change and thus to think widely enough about the context of their decision.

3.3 Banking Issue

It is possible to bank, borrow and loan money. However it is not possible to do this with all possible attributes: for example, it is not possible to borrow on next year's good will. Thus there is a need to treat some attributes differently from others. The ability to borrow from next year's attribute depends on the environment being considered. For example, within a commodities market it is possible to borrow money or grain on next year's harvest; however, within a single farm it is not possible to borrow grain from next year's harvest. Similarly, lives cannot be borrowed from one year to the next, and the treatment of situations involving lives cannot be treated in the same way as other attributes. Even in situations in which borrowing is possible, it may be constrained. Small firms may not have the credit rating to borrow all they need. In nuclear safety, governments may not have the ability to borrow more than a small fraction of the GNP; yet the clean up from a nuclear accident may cost much more than that. In many cases, discounting models assume that borrowing is both allowed and completely unconstrained (French, 1986).

3.4 Measuring Equity

Equity issues can further complicate intertemporal decision making. How to evaluate or measure the utility to individuals in a single time period is a problem in itself (see, e.g. French *et al*, 1995). When consequences accrue to populations over time, equity issues become more difficult still. Three approaches and examples of the difficulties which are encountered are outlined below.

Complete lives - this theory looks at the utility a person receives over the whole of their life and compares it with the total utility another person receives. Temkin (1992) gives the following example, if A and B are two people and their utility is:

$$A = 2,2,8,8 \qquad \text{and} \qquad B = 8,8,2,2$$

The theory would consider the two people equal, because their total utility is the same. However, there is inequity between A and B at every stage of their lives which is bad. Thus looking at total life consumption ignores the equity of consumption of one group in a population relative to another.

Simultaneous segments - This view divides history into segments of about 20 years, and compares the equity between people in the same segments. This considers the age of the person to be irrelevant and therefore may compare an 80 year old with an 18 year old. This is dubious because utility is often dependent on age.

Corresponding Segments - This view divides people's lives into segments of about 20 years and compares the utility people receive when they are the same age. This seems more logical as it compares like with like. However the simultaneous and corresponding segments views have another problem. Temkin (1992) shows that these views would see the inequality between A and B, and the inequality between C and D as equally bad.

A = 8,8,2,2 and B = 2,28,8

C = 8,8,8,8 and D = 2,2,2,2

This would seem objectionable, because the inequality between C and D may be considered worse than that between A and B. This approach can fall foul of the banking issue referred to above.

In the simultaneous segments view, inequality only matters between overlapping generations. Therefore inequality will matter within, but not between the past, present and future. On the other hand, the corresponding segments and whole lives views can deal with non overlapping generations. None of the views is ideal for all situations, and the view chosen will have important implications on how projects and equity is judged. It may be necessary to modify or combine the theories in order to give a better analysis of equity.

Analysing equity in populations is also difficult. If in two populations the absolute differences in utility between the rich and the poor are the same, and one population is larger than the other, then it seems logical to say that the larger population has more inequality. However, some would claim that the inequality is the same because the pattern of inequality in the two populations is the same. For an illustration, see Figure 1. Some would judge A and B to be the same because the absolute differences in inequality are the same though B has twice the population size.

This has significant implications for intergenerational equity as it is relevant in assessing equity between and within generations. Therefore, thought needs to go into how equity is to be measured with respect to the size of populations.

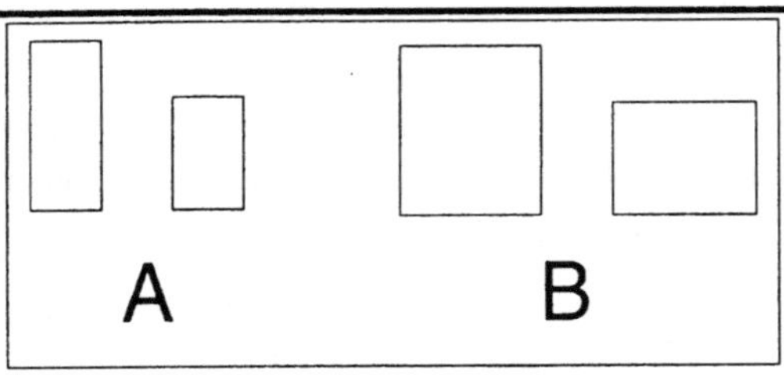

Figure 1: Population Utilities

4. Anomalies in Normative Theories of Intertemporal Choice

Normative theories of intertemporal choice try to describe how people should make decisions if they are to act in a rational manner. Discounted utility theory (DUT) is one of the most well known normative theories, it assumes decision makers choose between options by analysing the sum of discount factors based on time delays. However, it is based on axioms which people break when given certain choices. The following sections outline some of the main anomalies between DUT and how people actually make decisions.

4.1 Common Difference Effect

DUT says that preferences should depend on the absolute time interval between the delivery of outcomes. However, evidence has shown (Ainslie 1975 and 1985) that the impact of a constant time difference is less significant if outcomes are remote. This effect violates a stationarity property which is critical in the DUT.

For example: £20 in one month $\sim$ £25 in two months

£20 in 10 months $\prec$ £25 in 11 months

The result of the common difference effect is that preferences between two delayed outcomes may change if both delays are extended by a constant amount. It also implies that discount rates should decrease as a function of time delay, which has been shown to be true in empirical studies (Horowitz 1988). Therefore, increasing the outcome dates in a pair of temporal outcomes by a constant amount decreases the ratio between the time delays. This in turn implies a decrease in the relative importance of time because of the sensitivity to time ratios which can lead to changes in preference.

4.2 Immediacy Effect

Discontinuity has been found in preferences when outcomes occur immediately. This is because decision makers tend to be biased towards outcomes which occur immediately. Studies have shown (Thaler, 1981) this by discovering very high discount rates for short time delays.

For example: £20 immediately $\succ\succ$ £25 in one month

This effect also violates stationarity which discredits DUT. Therefore, people's preferences are biased by outcomes which occur immediately.

4.3 Sign Effect

The attractiveness of gains is discounted more steeply than the unattractiveness of losses. This has been shown in empirical studies (Thaler, 1981). In order for this to occur people must separate new choices from their existing assets and view the outcomes as gains or losses which they treat differently. This contradicts the assumption of integration in DUT. The sign effect means people are more eager to speed up a gain than they are to delay a loss which means that choices framed as gains and losses will be treated differently.

Kahneman and Tversky (1981) found that outcomes framed as losses induced risk seeking in subjects while outcomes framed as gains induced risk aversion. They constructed the following pair of problems, the number in brackets indicates the number of subjects in the study and the percentages indicate the proportion of subjects choosing that program.

Problem 1 : (N=152) Imagine the U.S. is preparing for the outbreak of an unusual Asian disease, which is expected to kill 600 people. Two alternative programmes to combat the disease have been proposed. Assume that the exact science estimates of the consequences of the programmes are as follows:

Program A: 200 people are saved. (72%)

Program B: There is a 1/3 probability that 600 people will be saved and a 2/3 probability that no people will be saved. (28%)

Which of the two programmes would you favour?

Problem 2 : (N=155) The same starting scenario is given but the options are:

Program C: 400 people will die. (22%)

Program D: There is a 1/3 probability that nobody will die and a 2/3 probability that 600 people will die. (78%)

Which of the two programmes would you favour?

In real terms programmes A and C are identical in their consequences, as are programmes B and D. However because the outcomes in problem 1 are framed as gains (lives saved) people act in a risk averse manner, but with problem 2 people act in a risk seeking manner because the programmes are framed as losses (lives lost).

The difference in risk attitude for gains and losses also stresses the importance of consistent and carefully planned framing of outcomes. The problem can also cause dominance violations in subjects (Kahneman and Tversky 1984). It is important to understand how subjects are viewing the outcomes of decisions, because if they view one outcome as a loss and another as a gain, their discount functions and risk aversion will be different which may be perceived as contradictory preferences when it is really a framing effect. One way to counteract this would be to state both the gains and losses incurred by a project. For example in the cases above the programmes would be defined as:

Program E: 200 people are saved and 400 people die.

Program F: There is a 1/3 probability that 600 people are saved and a 2/3 probability that 600 people die.

Presenting the information is this way may induce a risk neutral attitude, though McNeil *et al* (1986) found that mortality data was more prevalent than survival data. Therefore the choices made would be closer to those in problem 2. At least in this framing the subject is given all the information and not unduly biased by the analyst focusing their attention on one aspect of the data.

4.4 Magnitude Effect

Studies (Thaler, 1981) have shown a negative relationship between the absolute magnitude of an outcome and the rate at which it loses value when it is delayed. This contradicts the integration assumption made in DUT. This may occur because people are sensitive to relative and absolute differences in outcomes, which would change the desire to delay an outcome depending on its size. Thaler (1981) found the following:

$15 immediately ~ $60 in a year, discount rate = 0.25

$250 immediately ~ $350 in a year, discount rate = 0.71

Which suggests that discount rates depend on the magnitude of the outcome.

5. Framing Issues

Kahneman and Tversky (1984) found that the way a question is framed affects a decision maker's preference. One example is given above in Section 4.3. This can lead to choices which may seem irrational, because the decision maker perceives the problem in a different manner to that intended by the analyst. Here we discuss some issues which should be considered when framing questions to elicit preferences in order to avoid such biasing of the decision maker's responses. A major issue is the reference point to which the decision makers relate their preferences.

5.1 Delay - Speed-up Asymmetry

Generally the amount decision makers will pay to delay an outcome is larger than the amount they will pay to speed up its receipt by the same time. This effect has been shown empirically by Loewenstein (1988). A temporal shift which is framed as a delay has more significance than when it is framed as a speed up of receipt. Unless this asymmetry is recognised by noting the reference point when obtaining preferences, varying discount rates may be obtained for decision makers which will be due to the reference point and not actual preferences.

5.2 Buying and Selling Price

Asking how much people would be willing to pay to obtain an object will illicit a different value than asking how much they would need to be paid to give up an object. The first value is often called the buying price and the second the selling price. The selling price is usually much higher than the buying price. The difference in values is due to what is called the endowment effect, this is when people value an object more highly when they possess it than when they do not. The endowment effect has been shown in empirical studies by Loewenstein and Adler (1995) and implies that indifference curves shift when people obtain a good, which implies a taste change. Therefore, it is important when obtaining preferences to avoid the endowment effect and to note the differences between buying and selling prices, so that values are not biased and reflect the decision maker's real preferences.

5.3 Matching and Choice Questions

Preference questions can be stated in different ways. Matching questions are when subjects have to say which outcomes have the same value or give a monetary value to an outcome. Choice questions are when subjects have to choose which outcome they prefer. Matching questions involve the elicitation of discount rates and prices, while choice questions involve the elicitation of preferences. Albrecht and Weber (1995) found that the method of questioning determined the anomalies which occurred between subjects' answers and normative theories of choice. They found that in matching tasks certain outcomes were discounted more than risky ones and larger discount rates were used for initial waiting periods than for later ones. These anomalies were not found in the answers to choice questions. These results show that question frames can affect the discount rates obtained and that biases in decision makers' answers may occur when matching questions are used. Therefore it is important to know what type of question has been posed and the biases this may have on the results obtained.

5.4 Compatibility Effect

The weight of a stimulus attribute is increased by its compatibility with the response. This means that if a subject is asked to state which outcome has the highest value they will focus their attention on the value of the outcomes and give less weight to the other attributes of the outcomes. Alternatively if subjects are asked to rank the outcomes, they are more likely to evaluate the attributes fairly. This effect has been shown empirically by Griffin *et al* (1990). Therefore, preferences may be biased by the method of questioning. By asking a subject to state their preferences for outcomes, it may be possible to avoid causing them to put more emphasis on one of the attributes which may avoid biasing the answers. This effect is also a framing effect which biases people's preferences, and which must be taken into account when creating questions.

5.5 Sequences and Individual Outcomes

People react differently to sequences of outcomes than to individual outcomes. Prelec and Loewenstein (1993) have found that sequences tend to induce negative time preference while individual outcomes have positive time preferences. They found that two motives were important in time preference, impatience and preferences for improvement. Impatience dominates single outcome evaluations while preference for improvement dominates sequences of outcomes. Therefore it is important to know whether a subject views an outcome as a single event or a sequence of events as this will affect their discount rates and preferences which may lead to contradictory results.

6. Relating the Biases to Decision Support in the Event of a Nuclear Accident

The RODOS project is an European initiative to build a decision support system for emergency response before, during and after a nuclear accident. The system aims to support decision making in the short, medium and long term. That is from the moment an accident is imminent to years later. Many of the issues raised in this paper are important to RODOS, as the system presents information to the decision makers to help them to understand the issues involved in the decisions they have to make.

This section will use the issues raised in this paper to suggest methods of presenting information after a nuclear accident to aid decision making within the RODOS framework. Much of the information about the decisions faced after a nuclear accident comes from studies of the Chernobyl accident. However, it is not the aim of this paper to say how that situation should have been handled, the aim is to suggest ways that RODOS can present information to aid in the decision making processes.

The consequences of a nuclear accident stretch over 100s of years and the decisions which need to be made change with time. The decisions to be made after an accident can be categorised into three overlapping groups:

Short term :	Minutes to days after the accident
Medium term :	Weeks to months after the accident
Long term :	Months to years after the accident

Different issues affect the decisions in each time frame and different pressures apply.

Figure 2 shows some of the decisions which have to be made and the time frame they occur in. Because of the long time frame in which decisions have to be made it is essential that there is consistency in the presentation and interpretation of the information given to the decision makers.

The main concern that the decision makers face is the health risk to the public. There are regional, national and international guidelines which advise decision makers on the actions they should take depending on the radiation levels in an area. Usually these are based upon intervention levels. If the radiation is below the lower level no action is taken, if the radiation is above the upper level, certain actions should be considered seriously. Between the upper and lower levels matters are left to the decision makers' discretion.

Several studies have suggested attributes which need to be analysed; and, comfortingly, there is a degree of commonality between the results. Figure 3 is taken from the International Chernobyl Study: see French (1996b).

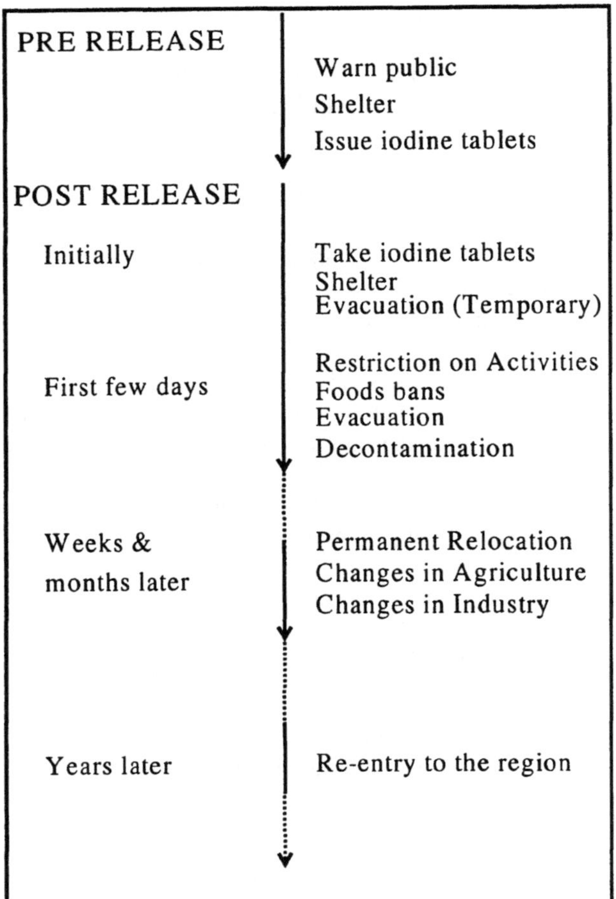

Figure 2: Time Frames of Decisions after a Nuclear Accident

RODOS must address the issue of how to present the information about these and related attributes of a strategy in order to obtain unbiased value judgements from the decision makers.

6.1 The Reference Point for Post Accident Decisions

As noted above, it is important to identify the decision makers' reference points. This is especially important in the context of emergency management as the current status quo is very different from the decision makers' normal position. It will be difficult for the decision makers to accept and understand this new status quo; therefore it is important to clearly define the situation and to remind them of it throughout the analysis.

There are several reference points from which the decision makers can make their decisions:

- The state of the world if no accident had occurred;
- The state of the world if all the countermeasures were applied;
- The state of the world if no countermeasures were applied.

As the first two positions are ideals which are unlikely to occur in reality they are unrealistic reference points. The most sensible reference point is the third, in this way all the options can be evaluated in terms of how they improve the current status quo. Therefore any action will be viewed as a gain from this reference point.

Section 4.3 showed that people treat gains and losses differently. People tend to be risk averse with respect to gains and risk seeking with respect to losses. As the gains and losses in the context of a nuclear accident are human lives it would be preferable for decision makers to be risk averse. This may be achieved by using the status quo as the reference point in this way any changes are evaluated as gains and will therefore be analysed in a risk aversive manner.

Although the concept of averted dose is useful for post accident decisions it is inapplicable for decisions when an accident threatens as there is no dose to avert. This also introduces the problem of *ex-post* and *ex-ante equity* issues: for further discussion see French *et al* (1995). The attributes which are important in decision making change depending on whether there is an accident or not. Before an accident stress may be more important as there is no dose to avert; but after an accident averted dose is more important: see French *et al* (1996) for further discussion.

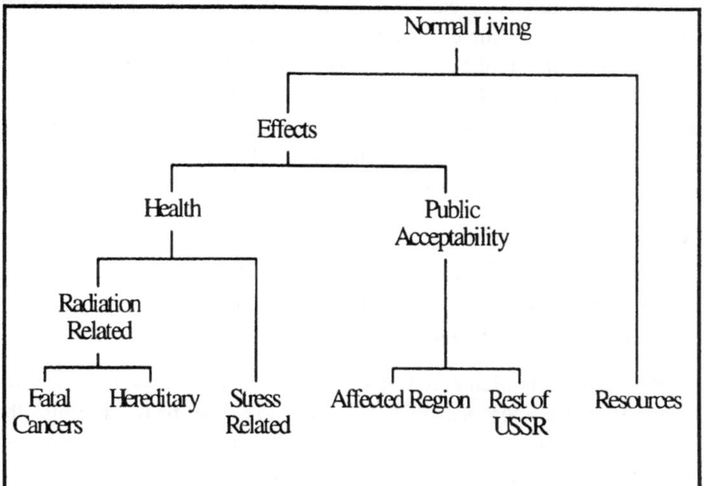

Figure 3: Attribute Hierarchy for Countermeasures

A problem which may occur if the outcomes are presented as gains is that people may discount the outcomes more steeply than if they had be framed as losses. Therefore long term gains may be perceived as much less attractive when compared to short term gains. This may result in the decision makers being biased towards short term programmes and rejecting long term strategies which are better. Therefore it will be important to indicate the long term implications of the strategies.

If the initial state of the environment is used as the reference point the decision makers may be affected by the a status quo bias (Section 3.2) as this state may not be the end stable state after the accident. The situation is dynamic and will be changing continuously depending on weather conditions and other issues. Therefore, it is important to update decision makers and make changes in the status quo very clear. In this way it may be possible to avoid the anchoring effect occurring towards the initial situation after the accident and stop the decision makers' preferences from being biased.

6.2 Intervention Level Biases

Intervention levels usually consist of two dose limits *a* and *b* which determine what action should be taken depending on the level of exposure to radiation. For example the intervention levels for evacuation may be:

Below *a*	- no action is taken
Between *a* and *b*	- protect the community (no evacuation)
Above *b*	- evacuate the community

Although these levels are useful guidelines they can bias the decision makers' actions. They create a compatibility effect (Section 5.4), which means that the decision makers may focus mainly on the radiation level in the situation and ignore the other attributes. This means they may act at the lower level of intervention which may be overcautious or cause more harm due to stress than the harm that would have been caused by the exposure to the radiation. To counteract this effect it will be important to emphasise the importance of the other attributes when strategies are being evaluated while using intervention levels. In this way it may be possible to avoid influencing decision makers into choosing strategies which are not health or cost effective due to the bias caused by intervention levels.

How should the radiation dose be measured?

- From the time of the accident to end of life;
- From the time of the decision to end of life;
- Average annual dose in the strategy period;
- Annual dose next year.

Any countermeasure implemented cannot affect any dose of radiation already received; therefore it is more sensible to look at the dose from the point that the decision is made and not before. The amount of radiation people are exposed to will change over time due to the decay of the radioactive isotopes therefore the average annual dose will change over time. Looking at the annual dose rate next year could therefore bias the decision makers' preferences because it may be unrealistically high in comparison to the levels of radiation people would receive in subsequent years.

The question of whether to look at the rest of someone's life or the period of the countermeasure strategy will depend on how long the strategy will continue. If the strategy is only for a very short period it would seem unreasonable to look at the rest of the person's life; however, if it is a permanent strategy such as relocation then looking at the rest of the person's life would be appropriate. The answer to this question will depend on the strategy's duration.

Should upper dose limits include countermeasures? Considering the use of the dose limits as defined earlier it would seem illogical to include countermeasures in the dose levels as the dose levels are being used to determine the countermeasures to apply. If countermeasures are included in the upper limit this may bias the decision makers' preferences. Yet in one study, decision makers applied the evacuation intervention levels to the radiological situation after sheltering had been implemented (Ahlbrecht *et al*, 1995).

6.3 Lives Lost or Decreased

In catastrophes there are usually a limited number of immediate deaths. In a nuclear accident there may be some deterministic effects which lead to early deaths, but we shall concentrate on the longer term stochastic effects arising from the contamination. Exposure to radiation affects the population's distribution of late life cancers which may only be detected years after the accident. In the years after an accident there is an increase in the percentage chance that people will contract cancer. People may also die early through a stress related illness. A person's life may be decreased for example from 60 to 30 years, or 82 to 80. The question is how this should be represented (assuming the distribution can be determined).

The health effects of 'low level' exposure to radiation are stochastic. There is a higher cancer risk and there are risks of genetic effects in future generations. The risks depend on the type of radiation and the exposure to it. It is possible to calculate a quantity called '*individual effective dose equivalent*' for an individual, from their exposure.

The main issue again is the reference point used for the evaluations. Is it the expected lifetime if no accident had occurred, or the expected lifetime if no action is taken? Though both values are important in assessing the options available to determine the usefulness of a strategy, the reference point should be the status quo, which would be the life expectancy if no action was taken.

This issue is compounded by the fact that medicine will advance in the future, and there is no way of estimating how these advances may affect the abilities to counteract the effects of cancer. This adds another element of uncertainty into the modelling which makes the long term decisions even more difficult to address.

Fischhoff *et al* (1978) found that people view the probability of loss as more important than the magnitude of loss. They found that when looking at individual car trips people did not wear seat belts because the probability of injury was very low. However, when the data was presented in a multi-trip dimension people said they would wear seat belts because the probability of an accident had increased from 0.01 to 0.33. This emphasises the need to guide decision makers towards analysing the long term health consequences of their strategies so that they avoid ineffective short term strategies.

McNeil *et al* (1986) found that presenting data on medical treatments as the life expectancy after the treatment encouraged the decision makers to look at the long term consequences of their decisions. Whereas presenting data as the cumulative probability of survival biased the decision makers towards short term strategies. As decision makers often exhibit an immediacy bias (Section 4.2) it may be better to express data about radiation exposure as the change in life expectancy to help focus decision makers on the long term consequences.

Medical research suggests that the health risks from radiation to the individual are linearly related to the individual dose. This is called the *linearity hypothesis*. This means that all the health effects resulting from radiation exposure are linearly related to the exposure. Therefore, when calculating the effects of countermeasures it is irrelevant which effect is measured for the purpose of calculations. However, the measure used can affect the decision makers' evaluation of the outcome.

6.4 Attribute Importance

Studies of Chernobyl have found that public acceptability and stress were very important factors in the success of countermeasure strategies. However, in the decision making process these were often only given nominal consideration. It is important therefore to focus the decision makers' attention on these attributes and the way they affect the countermeasure strategies chosen.

Policies such as evacuation may decrease the level of radiation considerably, but the levels of stress they induce may cause more health problems than might have occurred from radiation exposure. On the other hand not evacuating people but implementing restrictive countermeasures may reduce the quality of life so much that people suffer from stress and they would be in a better condition if they were evacuated. These stress effects extend into the distant future too. One might hope that they decrease with time and hence some form of discounting model might be applied in the evaluation. However, 10 years after Chernobyl the stress effects are still prevalent (Karaoglou *et al*, 1996).

6.5 Changes in Attribute Weight

Kornbluth (1992) investigated how people's attribute weights changed over time and how the volatility of their environment affected their attribute weights. He found that people change the weight they give to attributes as time progresses and they expect to change their weights. The more volatile the environment the more likely people are to change their attribute weights, and the more they focus on just a few of the attributes which they think are the most important.

Although these effects may impact more significantly in the years immediately after an accident, they are likely to continue at some level for many years. Therefore, intertemporal effects need to be considered. As the decisions to be made after a nuclear accident span over long time periods changes in attribute weights will affect the decisions people make. The situation itself is dynamic as the sort of radiation people are exposed to changes over time due to differences in the half lives of the radioactive isotopes. In the first few weeks radioactive iodine is the main threat, at the end of the first month it is beta radiation, then after the first few months caesium is the main cause of contamination. The way people are exposed to the radiation also changes initially they are under threat from the plume of radiation, then from contaminated surfaces and food. Therefore the countermeasures which are required to protect the public change over time, see

Figure 2. Therefore the decision support system must show the decision makers which issues are important in each time frame and help them to focus on the different issues.

History shows that the weight placed on "safety" has increased throughout the twentieth century. This has led to reductions in the "acceptable limits". This is often accelerated by accidents or the prospect of them. These changes may affect the strategies which should be taken.

The environment the post nuclear accident decisions have to be made in is very volatile and stressful, therefore, decision makers may focus on what they think are the main attributes of a decision as Kornbluth (1992) found. It is therefore important that the decision support system focuses the decision makers' attention on all the attributes of a decision even the ones they may think are unimportant.

6.6 Policy and Generation Changes

National and international policies on radiation intervention levels may change during the implementation of the countermeasure strategies. Therefore, it is important to get the decision makers to focus on the possibility of changes and how these may affect the strategies they have chosen. This may help to reduce the anchoring effect and help the strategies span governments and generations.

The people affected by the strategies may also change with time and it is important to focus decision makers attention how these changes may affect the effectiveness of their strategies. People's susceptibility to radiation exposure changes with their age and it is important to indicate to decision makers how strategies should develop over time to accommodate these changes. As the situation after a nuclear accident is dynamic it will be important to develop dynamic strategies which can be adapted to change and develop as the situation develops.

6.7 Immediacy Biases

Section 4.2 showed that people are biased towards outcomes that occur immediately, they can also suffer from the common difference effect (Section 4.1). Therefore it is important to stress the long term effects of strategies and their benefits. Decision makers will be biased towards immediate short term actions to reduce radiation risk now, however this may increase other harmful effects in the long run. For example stress in the case of relocation. To counteract this it will be important to emphasise the importance of all the attributes of the strategies and focus the decision makers' attention on the long term implications of their strategies.

6.8 Sequences or Single Event Strategies

The outcome of a protection strategy can be viewed as a single outcome or a series of outcomes. This depends on the countermeasure, for example taking iodine tablets is a one off action which gives an instant decrease in radiation intake; foods bans are ongoing procedures and decrease the intake over a period of time. The way the strategies are presented can affect the choices people make. People generally want sequences to give improving outcomes, but prefer a smaller single outcome immediately to a larger delayed single outcome (Section 5.5). It is therefore important to monitor if decision makers are treating strategies differently depending on if they are presented as single outcomes or sequences, and inform them if there are inconsistencies. As decision makers are usually biased towards immediate outcomes is may be better to present the outcomes of strategies as sequences to focus decision makers' attention on the long term consequences.

6.9 Equity Considerations

Equity is a significant issue in decision making after a nuclear accident. If the population feels they are being treated unfairly, they may suffer from stress and resentment which will compound the inequity. It is important to focus decision makers' attention on the equity issues as they can have a large effect on population moral and trust in the decision makers. The ability to treat people equally may be constrained by resources, and it may be necessary to balance the equity achieved against a decrease in the protection given.

The way decision makers view equity can be very different. In the decision conferences carried out to aid the design of RODOS (Ahlbrecht *et al*, 1995) the analysts found that different groups of decision makers considered the equal treatment of people differently. One group thought that equity was achieved by using strategies which would decrease the dose of individuals with the same exposure by the same amount. Any measure not giving the same dose decrease for all individuals was excluded. Another group viewed equal treatment as offering all individuals the same chance to decrease their doses, which may have resulted in different exposure to radiation for individuals. This small difference in wording caused large differences in the countermeasures implemented by the two groups.

Equity is often more an issue of public perception than an absolute measure and this must be taken into account when deciding on strategies. For example it may be the case that because the plume rose very high in the air the people nearest the plant were subjected to less radiation than those three miles away where the plume dropped. However, those nearest the plant would not accept that they were being treated equally if those further away from the plant were evacuated and they were not. To stop these inequitable scenarios from occurring RODOS has a filter system which removes strategies which would not be acceptable to the public. In this way the decision makers are only presented with strategies which have passed certain equity evaluations.

6.10 Discounting

The issue of discounting often depends on the object being discounted. Discounting money is usually accepted, though if the monetary sums equal a significant proportion of GDP one might argue against it. In Belarus the consequences of Chernobyl are claimed to cost on average 10-20 % GDP per annum, therefore discounting may not be appropriate.

To make judgements on the marginal cost per unit dose averted by countermeasures, decision makers use a reference value for the cost of the man-Sievert. This value is called α and creates an equivalence relationship between money and dose. Through α units of dose equals money, therefore discounting money is equivalent to discounting dose, which may occasionally be accepted. By the linearity hypothesis dose is equivalent to expected number of years of life expectancy lost. Discounting may be acceptable because individuals do discount the last years of their lives by, for example, smoking. However, not everyone will be comfortable with this idea, and if the life expectancy is reduced by a large amount people may be even more averse to discounting. The linearity hypothesis also means that dose equals number of deaths, and as society dislikes discounting lives this is unlikely to be acceptable to decision makers. Dose also equals number of genetic effects, by the linearity hypothesis. It is doubtful whether society would discount these unless there is a guarantee of medical advances.

The line of thought above shows that in radiation decisions discounting money has the effect of discounting many other factors because of the use of α and the linearity hypothesis. These result in discounting not being applicable to radiation decisions because the long term consequences reach into future generations, this however, does not answer the question of how to deal with the long term consequences and how the present should be balanced with the future.

7. Directions for Prescriptive Support of Intertemporal Choice

The anomalies outlined in this paper highlight the importance of question frames when obtaining preferences. The empirical evidence available shows that many anomalies are not due to people breaking normative axioms, but through the analyst obtaining the wrong answer due to the questions posed. This supports the view that normative theories should be used as the basis of how people 'should' make decisions, and be used as the ideal which decision makers are guided towards in prescriptive analysis.

The issues involved in the framing of questions highlights the thought and planning which must go into creating questions to obtain a decision maker's preferences. The evidence shows that what may initially appear to be inconsistent choices may in fact be due to framing effects and changes in the decision maker's reference point induced by the questions posed. If prescriptive analysis is to be successful more care will need to be applied when obtaining the decision maker's preferences.

The other issues highlighted in the paper require more theoretical analysis to determine suitable solutions. If discounting is unacceptable, what is acceptable? That is a question which philosophers and others should debate: there are major ethical issues to be addressed here. It is clear however, that there will not be one solution which will apply to all intertemporal decisions. The 'best' method of dealing with issues like equity will depend on the problem being faced.

8. Acknowledgements

Much of this work was carried out within the context of the RODOS project (EU contract no. F14P-CT95-0007). In addition, the support of the Frank Gott scholarship fund is gratefully acknowledged. We are grateful for many helpful discussions with others in the RODOS project and Theo Stewart.

9. References

G. Ainslie (1975) 'Specious Reward: A Behavioural Theory of Impulsiveness and Impulse Control'. *Psychological Bulletin,* **82**, 463-509.

G. Ainslie (1985) 'Beyond Microeconomics. Conflict among Interests in a Multiple Self as a Determinant of Value'. In J. Elster (Ed) *The Multiple Self.* Cambridge University Press.

M. Albrecht and M. Weber (1995) 'An Empirical Study on Intertemporal Choice Under Risk', Working Paper, University of Mannheim.

M. Ahlbrecht, J. Ehrhardt and S. French (1995) 'Designing the Evaluation Module in RODOS/RESY: Execution and Analysis of Elicitation Exercises with Emergency Management Teams'

D. Bell, H. Raiffa and A. Tversky (1988) 'Decision Making: Descriptive, Normative and Prescriptive Interactions'. Cambridge University Press.

CEC (1990) International Chernobyl Project Report. EUR 14543EN.

B. Fischhoff, S. Lichtenstein and P. Slovic (1978) 'Accident Probabilities and Seat Belt Usage: A Psychological Perspective', Accident Analysis and Prevention, **10**, 281-285.

B. Fischhoff, S. Lichtenstein and P. Slovic (1986) 'Informing the public about the risks from ionising radiation'. In H. Arkes and K Hammond (Eds) *Judgement and Decision Making*. Cambridge University Press.

S. French (1986) Decision Theory: an Introduction to the Mathematics of Rationality. Ellis Horwood, Chichester.

S. French (1996a) "The framing of statistical decision theory: a decision analytic view" in J. Berger, J.M. Bernardo, A.P. David and A.F.M. Smith (Eds.) *Bayesian Statistics 5: Proceedings of the Fifth Valencia International Meeting on Bayesian Statistics* Oxford University Press, 147-164.

S. French (1996b) 'Multi-Attribute Decision Support in the Event of a Nuclear Accident' *Journal of Multi-Criteria Decision Analysis*, **5**(1), 39-57.

S. French, E. Hall and D. Ranyard (1995) 'Equity and MCDA in the Event of a Nuclear Accident' Accident' School of Computer Studies, University of Leeds. RODOS(B)-RP(95)-04.

S. French, M. Harrison and D. Ranyard (1996) 'Event Conditional Attribute Modelling in Decision Making when there is a Threat of a Nuclear Accident' School of Computer Studies, University of Leeds. A shorter version of this paper was given at the XIIth International Conference on Multiple Criteria Decision Making in Hagen, June 19th to 23rd, 1995. RODOS(B)-RP(95)03.

D. Griffin, P. Slovic and A. Tversky (1990) 'Compatibility Effect in Judgement and Choice', in *Insights in Decision Making*, Ed. R. Hogarth, University of Chicago Press.

J. Horowitz (1988) 'Discounting Money Payoffs: An Experimental Analysis', *Working Paper*, Department of Agriculture and Resource Economics, University of Maryland.

D. Kahneman, J. Knetsch and R. Thaler (1991) 'The Endowment Effect, Loss Aversion, and the Status Quo Bias' *Journal of Economic Perspectives,* **5**(1), 193-206.

D. Kahneman and A. Tversky (1981) 'The Framing of Decisions and the Psychology of Choice', *Science*, **211**, 453-458.

D. Kahneman and A. Tversky (1984) 'Choice, Values, and Frames' *American Psychologist*, **39**(4), 341-350.

A. Karaoglou, G. Desmet, G.N. Kelly and H.G. Menzel (Eds) (1996) *The Radiological Consequences of the Chernobyl Accident.* EUR 16544 EN. Commission of the European Communities, Luxembourg.

R.L. Keeney and H. Raiffa (1976) Decisions with Multiple Objectives: Preferences and Value Trade-offs. John Wiley and Sons, New York.

J. Kornbluth (1992) 'Dynamic MCDM' Journal of Multi-Criteria Decision Analysis, **1**, 81-92.

G. Loewenstein (1988) 'Frames of Mind in Intertemporal Choice' *Management Science*, **34**, 200-214.

G. Loewenstein and D. Adler (1995) 'A bias in the Prediction of Tastes' *The Economic Journal*, **105**, 929-937.

G. Loewenstein and D. Prelec (1991) 'Decision Making Over Time and Under Uncertainty: a Common Approach' *Management Science*, **37**(7), 770-786.

G. Loewenstein and D. Prelec (1992) 'Anomalies in Intertemporal Choice: Evidence and an Interpretation' *Quarterly Journal of Economics*, **107**(2), 573-597

G. Loewenstein and D. Prelec (1993) 'Preferences for Sequences of Outcomes', *Psychological Review* **100**(1), 91-108.

G. Loewenstein and R. Thaler (1989) 'Intertemporal Choice' *Journal of Economic Perspectives*, **3**(4), 181-193.

B. McNeil, S. Pauker, H. Sox, A. Tversky, (1986) 'On Elicitation of Preferences for Alternative Therapies'. In H. Arkes and K. Hammond (Eds) *Judgement and Decision Making*. Cambridge University Press.

R.F. Meyer (1976) 'Preferences over time'. In Keeney and Raiffa (1976) 473-512.

L. Phillips (1982) 'Requisite Decision Modelling : A case study' Journal of the Operational Research Society, 33, 303-311.

L. Phillips (1984) 'A theory of Requisite Decision Models', Acta Psychologica, 56, 29-48.

L. Temkin (1992) 'Intergenerational Inequality'. In P. Laslett and J. Fishkin (Eds) Justice Between Age Groups and Generations. Yale University Press.

R. Thaler (1981) 'Some Empirical Evidence on Dynamic Inconsistency' Economic Letters, 8, 201-207.

Cancellation Conditions for Multiattribute Preferences on Finite Sets

Peter C. Fishburn

AT&T Labs – Research, Murray Hill, NJ 07974

Abstract. Applications of decision theory for multiple criteria or multiple attributes often assume a utility representation for preferences that is additive over criteria or attributes. Axiomatic theories for additive utilities are well developed but are not without gaps. A case in point arises with finite sets of alternatives, where two preference axioms are necessary and sufficient for additive utilities. One is weak ordering. The other is a cancellation axiom that consists of an infinite scheme of cancellation conditions, one for each positive integer $K \geq 2$. It is known that the infinite scheme can be truncated to a finite scheme for $K \leq K^*$ that depends on the size of the set of alternatives, but very little is known about the value of K^* which ensures additivity for all finite sets of that size. The present paper contributes to the determination of K^*. A fundamental result is that if there are m attributes, the jth of which has n_j elements in its attribute set, then $\sum n_j - (m-1)$ is an upper bound on K^*. Lower bounds on K^* that are near to this upper bound are obtained for special cases of $(n_1, n_2, \ldots, n_m)$.

1 Introduction

This paper investigates the extent to which finite truncations of a denumerable scheme of necessary cancellation conditions imply the existence of a real-valued, order-preserving and additive representation of a weak order $\precsim$ on a finite product set $X = X_1 \times X_2 \times \cdots \times X_m$ with $m \geq 2$. Our broad context is the theory of additive conjoint measurement [13]. For interpretive purposes we view X as a set of decision alternatives or outcomes and each factor X_i as a finite set of levels of a particular criterion, attribute or commodity. Let $n_i = |X_i|$, the cardinality of X_i, so that X contains $n_1 n_2 \cdots n_m$ m-tuples $x = (x_1, x_2, \ldots, x_m)$ with $x_i \in X_i$ for every i. The *size* of X is defined as the m-tuple $(n_1, n_2, \ldots, n_m)$. We are interested in determining the most restrictive finite truncation of our denumerable cancellation scheme that implies the existence of an additive order-preserving representation for every weak-ordered set of a particular size. The most restrictive truncation sufficient for additivity is known precisely for only a few special sizes. In several other cases we obtain tight bounds on the critical truncation parameter, and for all sizes we derive an upper bound on this parameter that seems likely to be close to the most restrictive sufficient truncation value.

We say that x is *not preferred to* y when $x \precsim y$ in our multicriteria or multiattribute setting. The induced strictly less preferred than relation $\prec$ and indifference relation $\sim$ are defined on X from $\precsim$ by

$$x \prec y \quad \text{if} \quad x \precsim y \quad \text{and not}(y \precsim x) \ ,$$

$$x \sim y \quad \text{if} \quad x \precsim y \quad \text{and } y \precsim x \ .$$

It is assumed throughout that $\precsim$ on X is a *weak order*, i.e., it is transitive ($x \precsim y$ & $y \precsim z \Rightarrow x \precsim z$) and complete (for all $x, y \in X$, $x \precsim y$ or $y \precsim x$). We refer to $\precsim$ as a *linear order* when it is a weak order for which $x \sim y$ if and only if $x = y$. It is also common to call $\prec$ a linear order. Our weak order assumption implies that $\sim$ is an equivalence relation (reflexive: $x \sim x$; symmetric: $x \sim y \Rightarrow y \sim x$; transitive) that partitions X into equivalence classes or indifference classes that are linearly ordered by the natural extension of $\prec$ to these classes. If $\prec$ itself is linear, then each indifference class is a singleton subset of X.

Because X is finite, the weak order assumption implies the existence of an order-preserving utility function $u : X \to \mathbb{R}$ such that, for all $x, y \in X$,

$$x \precsim y \Leftrightarrow u(x) \leq u(y) \ . \tag{1}$$

The same representation holds for infinite sets of alternatives or outcomes if and only if weak order is augmented by a suitable Archimedean or continuity condition, but such conditions play no role here.

Since about 1960, investigators have considered various axioms or restrictions on $\precsim$ that lead to special forms for u in (1). These include a few models that accommodate preference interdependencies among attributes or criteria [1, 3, 10], but the bulk of the research focuses on models that satisfy the first-order independence condition

(2) $x \precsim y \Leftrightarrow z \precsim w$ when there is a nonempty proper subset M of $\{1, 2, \ldots, m\}$ such that $x_i = z_i$ and $y_i = w_i$ for all $i \in M$, and $x_i = y_i$ and $z_i = w_i$ for all other i.

When $M = \{i\}$, (2) induces the weak order $\precsim_i$ on X_i defined by

$$x_i \precsim_i y_i \quad \text{if} \quad x \precsim y \quad \text{when} \quad x_j = y_j \quad \text{for all} \quad j \neq i \ . \tag{3}$$

We define $x_i \prec_i y_i$ if $x_i \precsim_i y_i$ and not($y_i \precsim_i x_i$).

Representations that satisfy (2) include lexicographic utility models [4, 5] and the closely-related additive utility model

$$x \precsim y \Leftrightarrow \sum_{i=1}^{m} u_i(x_i) \leq \sum_{i=1}^{m} u_i(y_i) \ , \tag{4}$$

in which $u_i : X_i \to \mathbb{R}$ for every i. The popularity of (4) seems attributable to its simplicity, ease of assessment, approximate fidelity to more complex preference structures in many situations, and the elegance of its axiomatic foundations. Accounts of its foundations for both finite and infinite sets are available in Fishburn [2, 7], Krantz et al. [13], Keeney and Raiffa [11] and Wakker [17].

Although first-order independence is necessary but not generally sufficient for (4) in our finite weak order setting, it is sufficient for a limited set of sizes [13, pp. 427–428].

THEOREM 1. *If* $m = 2$ *and* $\min\{n_1, n_2\} = 2$, *then* (2) *holds if and only if there are real-valued functions* u_1 *on* X_1 *and* u_2 *on* X_2 *that satisfy* (4) *for all* $x, y \in X_1 \times X_2$.

The smallest sizes that satisfy (2) but not (4) are $(2,2,2,2,2)$ for $m = 5$ and $(3,3)$ for $m = 2$. An example for the former size was first given in [12]. A linear order of size $(3,3)$ for $X = \{1,2,3\} \times \{a,b,c\}$ that satisfies (2) but not (4) is

$$1a \prec 1b \prec 2a \prec 2b \prec 3a \prec 1c \prec 2c \prec 3b \prec 3c\ ,$$

where (x_1, x_2) is written as $x_1 x_2$. Here, (2) holds with $1 \prec_1 2 \prec_1 3$ and $a \prec_2 b \prec_2 c$ for (3), but (4) requires

$$u_1(1) + u_2(b) < u_1(2) + u_2(a) \quad \text{for} \quad 1b \prec 2a$$

$$u_1(3) + u_2(a) < u_1(1) + u_2(c) \quad \text{for} \quad 3a \prec 1c$$

$$u_1(2) + u_2(c) < u_1(3) + u_2(b) \quad \text{for} \quad 2c \prec 3b\ ,$$

which are impossible because addition and cancellation of the three inequalities leaves $0 < 0$.

An extension of (2) that is sufficient as well as necessary for (4) for all sizes was first described by Kraft et al. [12] in the context of subjective probability and by Scott [15] and others for multiattribute preferences. It is the following general independence axiom or denumerable cancellation scheme:

$$\text{(5)}\quad \begin{array}{l} \text{for every } J \geq 2 \text{ and all } x^j = (x_1^j, \ldots, x_m^j) \text{ and } y^j = (y_1^j, \ldots, y_m^j) \\ \text{in } X \text{ for } j = 1, 2, \ldots, J\text{: if } (y_i^1, y_i^2, \ldots, y_i^J) \text{ is a permutation} \\ \text{of } (x_i^1, x_i^2, \ldots, x_i^J) \text{ for } i = 1, \ldots, m, \text{ then it is false that} \\ x^j \precsim y^j \text{ for } j = 1, 2, \ldots, J \text{ and } x^j \prec y^j \text{ for some } j \in \{1, \ldots, J\}\ . \end{array}$$

The first-order condition (2) is tantamount to (5) for $J = 2$. The impossibility of (4) in the preceding example arises from the failure of (5) for $J = 3$ that has

$$(1,b) \prec (2,a),\ (3,a) \prec (1,c),\ (2,c) \prec (3,b)\ .$$

Here $[x^1, x^2, x^3] = [(1,b), (3,a), (2,c)]$ and $[y^1, y^2, y^3] = [(2,a), (1,c), (3,b)]$.

The necessity of (5) for (4) is easily demonstrated. If (5) fails for specific $x^j \precsim y^j$ and $x^j \prec y^j$, and (4) is presumed to hold, then addition and cancellation of the u_i inequalities in (4) implied by those x^j and y^j leaves $0 < 0$.

Sufficiency of (5) for the existence of additive utilities that satisfy (4) follows from solution theory for finite systems of linear inequalities. Sufficiency proofs are included in Scott [15], Fishburn [2] and Krantz et al. [13].

Because (5) is a denumerable scheme of cancellation conditions for $J = 2, 3, \ldots$, it is of theoretical as well as practical interest to consider the minimum value of J for each possible size, say $J^*(n_1, \ldots, n_m)$, so that if (5) holds for every $J \leq J^*(n_1, \ldots, n_m)$ for a weak order of the designated size, then (5) holds for *all* J. When they mentioned the result of Theorem 1, that $J^*(2, n) = 2$ for all $n \geq 2$, Krantz et al. [13] said that little more is known about J^*. This remained so over the next 25 years, but the situation is changing. In two recent papers [8, 9], significant progress is made on a similar problem for additive subjective probability, and these results apply to the present setting when $(n_1, \ldots, n_m) = (2, \ldots, 2)$. The present paper includes new results for larger n_i.

We summarize our findings in the next section after (5) is reformulated to account for failures that require repetitions of specific (x^j, y^j) pairs. An example is shown by Figure 1 in [8] with a failure of (5) for $J = 6$ that requires two of the six (x^j, y^j) to be identical. Our reformulation focuses on the number of *distinct* (x^j, y^j) in potential failures of (5).

We highlight two new results here. First, if (5) is false for a weak order of size $(n_1, \ldots, n_m)$, then it fails for an example of the $x^j \precsim y^j$ and $x^j \prec y^j$ type that uses no more than $\sum n_i - (m-1)$ distinct (x^j, y^j) pairs. Second, there is a linear order of size $(3, n)$, $n \geq 3$, for which (5) fails only when a violating example uses at least $n - 1$ distinct (x^j, y^j) pairs. We prove the first result in the next section and the result for $(3, n)$ in Section 3.

Section 4 concludes with a brief summary and a discussion of open problems.

2 Cancellation conditions

We now formulate a sequence of cancellation conditions based on (5) that involve fixed numbers of different (x^k, y^k). There is one condition for each $K \in \{2, 3, \ldots\}$ that we refer to as $C(K)$. Condition $C(K)$ considers precisely K distinct (x^k, y^k) in $X \times X$. If K exceeds $|X|^2$ then $C(K)$ holds trivially.

For positive integers $a_1, a_2, \ldots, a_K$ and distinct (x^1, y^1), $(x^2, y^2), \ldots$, (x^K, y^K) in $X \times X$, let $\sum_K a_k(x^k, y^k) \in C_K$ mean that for all $i \in \{1, 2, \ldots, m\}$, the sequence $a_1 x_i^1$ [x_i^1 repeated a_1 times], $a_2 x_i^2, \ldots, a_K x_i^K$ of $\sum a_k$ terms from X_i is a permutation of the sequence $a_1 y_i^1$, $a_2 y_i^2, \ldots, a_K y_i^K$. Thus $\sum_K a_k(x^k, y^k) \in C_K$ is an instance of $J = \sum a_k$ in (5).

$C(K)$. For all $\sum_K a_k(x^k, y^k) \in C_K$, it is false that $x^k \precsim y^k$ for $k = 1, 2, \ldots, K$ and $x^k \prec y^k$ for some $k \in \{1, \ldots, K\}$.

We usually assume that the a_k for an instance of C_K have no common divisor greater than 1. In many instances each (x^k, y^k) is used only once, so $a_1 = a_2 = \cdots = a_m = 1$ for these instances. Moreover, the only relevant cases for $C(2)$ have $a_1 = a_2 = 1$, and $C(2)$ is equivalent to the nontrivial part of (2).

Our focus on the number of distinct preference comparisons that may be needed to detect a violation of additivity, as in (4), for each size of X is made clear by the following definition of f on the set of sizes. We assume $m \geq 2$ and $n_i \geq 2$ for all i.

DEFINITION 1. $f(n_1, n_2, \ldots, n_m)$ is the smallest positive integer K^* such that every weak order on X of size $(n_1, n_2, \ldots, n_m)$ that violates (5) violates condition $C(K)$ for some $K \leq K^*$.

In other words, if $f(n_1, n_2, \ldots, n_m) = K^*$ then

(i) there is a weak order $\precsim$ on X of size $(n_1, n_2, \ldots, n_m)$ that violates $C(K^*)$ but satisfies $C(K)$ for all $K \in \{2, \ldots, K^* - 1\}$;

(ii) every weak order on X of size $(n_1, n_2, \ldots, n_m)$ that satisfies $C(K)$ for $K = 2, \ldots, K^*$ also satisfies $C(K)$ for all $K > K^*$ and therefore has an additive utility representation as in (4).

Theorem 1 is rewritten in terms of f as follows.

THEOREM 1. *$f(2, n) = 2$ for all $n \geq 2$.*

Our first main result is an upper bound on f for all sizes. Although the bound is far above $f(2, n)$ for large n, it is very close to f in other cases.

THEOREM 2. $f(n_1, n_2, \ldots, n_m) \leq \sum_{i=1}^{m} n_i - (m-1)$.

Proof. Assume without loss of generality that the X_i are mutually disjoint and that X has size $(n_1, n_2, \ldots, n_m)$. Let $X_i = \{x_{i1}, x_{i2}, \ldots, x_{in_i}\}$ for every i, let

$$Y = \bigcup_{i=1}^{m} (X_i \setminus \{x_{in_i}\}) \quad \text{and} \quad N = \sum_{i=1}^{m} (n_i - 1) = \sum_{i=1}^{m} n_i - m \,,$$

and let d be a bijection from Y onto $\{1, 2, \ldots, N\}$. We view $d(x_{ij})$ as the dimension or coordinate of $\mathbb{R}^N$ assigned to x_{ij}. Also define $s : X \times X \to \{1, 0, -1\}^N$ by $s(x, y) = (s(x, y)_1, \ldots, s(x, y)_N)$ with

$$s(x, y)_{d(x_{ij})} = \begin{cases} 1 & \text{if } x_i = x_{ij} \text{ and } y_i \neq x_{ij} \\ -1 & \text{if } x_i \neq x_{ij} \text{ and } y_i = x_{ij} \\ 0 & \text{otherwise} \,, \end{cases}$$

and let $\mathbf{0}$ denote the zero vector in $\mathbb{R}^N$. It follows from our definitions that, for positive integers $a_1, \ldots, a_K$ and $(x^k, y^k) \in X \times X$ for $k = 1, \ldots, K$,

$$\sum_K a_k(x^k, y^k) \in C_k \Leftrightarrow \sum_{k=1}^{K} a_k s(x^k, y^k) = \mathbf{0} . \tag{6}$$

Suppose, for example, that $X_1 = \{a, b, c\}$ with $d(a) = 1$ and $d(b) = 2$. Let the nine (x_1^k, y_1^k) in $X_1 \times X_1$ have summed a_k coefficients

α_1	for	(a, a)	α_4	for	(a, b)	α_7	for	(c, b)
α_2	for	(b, b)	α_5	for	(b, a)	α_8	for	(a, c)
α_3	for	(c, c)	α_6	for	(b, c)	α_9	for	(c, a)

with $\sum \alpha_j = \sum a_k$. Then $\sum_K a_k(x^k, y^k)$ satisfies C_K for X_1 if and only if

$$\begin{aligned} \alpha_1 + \alpha_4 + \alpha_8 &= \alpha_1 + \alpha_5 + \alpha_9 \quad (\text{for } a) \\ \alpha_2 + \alpha_5 + \alpha_6 &= \alpha_2 + \alpha_4 + \alpha_7 \quad (\text{for } b) \\ \alpha_3 + \alpha_7 + \alpha_9 &= \alpha_3 + \alpha_6 + \alpha_8 \quad (\text{for } c) . \end{aligned}$$

These three equations hold if and only if $\alpha_4 + \alpha_8 = \alpha_5 + \alpha_9$ and $\alpha_5 + \alpha_6 = \alpha_4 + \alpha_7$. On the other hand, the values of $\sum \alpha_k s(x^k, y^k)$ for the first and second dimensions of $\mathbb{R}^N$ are $(\alpha_4 + \alpha_8) - (\alpha_5 + \alpha_9)$ and $(\alpha_5 + \alpha_6) - (\alpha_4 + \alpha_7)$ respectively. Hence $\sum \alpha_k s(x^k, y^k)$ vanishes on the two dimensions for X_1 if and only if $\sum_K \alpha_k(x^k, y^k)$ satisfies the X_1 part of C_K.

Contrary to Theorem 2, suppose there is a $K > N + 1 = \sum n_i - (m - 1)$ and a weak order $\precsim$ on X that satisfies $C(K')$ for all $K' < K$ but violates $C(K)$ with $\sum_K a_k(x^k, y^k) \in C_K$, $x^k \precsim y^k$ for all k, and $x^k \prec y^k$ for some k. Take $x^K \prec y^K$ without loss of generality. Because $C(2)$ holds, every instance of $x_i^k = y_i^k$ can be replaced by $x_i^k = y_i^k = x_{i1}$ without affecting $\precsim$ or $\prec$ or the permutation balance of $\sum_K a_k(x^k, y^k)$. Moreover, because $C(K')$ is assumed to hold for all $K' < K$, the (x^k, y^k) remain distinct after the replacements, their corresponding $s(x^k, y^k)$ vectors will also be distinct, and $s(x^k, y^k) \neq \mathbf{0}$. By the formulation of the preceding paragraph,

$$\sum_{k=1}^{K} a_k s(x^k, y^k) = \mathbf{0} .$$

Moreover, the first $N + 1$ $s(x^k, y^k)$ can not be linearly independent, so there are rational numbers $b_1, \ldots, b_{N+1}$, not all 0, such that

$$\sum_{k=1}^{N+1} b_k s(x^k, y^k) = \mathbf{0} .$$

It follows for every real λ that

$$\sum_{k=1}^{N+1} (a_k - \lambda b_k) s(x^k, y^k) + \sum_{k=N+2}^{K} a_k s(x^k, y^k) = \mathbf{0} .$$

As λ moves away from 0, we obtain a first instance of $a_k - \lambda b_k = 0$. When λ is fixed at that value and coefficient fractions are cleared, translation by (6) back into the original formulation yields the contradiction that $C(K')$ fails for some $K' < K$. Consequently, if a weak order on X is not representable as in (4), it must violate a cancellation condition $C(K')$ with $K' \leq N+1$. ∎

The next two theorems include lower bounds for two special size classes: $n_i = 2$ for $i = 1, \ldots, m$; $m = 2$ with $3 = n_1 < n_2$. My lower bound constructions are based on linear orders, and to note this we define $f_L(n_1, n_2, \ldots, n_m)$ by Definition 1 with "linear order" in place of "weak order". Because every linear order is a weak order, $f_L \leq f$. I suspect that $f_L = f$, but have no proof and offer it as

CONJECTURE 1. $f_L = f$.

In the following theorem and later, 2_m denotes $(2, 2, \ldots, 2)$ with m entries.

THEOREM 3. $m - 1 \leq f_L(2_m) \leq f(2_m) \leq m + 1$ *for* $m = 5, 6, \ldots$.

The upper bound, valid for all $m \geq 2$, is from Theorem 2. The lower bound is proved in [9]. The exact values of $f_L(2_m)$ for $m = 2, 3, 4, 5$ are $2, 2, 2, 4$ respectively. These are verified in [8]. It also seems likely that $f_L(2_6) = 5$ and $f_L(4, 4) = 4$. Of wider concern is

CONJECTURE 2. $f_L(2_m) = m - 1$ *for all* $m \geq 5$.

A proof of this could lead to lower upper bounds for other cases and, best of all, to replacement in Theorem 2 of $-(m-1)$ by $-(m+1)$ for all sizes except $(2, 2)$.

Our second lower-bound result is

THEOREM 4. $f_L(3, n) \geq n$ *for all even* $n \geq 4$; $f_L(3, n) \geq n - 1$ *for all odd* $n \geq 5$.

Theorem 2's upper bound is $f(3, n) \leq n+2$, so our bounds for Theorems 3 and 4 are very tight. We prove the first part of Theorem 4 in the next section. Its second part is a trivial extension of the first part.

3 Proof for $(3, n)$

Assume that X has size $(3, n)$ with n even and $n \geq 4$, and let $\prec$ denote a linear order on X that satisfies condition $C(2)$. For definiteness, take

$$\begin{aligned} X_1 &= \{1 \prec_1 2 \prec_1 3\} \\ X_2 &= \{b_1 \prec_2 b_2 \prec_2 \cdots \prec_2 b_n\}. \end{aligned}$$

We specify such a $\prec$ that violates $C(n)$ but satisfies $C(K)$ for all $K < n$, so $f_L(3, n) \geq n$.

Our violation of $C(n)$ is given by

$$\begin{array}{c} 2b_1 \prec 1b_{n-1} \\ 1b_n \prec 2b_2 \\ \hline 2b_3 \prec 3b_1 \\ 3b_2 \prec 2b_4 \\ 2b_5 \prec 3b_3 \\ 3b_4 \prec 2b_6 \\ \vdots \\ 2b_{n-1} \prec 3b_{n-3} \\ 3b_{n-2} \prec 2b_n \ . \end{array}$$

Note that point $1 \in X_1$ is used only in the first two comparisons and each b_j appears exactly once on each side of $\prec$. Let $\lambda = (n-2)^2/2 - 1$ and $\epsilon > 0$. The following u_i preserve all $\prec$ additively in the preceding list, except for $3b_{n-2} \prec 2b_n$:

$$\begin{aligned} u_1(1) &= 0 \\ u_1(2) &= n - 2 + \lambda\epsilon \\ u_1(3) &= n + (\lambda + n - 1)\epsilon \end{aligned}$$

$$\begin{aligned} u_2(b_1) &= 0 \\ u_2(b_{2j}) &= 2j - 1 + [(j-1)n + 2]\epsilon \quad \text{for} \quad j = 1, 2, \ldots, (n-4)/2 \\ u_2(b_{2j+1}) &= 2j + j(n-2)\epsilon \quad \text{for} \quad j = 1, 2, \ldots, (n-4)/2 \\ u_2(b_{2n-2}) &= n - 3 + (\lambda + 1)\epsilon \\ u_2(b_{2n-1}) &= n - 2 + (\lambda + 1)\epsilon \\ u_2(b_n) &= n - 1 + (\lambda + 1)\epsilon \ . \end{aligned}$$

For each of the first $n-1$ preference comparisons in the preceding list, $u_1 + u_2$ for the right side minus $u_1 + u_2$ for the left side equals ϵ. For the final comparison, $[u_1(2) + u_2(b_n)] - [u_1(3) + u_2(b_{n-2})] = -(n-1)\epsilon$. Moreover, the additive utility difference for *every* other $\prec$ comparison has an absolute value on the order of 1 or more when ϵ is near zero, for when we set $\epsilon = 0$ the $u_1 + u_2$ sums are as follows:

	b_1	b_2	b_3	b_4	$\cdots$	b_{n-2}	b_{n-1}	b_n
1	0	1	2	3		$n-3$	$n-2$	$n-1$
2	$n-2$	$n-1$	n	$n+1$		$2n-5$	$2n-4$	$\underline{2n-3}$
3	n	$n+1$	$n+2$	$n+3$		$\underline{2n-3}$	$2n-2$	$2n-1$

The full $\prec$ on X is specified by additive-utility preservation as in (4) with ϵ near 0, except for $3b_{n-2} \prec 2b_n$, which reverses the additive order on these two adjacent terms. It follows for ϵ near 0 that every violation of a cancellation condition must contain $3b_{n-2} \prec 2b_n$ as one of its comparisons. Moreover, no violation of cancellation can have a $\prec$ comparison not in our initial list for $C(n)$. Otherwise, say with $a_1(3b_{n-2}, 2b_n) + \sum_{k\geq 2} a_k(y^k, x^k)$ in C_K with $y^k \prec x^k$ for all $k \geq 2$ and some $y^k \prec x^k$ not in the initial list, ϵ near 0 would give

$$a_1(n-1)\epsilon + \sum_{k\geq 2} a_k[u_1(y_1^k) + u_2(y_2^k) - u_1(x_1^k) - u_2(x_2^k)] < 0 ,$$

in contradiction to the total balance on each dimension required by membership in C_K.

Hence, besides $3b_{n-2} \prec 2b_n$, only $\prec$ comparisons in our initial list can appear in a violation of a cancellation condition, and the structure of that list shows that all n comparisons are needed to satisfy the dimensional balance required for membership in some C_K. Thus every $C(K)$ for $K < n$ holds, and we conclude that $f_L(3, n) \geq n$.

4 Discussion

For a given size $(n_1, n_2, \ldots, n_m) = (|X_1|, |X_2|, \ldots, |X_m|)$ of a finite multicriteria or multiattribute set $X = X_1 \times X_2 \times \cdots \times X_m$, we have considered the smallest positive integer $K^* = f(n_1, n_2, \ldots, n_m)$ such that, for every weak order on X, the weak order is representable by order-preserving additive utilities if and only if it satisfies cancellation condition $C(K)$ for $K = 2, \ldots, K^*$. We proved that

$$K^* \leq \sum_{i=1}^{m} n_i - (m-1)$$

and presented lower bounds for sizes $(2, 2, \ldots, 2)$ and $(3, n)$ which suggest that K^* for most sizes except $(2, n)$ is near to $\sum n_i - m$.

This has mixed practical implications. The good news is that only about $\sum n_i - m$ preference comparisons out of the total $\binom{n_1 n_2 \cdots n_m}{2}$ are ever needed to demonstrate a failure of additivity. The bad news is that there is usually a huge number of subsets of $\sum n_i - m$ preference comparisons, and it may be impossible to test very many for failures of cancellation. An efficient algorithm to deal with this problem is described in Sherman [16]. Many situations that exhibit preference interdependencies will have easily demonstrated failures of $C(2)$ or $C(3)$, but others will require careful scrutiny. Of related interest is the degree of inaccuracy of an additive-utility representation when $C(2)$ holds or $C(2)$ and $C(3)$ hold but $C(K)$ fails for a larger K. I am aware of little theoretical work on the additive approximations issue apart from things mentioned in [6, 14].

Many other problems remain unresolved. One is verification of good lower bounds on f or f_L for sizes other than those in Theorems 1, 3 and 4. Another is the determination of exact f and f_L values for small sizes. A third asks for good bounds on J^* defined after (5). Several other theoretical issues, posed as questions, conclude the paper.

1. Are the f values for weak orders and linear orders identical? (Conjecture 1)

2. Is $f(2_m) = m - 1$ for all $m \geq 5$? (Conjecture 2)

3. Is there an argument that decreases the upper bound noted above for K^* by 1 or 2?

4. If a weak order of size $(n_1, \ldots, n_m)$ violates $C(K^*)$ with $K^* = f(n_1, \ldots, n_m)$ but satisfies $C(K)$ for all $K < K^*$, is there an instance of the failure of $C(K^*)$ in which all $a_k = 1$?

5. What is the largest value of $\sum a_k$ that ever occurs in a failure of $C(K)$ when $C(K')$ holds for all $K' < K$? (We assume that the a_k have no common divisor greater than 1.)

6. For each $N \geq 4$, what is $\max\{f(n_1, \ldots, n_m) : \sum n_i = N\}$, and what sizes with sum N achieve the maximum?

References

[1] Farquhar, P. H. and V. R. Rao, A balance model for evaluating subsets of multiattributed items, *Management Science* **22** (1976), 528–539.

[2] Fishburn, P. C., *Utility Theory for Decision Making*, Wiley, New York, 1970.

[3] Fishburn, P. C., Interdependent preferences on finite sets, *Journal of Mathematical Psychology* **9** (1972), 225–236.

[4] Fishburn, P. C., Lexicographic orders, utilities, and decision rules: A survey, *Management Science* **20** (1974), 1442–1471.

[5] Fishburn, P. C., Axioms for lexicographic preferences, *Review of Economic Studies* **42** (1975), 415–419.

[6] Fishburn, P. C., Approximations of two-attribute utility functions, *Mathematics of Operations Research* **2** (1977), 30–44.

[7] Fishburn, P. C., A general axiomatization of additive measurement with applications, *Naval Research Logistics* **39** (1992), 741–755.

[8] Fishburn, P. C., Finite linear qualitative probability, *Journal of Mathematical Psychology* **40** (1996), 64–77.

[9] Fishburn, P. C., Failures of cancellation conditions for additive linear orders, AT&T Bell Laboratories, Murray Hill, NJ, 1995.

[10] Fishburn, P. C. and I. H. LaValle, Binary interactions and subset choice, *European Journal of Operational Research* **92** (1996), 182–192.

[11] Keeney, R. L. and H. Raiffa, *Decisions with Multiple Objectives*, Wiley, New York, 1976.

[12] Kraft, C. H., J. W. Pratt and A. Seidenberg, Intuitive probability on finite sets, *Annals of Mathematical Statistics* **30** (1959), 408–419.

[13] Krantz, D. H., R. D. Luce, P. Suppes and A. Tversky, *Foundations of Measurement, Volume I*, Academic Press, New York, 1971.

[14] Leung, P., Sensitivity analysis of the effect of variations in the form and parameters of a multiattribute utility model: A survey, *Behavioral Science* **23** (1978), 478–485.

[15] Scott, D., Measurement structures and linear inequalities, *Journal of Mathematical Psychology* **1** (1964), 233–247.

[16] Sherman, B. F., An algorithm for finite conjoint additivity, *Journal of Mathematical Psychology* **16** (1977), 204–218.

[17] Wakker, P. P., *Additive Representations of Preferences*, Kluwer Academic, Dordrecht, 1989.

Measurement of Differences in Preferences Expressed by a Simple Additive Rule

O. I. Larichev[1]

[1]Institute for Systems Analysis, Moscow, 117312, pr.60-let Octyabrya, 9, Russia

Abstract

The Method of Weighted Sum of Criteria Estimates (WSCE) is popular in multicriteria decision making. The question arises, how to measure differences in the preferences of different decision makers or the same decision maker (DM) over time when using the WSCE method. A new measure is suggested:the number of alternative pairs in complete order given by WSCE for which the superiority of one alternative upon the other depends on DM's preferences.They are called as potentially contradictory pairs (PCA). It is shown that PCA created from the elements of the first Pareto-layer very reliably represent PCA in a complete order in the case of additivity.This makes it possible to only use PCA to measure differences in DM's preferences.

1. Introduction

A popular version of MAUT (multiattribute utility theory) is the method of Weighted Sum of Criteria Estimates (WSCE) (Von Winterfeld and Edwards [1986]). In the WSCE method the decision maker (DM) estimates the quantitative criterion weights and defines the estimates of the alternatives on the criterion scales. The utility of alternative A is defined by:

$$U(A_j) = \sum_{i=1}^{N} w_i x_i^j \tag{1}$$

where: w_i= the weight of the i-th criterion; $\sum_{i=1}^{N} w_i = 1$;

N = the number of criteria;

x_i^j = the estimates of A_j on the i-th criterion.

The reason of the popularity of the WSCE method stems from its ease of use (Edwards [1977]). In many cases it is necessary to define the difference in the preferences of DMs given by the WSCE method. If every DM in the group gives the information needed for the utilization of the WSCE method, one can construct the complete order of all hypothetical alternatives using formula (1). The question arises: how to define the difference between two complete orders given by different DMs using theWSCE method. This paper is devoted to this problem.

2. The Statement of the Problem

Let us consider the following simple multicriteria problem.

Assume N criteria having two quality levels (1 and 2, 1 better) on an ordinal scale. The relation between the multicriteria alternatives can be represented by a quasi-order.

Two alternatives in the quasiorder can have one from the following relations:

$$A_i\,R\,A_j;\;\; A_j\,R\,A_i;\;\; A_i\,IN\,A_j;\;\; A_i\,I\,A_j$$

where:
R = the relation of preference (it includes the Pareto dominance relations),
I = the relation of equivalence,
IN = the relation of incomparability.

If the DM defines the weights of the criteria and the quantitative criterion scales, one can use (1) and construct the complete order of all possible alternatives (all possible combinations of criterion estimates). The number of possible alternatives is equal to 2^N.

If two DMs give the weights and the quantitative scales, one has two different policies, expressed via complete preference orders. The question that we investigate is: how to measure the difference between two complete orders of alternatives in terms of relations between the alternatives.

3. The Properties of Additive Structures

The WSCE method is based on additive preference structures.

Let us review the properties of additive structures (Edwards [1977]). We shall use them in a way appropriate for ordinal binary criterion scales.

Let: $U_i(x_k^i)$ = the utility of i-th alternative.

$k = 1,\dots,N;\ x_k^i \in \{1,2\}$

1) $U_i + U_j = U_j + U_i$

2) If $U_i \geq U_j$ by best estimates on some criteria, then $U_i \geq U_j$ by worst estimates on the same criteria.

3) If $U_i \geq U_j$ and $U_j \geq U_y$, then $U_i \geq U_y$.

4) If $U_i \geq U_y$ and $U_j \geq U_w$, then $U_i + U_j \geq U_y + U_w$.

5) Let $U_i(x_k^i) = U'$ and $x_t^i = 2;\ x_k^i = 1,\ k = 1,\dots,N;\ k \neq t$.

Let $U_j(x_f^j) = U''$ and $x_p^j = 2,\ x_f^j = 1,\ f = 1,\dots,N;\ f \neq p,\ f \neq t$.

Then $U_i(x_k^i) + U_f^j(x_f^j) = U_z(x_g^m) = U' + U''$; where $x_t^m = x_p^m = 2;\ g = 1,\dots,N;$ $g \neq t, g \neq p$.

4. A Simple Example

For any quasiorder it is possible to construct the Pareto layers. The elements of the j-th Pareto layer are in I or IN relations and for each element of the j-th layer there is at least one element in (j-1)-th Pareto layer which is in R relation with it.

If we shall not take into account the first (the combination of all best estimates on criterion scales) and the ast (the combination of all worst estimates on criterion scales) elements of the quasiorder there will be (N-1) Pareto layers for quasiorder in a multicriteria problem.

If we use the properties of additive structures (Edwards [1977]), the R or I relations in some alternative pairs (given by DM) can be transferred into different alternative pairs. If we define the DM's preferences for some pairs, the numbers of alternative pairs remaining in IN relation (we call them remaining pairs) can be small.

Let us consider a simple example: 3 criteria with binary scales. The existing alternatives are given in Fig 1 (1 - best estimate)

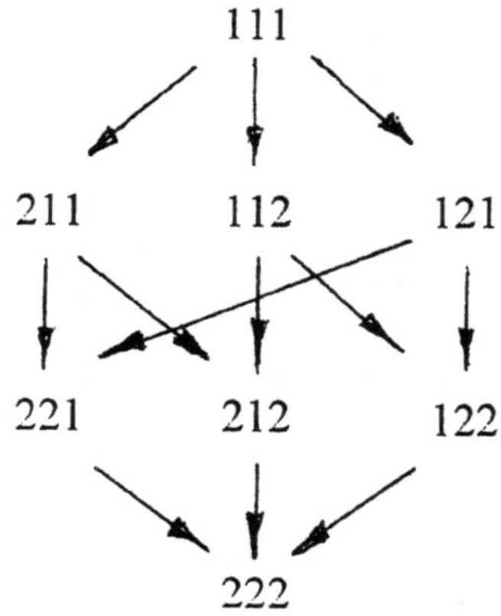

Fig. 1. A Simple Example of Quasiorder

By passing from the quasi-order (Fig.1) to the complete order, it is necessary to define some relations for alternative pairs. In this example we have 28 alternative pairs, from which 19 pairs are in the R relation independent of the complete order construction. The remaining 9 pairs represent 3 IN relations between the elements of the first Pareto layer (211, 112, 121), 3 IN relations between the elements of the second Pareto layer (221,212,122). The remaining 3 IN pairs are the IN relations between the elements of the two layers (each element of the first layer has one incomparable element in the second one).

If we use the additivity properties for complete orders, the number of independent incomparable pairs is smaller. It is easy to demonstrate that the relations between the elements of the first Pareto layer define the relations between the elements of the second Pareto layer. One can say that the second Pareto layer has not independent pairs. After ordering the elements of the two layers we have only one remaining alternative pair between the last element in the order for the elements of the first layer and the first element in the order for the elements of the second layer. Hence we have only 4 alternative pairs which can be in different relations (R,I) in dependence on the DM's preferences.

Let us consider two orders in which the relations for these 4 pairs are contradictory (Fig.2).

1 1 1	1 1 1
1 2 1	2 1 1
1 1 2	1 1 2
1 2 2	2 1 2
2 1 1	1 2 1
2 2 1	2 2 1
2 1 2	1 2 2
2 2 2	2 2 2

Fig.2. Two Orders with Opposite DM Preferences

Using the Kendall coefficient of concordance we have: T=0.43. This coefficient does not represent the elations between two orders. Really, in the second order the DM's preferences are opposite. Hence it is necessary to come up with a different indicator showing the degree of differences in the DM's preferences for two orders having additive structures.

5. A New Indicator

We suggest to use the following indicator:

$$L = 1 - \frac{P}{P_{max}} , \tag{2}$$

where P_{max} is the maximum number of alternative pairs which can change relations in accordance with the DM policy. P is the number of such alternative pairs which do change the relations in the second order comparison with the first one. If we use this indicator for the above example (Fig.2), we receive $L = 0$; it corresponds to a real difference in the preferences.

To be able to use this indicator we must define how it is calculated. Let us recall that the number of alternative pairs increases very rapidly when the number of criteria (N) is increased.

6. Base Potentionally Contradictory Alternatives Pairs (PCA)

Let us consider two orders generated by additivity rules and two alternatives A_i and A_j which are not in dominance relation. We call this pair 'potentially contradictory alternatives' (PCA), if A_i and A_j can change the relation in dependence on the DM's preferences. Let U_i and U_j be the utilities of those alternatives in the first order and $U_i^{'}$ and $U_j^{'}$ the utilities of those alternatives in the second order. If PCA has in the second order the opposite relation, then:

$$\begin{matrix} U_i > U_j \\ U_i' \le U_j' \end{matrix} \text{ or } \begin{matrix} U_i < U_j \\ U_i' \ge U_j' \end{matrix} \tag{3}$$

Hence:

$$(U_i - U_j)(U_i' - U_j') \le 0 \tag{4}$$

The condition (4) is true for contradictory alternative pairs in two orders. Our problem is to find the contradictory pairs. Among different PCA we select base PCA which have the following property: both alternatives belong to the first Pareto layer.

We shall show that in our case (the additive structures) it is possible to construct an almost complete order (the number of remaining alternatives pairs is small) using only the comparisons for the base PCA. In other words, the comparison of the elements of the first Pareto layers define an almost complete order. Let us show how to construct such an order . We shall call PCA which cannot be defined using the comparisons for the base PCA and additivity rules as remaining independent PCA. We shall also give the estimation for the number of remaining independent PCA. All results are given for W_i=2 (binary scales).

The general approach consists of the following. We construct the order for base PCA comparing the elements of the first Pareto layer. Using the results of comparisons we shall construct quasiorders for all Pareto layers until 10 and we estimate the number of remaining PCA . We shall see that the number is relatively small.

We compare the elements of the Pareto layers between themselves and estimate the number of remaining PCA.

7. The Pareto Layers, their Properties and the Transfer of the Relations

In the general case we have Q different alternative pairs:

$Q = C_E^2$, where $E = 2^N$ or $Q = 0.5 \cdot 2^N (2^N - 1)$.

The number of alternative pairs which always belong to the R-relation is defined by the formula:

$$D = 2^N - 1 + \sum_{j=1}^{N-1} C_N^j (2^N - 1). \tag{5}$$

The proof of (5) is given in Appendix 1.

Table 1 characterizes the relation between Q and D.

N	3	4	5	6	7	8	9	10
Q	28	120	496	2016	8128	32640	130816	523776
D	19	65	211	665	2059	6305	19171	58025
D/Q	0.68	0.54	0.425	0.33	0.25	0.19	0.15	0.11

Table 1. The Ratio of Number D of Alternative Pairs which are in Dominance Relation and the Number Q of All Possible Alternative Pairs.

For the j-th Pareto layer we have T_j elements:

$$T_j = C_N^j. \tag{6}$$

The number of PCA in the j-th Pareto layer is equal:

$$S_j = 0,5 C_N^j (C_N^j - 1). \tag{7}$$

Let us recall that by construction the Pareto layers are symmetric:

$$C_N^m = C_N^{N-m}.$$

For any odd N the two biggest layers have an equal number of elements; for an even N one central layer has he biggest number of elements. So, for N = 5 there are 4 layers (5, 10, 10, 5 elements), for $N = 6$ there are 5 layers (6, 15, 20, 15, 6 elements).

Lemma 1. For the layers having an equal number of elements (the symmetric ones) all PCA of the second symmetric layer can be obtained from the PCA of the first symmetric level.

Proof. According to the construction, the elements of the first symmetric layer have m second estimates on the criteria, the elements of the second layer have

$(N\text{-}m)$ second estimates. It is evident that m<0,5N and $N\text{-}m>0.5N$. Let us take one PCA from the first symmetric layer with the alternatives A and B, which have not second estimates on the same criteria. Let us replace in A, B the second estimates on criteria by the first and the first estimates by the second. We obtain two elements of the second symmetric layer C and D. Let us replace in C, D the second estimates on the same criteria by the first $(N\text{-}2m$ common estimates). We obtain the elements A and B of the first symmetric layer. **Q.E.D.**

Corollary. The second symmetric layer has no PCA which are different from PCA of the first symmetric level.

Lemma 2. The quasi-order for the elements of every Pareto layer can be received by only using comparisons for the base elements.

The number of remaining independent PCA for $N \leq 10$ is given by formula

$$Z_j = 0.5C_N^j(C_N^j - 1) - 0.5N(N-1)C_{N-2}^{j-1} - F(N)C_{N-4}^{j-2} - G(N)C_{N-8}^{j-4}, \quad (8)$$

where:

$$F(N) = 2C_{N-1}^3 + F(N-1); F(4) = 2; \; G(N) = 5C_{N-1}^5 + G(N-1); G(6) = 5;$$
$$J(N) = 14C_{N-1}^7 + J(N-1); J(8)=14$$

The proof is given in Appendix 2.

Let us now consider the comparisons between the layers. For $N<10$ it is sufficient to compare p layers among themselves ($p=N/2$ for even N and $p=0.5(N\text{-}1)$ for odd N). It is easy to see that comparison of the lements from the $(N\text{-}m)$-th and the $(N\text{-}m\text{-}i)$-th layers are equivalent to the comparison of the elements of the m-th and the $(m+i)$-th layers. Hence for $N<10$ it is enough to generate the results for the five first layers.

Lemma 3. The number of remaining incomparable pairs obtained in the comparison of the first and the i-th Pareto-layers is equal to:

$$H_{1,i}^N = C_{N-1}^i + H_{1,i}^{N-1} \quad (9)$$

$$N \geq 3; \; i \leq N\text{-}1; \; H_{1,2}^3 = 1.$$

The proof is given in Apendix 3.

Lemma 4. For $N=j+i$ (j=2, 3, 4) the number of remaining incomporable pairs is equal to:

a) in the comparison of the second and the i-th Pareto layer:

$$H_{2,i} = C^i_{N-1} + 1 \tag{10}$$

b) in the comparison of the third and the i -th Pareto layer

$$H_{3,i} = C^i_{N-1} + C^i_{N-2} + 2; \tag{11}$$

c) in the comparison of the fourth and the i -th Pareto layer:

$$H_{4,i} = C^i_{N-1} + C^i_{N-2} + 2C^i_{N-3} + 4 \tag{12}$$

The proof is given in Appendix 4.

In the general case for the comparisons between the layers the following is true.

Lemma 5. The number of remaining incomporable PCA obtained in the comparison of the second and the i-th Pareto layers is no more than:

$$H_{2,i} < C_N C_{N-2} - \left[\frac{(C_N^{i-2} - 1)F(N-i+2)}{C_i^2}\right] - F(N-i+2) \tag{13}$$

where [expression] =- the integer part of the expression.

The proof is given in Appendix 5.

Lemma 6. The number of remaining incomparable PCA obtained in the comparison of the third and the i-th Pareto-layers is no more than:

$$H_{3,i} < C_N^3 C^i_{N-3} - \left[\frac{G(N-i+3)(C_N^{i-3} - 1)}{C_i^3}\right] - G(N-i+3) \tag{14}$$

where: [expression] = the integer part of the expression. The proof is analogous to the proof of Lemma 5.

Lemma 7. The number of remaining incomparable PCA obtained in the comparison of the fourth and the i-th Pareto layers is no more than:

$$H_{4,i} < C_N C_{N-4} - \left[\frac{(C_N^{i-4} - 1)J(N-i+4)}{C_i^4}\right] - J(N-i+4) \tag{15}$$

The proof is analogous to the proof of Lemma 5.

The application of (8) allows one to obtain the number of remaining PCA in the Pareto-layer (Table 2).

N	The number of remaining independent PCA in layers. Number of layers				The general number of remaining PCA
	2	3	4	5	
3	0				0
4	1				1
5	5				5
6	15	35			50
7	35	140			175
8	70	448	721		1239
9	126	1050	2709		3885
10	210	2310	8295	13516	24331

Table 2. The Number of Remaining PCA in Pareto-Layers

The application of (9)-(15) allows one to obtain the number of remaining PCA in the comparisons between the layers (Table3).

N	$H_{1,i}$	$H_{2,i}$	$H_{3,i}$	$H_{4,i}$	
3	1				
4	5				5
5	16	5			21
6	42	40			82
7	99	189	22		310
8	219	692	213		1124
9	502	2158	1223	93	3976
10	1013	6132	5226	803	13174

Table 3. The Number of Remaining PCA in the Comparisons Between the Layers

Thus, the above method of the quasi-order construction allows one to obtain the quasi-order by only using comparisons of base elements. The estimation of its exactness is given by Table 4, where the general number B of potentionally contradictory alternatives is also given. (S is obtained by summing the date from Tables 2 and 3.)

N	3	4	5	6	7	8	9	10
S	1	6	26	132	485	2363	7861	37505
$B=Q-D$	9	55	285	1351	6069	26335	111645	465751
S/B	0.11	0.10	0.09	0.10	0.08	0.09	0.07	0.08

Table 4. The Estimation of the General Number of Remaining PCA

8. An Approximate Method for Calculating the New Indicator

Table 4 shows that the comparisons between base elements only allow us to obtain information about 90-93% of all PCA. So, it is possible to judge the contradictions between two orders by the contradictions between the base PCA. Our simple method of calculation is as it follows. Let us take from the first and second order (obtained by the WSCE method) the values for base elements. Let us order the elements according to the utility in the first order. After that the number of pairwise permutations which gives the order for the utilities from the second order define the number of contradictory pairs of alternatives for the two orders. Each contradictory pair of base elements gives (as a minimum) 2^{N-2}contradictions in the order.

We suggest to calculate the approximate indicator of concordance between the two orders by formula

$$L_{app.} = 1 - \frac{P_B}{C_N^2} \tag{16}$$

where P_B = number of base PCA which changes the relation in the second order (in comparison with the first).

9. An Example

Let us illustrate the calculation of L for two orders. In Fig.3 the two orders are given for N=5 and $W_i = 2$.

The base elements in the orders have the numbers 2, 3, 5, 9, 17.

Let us write in Table 5 the utility values for the base elements.

Order I	Order II
estim [1] → 5.00	estim [1] → 5.00
estim [2] → 4.69	estim [2] → 4.55
estim [3] → 4.77	estim [3] → 4.09
estim [4] → 4.46	estim [4] → 3.64
estim [5] → 4.38	estim [5] → 4.36
estim [6] → 4.08	estim [6] → 3.91
estim [7] → 4.15	estim [7] → 3.45
estim [8] → 3.85	estim [8] → 3.00
estim [9] → 3.92	estim [9] → 4.00
estim [10] → 3.62	estim [10] → 3.55
estim [11] → 3.69	estim [11] → 3.09
estim [12] → 3.38	estim [12] → 2.64
estim [13] → 3.31	estim [13] → 3.36
estim [14] → 3.00	estim [14] → 2.91
estim [15] → 3.08	estim [15] → 2.45
estim [16] → 2.77	estim [16] → 2.00
estim [17] → 3.46	estim [17] → 4.00
estim [18] → 3.15	estim [18] → 3.55
estim [19] → 3.23	estim [19] → 3.09
estim [20] → 3.92	estim [20] → 2.64
estim [21] → 2.85	estim [21] → 3.36
estim [22] → 2.54	estim [22] → 2.91
estim [23] → 2.62	estim [23] → 2.45
estim [24] → 2.31	estim [24] → 2.00
estim [25] → 2.38	estim [25] → 3.00
estim [26] → 2.08	estim [26] → 2.55
estim [27] → 2.15	estim [27] → 2.09
estim [28] → 1.85	estim [28] → 1.64
estim [29] → 1.77	estim [29] → 2.36
estim [30] → 1.46	estim [30] → 1.91
estim [31] → 1.54	estim [31] → 1.45
estim [32] → 1.23	estim [32] → 1.00

Fig. 3. Two Different Orders of Alternatives for N=5 and w_i=2.

N	Order I	Order II
2	4.69	4.55
3	4.77	4.09
5	4.38	4.36
9	3.92	4.00
17	3.46	4.00

Table 5. The Utility Values for Base Elements in Two Orders given by Fig. 3.

N	Order I	Order II
3	4.77	4.09
3	4.69	4.55
5	4.38	4.36
9	3.92	4.00
17	3.46	4.00

Table 6. The base elements ordered by utility values in first order given by Fig. 3.

In Table 6 there are the same elements but ordered according to the utilities in the first order. It is easy to see that it is necessary to make two permutations in the right column of Table 6 (the pairs 3-2, 3-5) to receive the correct order. Hence $L_{app.} = 1 - 2/10 = 0{,}8$.

10. Conclusion

In the conclusion we suggest some generalizations.

1) The above results have been obtained for $W_i = 2$ (binary scales). But we can on an empirical basis suggest to use this simple method for problems with any structure. What is the base for such a suggestion? When one uses the additive rules, a contradiction in one base PCA creates a larger number of contradictions in one non-base alternative pair. The contradictions in base alternative pairs are much more important for any complete order. That is why, the comparisons in base pairs allow one to judge the concordance of the orders in a reliable way.

2) All results are obtained above for $N<10$. But one can use the same ideas to obtain the results for any N.

3) To calculate the new indicator (12) it is sufficient to use only the ranks of the alternatives in the order (as well as in the concordance measure). It means that this indicator can be applied also for the quasiorder constructed on the basis of qualitative measurements (Larichev et al. [1995]).

Acknowledgements

The author expresses the gratitude to Professor L.Sholomov (Institute for System Analysis, Moscow, Russia) and to Professor K.Borcherding (University of Darmstadt,Germany) for valuable comments and suggestions. He is thankful to Miss A. Pieritz and Miss V. Kostina for their help in the preparation of this manuscript.

References

Von Winterfeldt, D. and Edwards,W. (1986) *Decision Analysis and Behavioral Research*, Cambridge University Press, Cambridge.

Edwards, W. (1977) "How to Use Multiattribute Utility Measurement for Social Decision Making". *IEEE Transactions on System, Man and Cybernetics*, SMC-7, 326-40.

Larichev, O. and Moshkovich, H. (1995) "Zapros LM-A Method and System for Ordering Multiattribute Alternatives". *European Journal of Operational Research* 82, 503-521.

Appendix 1

The formula for the calculation of D has been obtained in the following way. The first element in the order dominates all remaining elements $(2^N - 1)$. The one element of the first Pareto layer dominates all possible combinations of estimates upon (N-1) criteria except one: $(2^{N-1} - 1)$. The number of the elements of this layer is C_N^1. In an analogous way we can obtain the number of dominated alternatives for each element of each layer.

Appendix 2

We know that the number of PCA in the j-th Pareto layer is $0.5(C_N^j - 1)C_N^j$. By using the additivity rule, we can add to every alternative in a pair of the first layer (the base PCA pair) (j-1) second estimations on criteria and can obtain one pair of the j-th layer.

The number of such pairs is $0.5C_N^1(C_N^1 - 1)C_{N-2}^{j-1}$,

where: $0.5C_N^1(C_N^1 - 1)$ = the number of base PCA pairs and C_{N-2}^{j-1} = the number of possible ways of adding to each base PCA pair (j-1) second estimations on criteria.

We can, however, utilize the order of the base elements in a different way.We can use the sums of two elements from the order of the base elements which dominate a different sum of two elements. (By the sum we mean the element of the second layer, two second estimates of which correspond to two second estimates in two base elements.) We can construct such sums in the following way. Let us order the second estimates of an N-dimensional vector in accordance with the order of the base element from right to left. Let us fix the first element from the right. Its second estimates must belong to the first (dominating) sum. Let us take any three consequent elements in the order. From those three elements the left one must belong to the second (dominated) sum. One from other two elements can belong to the first or to the second sum. Hence for the first element and those three elements we have the following number of such pairwise sums (the first dominates the second): $F(4) = C_2^1 = 2$.

For N criteria we must multiply $F(4)$ by the number of combinations from (N-1) by 3. So, the general number of pairwise sums is equal to : $F(N) = 2C_{N-1}^3 + F(N-1)$.

Analogously, we can obtain the sums of three elements dominating other three elements. Let us take N=6. The first element from the right belongs to the first sum, the last from the right - to the second sum. From four elements located in the middle we can create the combinations of size two and all these combinations except one can be added to the first element from the right to obtain the dominating sum. As the result we have: $G(6) = C_4^2 - 1 = 5$.

For N>6 we must take all possible combinations of size 5 from (N-1) elements: $G(N) = 5C_{N-1}^5 + G(N-1)$.

Analogously, we can obtain the sums of four elements dominating other four elements. Let us take N=8. $J(N) = (C_5^2 - 1) + (C_4^2 - 1) = 14$.

In the general case: $J(N) = 14C_{N-1}^7 + J(N-1)$.

It is possible to add to each sum ($F(N)$, $G(N)$, $J(N)$) the same second estimations on criteria and also to use those pairs of elements. **Q.E.D.**

Appendix 3

Let us order the second estimates of an N-dimensional vector in accordance with the order of base elements from right to left. Let us take N=i+1. In this case we have one incomparable PCA of such type:

$$\begin{array}{c} 211\ldots11 \\ \underbrace{122\ldots22}_{i} \end{array}$$

In the general case, when N>i+1, we shall have for the same first element in PCA several (C_{N-1}^i) second elements which create C_{N-1}^i incomparable PCA. But we have some incomparable PCA for the first element of type:1211....1. It means that we must take into account all incamporable PCA of dimension (N-1). The resulting formula for remaining incomporable PCA obtained in the comparison between the first and the *i*-th Pareto layers is the following: $H_{1,i}^{N} = C_{N-1}^{i} + H_{1,i}^{N-1}$, $N \geq 3$; $i \leq N-1$; $H_{1,2}^{3} = 1$.

Appendix 4

The proof is analogous to the proof of App.3. Let us take the same order of the second estimates in the vector, as was taken in App.3. Let us consider the second layer and *i*=*N*-2. In this case we have (*i*+1) incomparable PCA, for which the first element has one second estimate in the left dimension and one on any other dimension. But we have one additional incomparable PCA with the following first element: 1221....1. So for *i*=*N*-2 we have $C_{i+1}^{i} + C_{i}^{i}$ incomparable PCA.

$$H_{2,i}^{N} = C_{N-1}^{i} + 1 = i + 2..$$

Let us take the third layer and *i*=*N*-3. In this case, incomparable PCA have the following first elements:

21.....; 122....; 12122..; 11222...

$$H_{3,i}^{N} = C_{N-1}^{i} + C_{N-2}^{i} + 2.$$

Let us take the fourth layer and *i*=*N*-4. In this case incomparable PCA have following first elements:

21...........; 122.........; 11222....; 1112222..;
12122....; 1212122..; 1121222..; 1122122...

$$H_{4,i}^{N} = C_{N-1}^{i} + C_{N-2}^{i} + 2C_{N-3}^{i} + 4.$$

Appendix 5

Let us consider the second and the *i*-th Pareto-layers. If two elements in PCA created by one element from the second and one from the *i*-th layer have one common second estimate, we can transform this PCA into one between the first and the (*i*-1)-th layers. That is why it is enough to investigate the PCA, the elements of which have no common (second) estimates.

The number of such PCA in which the elements are different by (2+*i*) estimates equals: $C_{N}^{2} C_{N-2}^{i}$. Let us call them C-PCA. Among them we have comparable (D-PCA) and incomparable PCA. Each D-PCA can be represented as the sum of one D-PCA (2x2) of the second Pareto layer (D-

PCA different on 4 criteria) and one E-PCA for which one element has only best (first) estimates and second one has worst (second) estimates by (i-2) criteria.

Let us obtain all D-PCA of such kind in the following way. One can take (i-2) different E-PCA of dimension one for which the first element has all best estimates and the second one has one worst estimate. One can create all possible combinationsof size (i-2) from N elements and allocate E-PCA of dimension one to different places in the N-dimentional vector. For each combination one can use the remaining (N-i+2) places in the vector by inserting all possible D-PCA from the second Pareto layer.The number of such D-PCA is equal to F(N-i+2). In this way, one can obtain F(N-i+2) PCA. Among them there are non-repeated PCA and repeated PCA.

Let us allocate the components of an N-dimensional vector according to the order of the base elements from right to left. In this case for each C-PCA the second estimation in the first element in the i-th place from the right dominates the second estimate in the second element located in the (i+k)-th place (k=1, 2, ...).Let us put all E-PCA (the number is (i -2)) in the rightmost (i-2) places. Let us add D-PCA taken from the second layer to it. We obtain $F(N$-i+2) D-PCA. It is easy to see that they are not repeated ones, because not one from the second estimates of D-PCA (2x2) can be replaced by the second estimate from E-PCA.

Let us put all E-PCA into the leftmost (i-2) places and add all possible D-PCA (2x2) taken from the second layer. It is easy to see that they are repeated ones, because every second estimation from the second element of D-PCA (2x2) can be replaced by second estimates taken from E-PCA. The maximum number of dublications is equal to C_i^2. That is why the real number of non-dublicated D-PCA does not exceed: $\left[\frac{F(N-i+2)C_N^{i-2}-F(N-i+2)}{C_i^2}\right]$, where [expression] is the integer part of the expression.

The number of remaining incomparable pairs does not exceed:

$$H_{2,i} < C_N^2 C_{N-2}^i - \left[\frac{F(N-i+2)C_N^{i-2}-F(N-i+2)}{C_i^2}\right] - F(N-i+2).$$

A Foundation of Principles for Expanding Habitual Domains†

Po-Lung Yu[1] and Liping Liu[2]

[1] School of Business, University of Kansas, Lawrence, KS 66045, USA

[2] School of Business, Albany State University, Albany, GA 31705, USA

Abstract. A human brain is like a super computer (hardware) and has practically unlimited potential capacity. However, because of their habitual patterns of thinking (software), people use only a very small percentage of the capacity. Their ability for effective problem-solving very much relies on how they upgrade the software and break undesirable and inefficient habits. Habitual domain studies aim at providing general principles, techniques, and theoretical foundations of how to expand and enrich habitual domains. In this paper, we show the importance of studying habitual domains using familiar examples. We introduce basic concepts of the habitual domain theory. We present three of the most commonly used principles of expanding habitual domains. Finally, as a new contribution of this paper, we present a theoretical interpretation of the principles using the notion of activation probabilities and attention spectra.

Keywords. Habitual domain, MCDM, Mental Focus Control

1 Introduction

Each normal human brain has approximately 100 billion neurons. If each neuron is regarded as a micro-computer (see Fischbach [1992]), then a human brain would be a super computer with countless parallel micro-computers networked with each other. We can hardly imagine the potential power of that "computer." Unfortunately, people use less than ten percent of the potential power in their daily life and work (The Diagram Group [1982]). This may be due to the restrictions of their habitual domains (software). The constraining force of habitual domains is invisible and unconscious. However, its power can constrain the behavior of giant creatures and hinder the prosperity of an enterprise.

Case 1. In India or circus troupes, people often see that a giant elephant is tied to a small post by a rope. In terms of its potential power, an elephant can easily up-root the post and do whatever it wants to do. But why does the elephant remain obediently at the post? The elephant does not yield to the strength of the post due to its trained habits.

---

† This research is partially supported by Chiang Ching-Kuo Foundation for International Scholarly Exchange, Taipei, Taiwan to the first author.

[0] *Volume in Honor of Stanley Zionts—on the Occasion of His 60th Birthday*. Edited by M. Karwan, J. Spronk and J. Wallenius.

A trained elephant loses its potential power when facing a post, and so does a 'tamed' human being in solving problems. People are tamed by education, law and order, cultural traditions and religion. With the development of machines, we have learned to use the machine as a metaphor to mold the internal and external world in accordance with mechanical principles. People are trained to behave as if they were parts of machines. Human brains are programmed and conditioned by rules, laws, and procedures of thinking and are expected to work as machines: in a routinized, efficient, reliable, and predictable way. Gradually, a habitual domain is being developed and stabilized. Habitual domains help people think, act, judge and respond efficiently when facing routine problems. They also help stabilize society and improve the productivity of organizations. However, they hinder the development of human mental capacity. Habitual domains tend to restrict rather than mobilize the function of the human brain and constrain our behavior and thinking. Because of their inhibition, our full potential capacity is rarely reached; our brains are degenerating; and idiosyncrasies are decaying. As a result, people are puzzled or even defeated when facing non-routine problems that require newfound intelligence beyond our habitual ways of information processing.

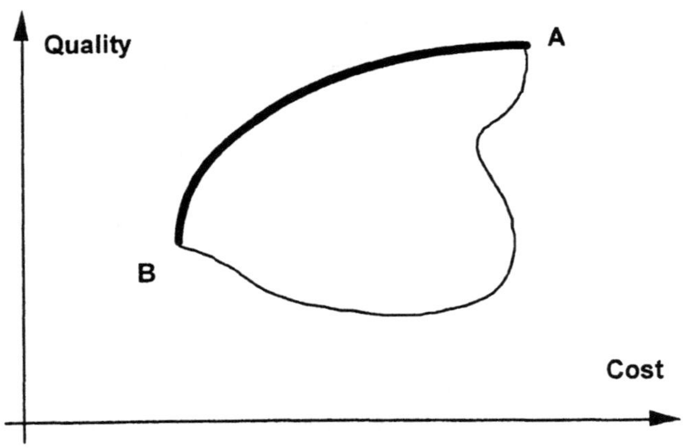

Fig. 1 Trade-off between quality and cost

In human history many unusual successes in problem-solving and scientific discovery are due to non-routine inquiries of problems and the expansion of habitual domains. As an example of habitual domains, literally millions of people take it for granted that an apple will fall down to the ground when detached from a twig. Breaking this habitual response. Newton asked why does an apple fall down but not fly away. It is well known that this inquiry eventually led to the discovery of the law of universal gravitation.

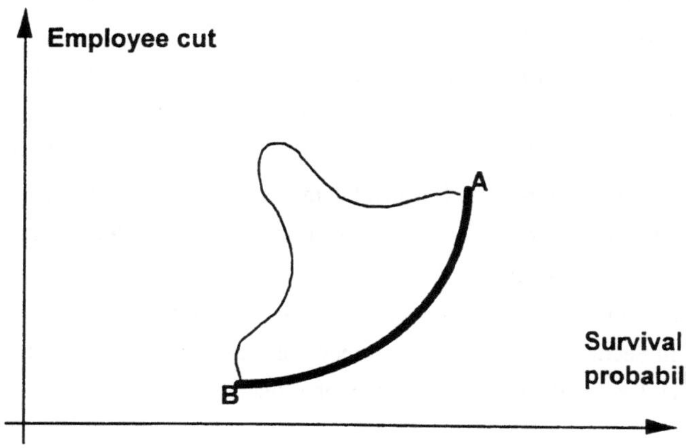

Fig. 2. Trade-off between employee cut and survivability

Case 2. It is commonly believed that the better the quality, the higher the cost. In terms of multicriteria decision making, people must make a trade-off along the non-inferior frontier AB (see Fig. 1): higher cost exchanges for higher quality or vise versa. However, Henry Ford came along with a different response: Why didn't we have both high quality and low cost cars so that more people could enjoy automobiles? He broke the common belief and successfully introduced mass production and standardization techniques into the auto-industry. His unusual inquiry eventually resulted in a revolutionary change in manufacturing, transportation, as well as many aspects of human life.

Case 3. Corporate downsizing and laying off employees tends to be a natural response to economic depression in order to maintain survival and/or profit-maximization. In terms of multicriterion decision making, firms struggle on the Pareto-efficient frontier shown in Fig. 2 and make trade-offs between the number of employees cut and survivability, because it is believed that the more employees it lays off, the more likely a company will survive and vice versa. In 1929, Matsushita Company faced such a dilemma. It had only 50 percent of regular sales and had accumulated a large stock of inventory. Firing employees seemed to be imperative for its survival. At this time. Mr. Matsushita asked why it could not have 100 percent survival probability without firing its employees and/or reducing their benefits. He had an unusual solution: no lay-offs and no bankruptcy. The vi-able strategy was to reschedule operations such that the production department produced new products only half a day and sold the products in the other half of the day. Employees became excited and happy because they were able to keep their jobs. They worked hard with enthusiasm. In less than one half year, they solved the overstocked inventory problem and rescued the company.

The above cases and numerous more examples suggest that to solve challenging problems effectively, one needs to break his habitual constraints and expand and enrich a new domain of intelligence. The habitual domain theory introduced in Yu [1985, 1990, 1995] describes a conceptual model of habitual domains and general principles and techniques for expanding and enriching habitual domains in order to empower people to solve problems more effectively and efficiently. In this paper, we first sketch a basic framework of the habitual domain theory. In Section 3, we discuss three principles of expanding habitual domains. In Section 4, we present a theoretic foundation to justify the effectiveness of the three principles. Final remarks are given in the conclusion.

2 Concepts of Habitual Domains

Human behavior is both dynamic and stable. Our daily observations and experiments show that human behavior has many varieties. For example, Alinsky [1971, pp. 86-87], looking tough and without a tie, once walked down a busy Los Angeles street, offering a ten-dollar bill to anyone who came along, saying " Here, take this." He got responses like these:

"I'm sorry, I don't have any change."

"I'm sorry, I don't have any money on me right now."

"I'm not that kind of girl."

"I don't come that cheap."

"What kind of con game is this?"

Although different people or one individual at different times may behave very differently, it has been recognized that each person has his own habitual ways of responding to stimuli, sometimes called conditioned or programmed behavior patterns. For example, each of us has habitual ways of eating, dressing, speaking and thinking.

Conceptually, the collection of one's habitual ways of perceiving, thinking, responding, and acting, together with their formation, dynamics, and basis in experience and knowledge, is called one's habitual domain (HD). A habitual domain consists of four elements:

(i) The *potential domain* (PD)—the collection of ideas and actions that can be potentially activated.

(ii) The *actual domain* (AD)—the set of ideas and actions that are actually activated.

(iii) The *activation probabilities* (AP)—the probabilities that ideas and actions in PD that also belong to AD.

(iv) The *reachable domain* (RD)—the set of ideas and actions that can be attained from a given activated set in PD.

The relation between the AD and the PD is similar to that between a sample and the sample space in statistics. The concept of the RD arises from the fact that ideas can generate other ideas and that actions can prompt other actions by using operators (see Yu [1985, 1990, 1995]). The notion of the RD plays an important role in human cognition and the recognition of suggestions.

Let us use an example to illustrate the above concepts. Consider a case in which someone is frustrated because he has failed unexpectedly to meet a goal. Many possible ideas (members of his PD) may be activated in his mind. One possibility is a self inquiry "Why did I fail?" Once activated, it is a member of his AD. This thought can lead to a number of other ideas (members of his RD), such as "Because someone did not support me," or "Someone doesn't like me." On one hand, these newly activated ideas can affect both his mood and his judgment of the experience; on the other hand, they may activate other related ideas.

Now suppose that, instead of asking "Why did I fail?" he asks "What can I learn from this experience?" or "How do I make it work?" (new elements of his AD). Then he will reach a new set of ideas (in his RD), and his mood and judgment of the experience will be different.

In Case 2 of the previous section, most people believed that the better the quality, the higher the cost. This thought when activated leads to the concept of making a trade-off on the noninferior frontier AB (RD). Ford activated a different thought: why not better quality with lower cost, which led to the discovery of mass production and standardization (RD). Similarly, in Case 3, because Matsushita activated a different thought (AD), he reached a different set of concepts (RD) to solve his dilemma. We note that by controlling AD we could control RD and thus control our emotion and judgment.

Since an HD can change over time, it is dynamic, as are its four elements, PD, AD, AP, and RD. But it will stabilize most of time except during some transient periods. There are many causes that contribute to the stabilization. For example, the more one learns, the less likely that arriving messages or events are new. For another example, since one often uses his pre-existing models to interpret arriving messages and events, there is a tendency toward consistency. For still another example, environment may converge toward a regular rhythm and result in regularity of the input of information. Therefore, in the absence of extraordinary events, an individual PD and AP may stabilize. For a mathematical formulation and the proof of the stability of the HD, the reader is referred to Chan and Yu [1985] or Yu [1985, Chapter 9].

Observe that, for a given initial stimulus *E*, using an operator *O* (a thinking process) can produce multiple outputs—retrieved ideas, responses, judgments etc. Usually, *O* is a set-valued mapping from AD to PD.

If PD and AP are stable, it can be shown that the set *O*(*E*) will be stable. It implies that one will have a stabilized set of responses to a given stimulus *E*. Of course, *O*(*E*) may not be a singleton, i.e., one's actual response to *E* may not be deterministic and fixed. When multiple operators *O* are applied, the reachable domain from stimulus *E* is the union of all the *O*(*E*), i.e., $\bigcup_{\text{all posible } O} O(E)$, which is larger than when a single operator is applied. Thus, there is a possibility that the set of one's responses to a given stimulus *E* may be quite large even though his HD is stabilized.

However, this possibility is very small according to many daily observations and laboratory experiments. What we find is that how people think, judge and make decisions (i.e., apply operators) are fairly rigid. One often works under fairly fixed paradigms to observe and influence reality. For example, Churchman [1979, p.50] describes the analyst as someone who saw the world as a mathematical program. Maslow put it more vividly, "If the only tool one has is a hammer, he tends to see every problem a nail."

Through a number of laboratory and field studies, Mackenzie [1988] found that what people do is less stable than how they do it and that the processes of change are more stable than that to which the change is directed. This finding indicates that operators (how) people are using are habitually more rigid and stable than the resulting ideas--the activation of elements in *O*(*E*) (what).

The above discussion suggests the following important observations of habitual domain studies:

(i) How one views reality and responds to a stimulus is fairly inflexible or rigid and it often exhibits some programmed patterns. Consequently, the reachable domain from a stimulus tends to be small and the actual domain tends to be predictable;

(ii) A limited habitual domain is appropriate only for solving routine problems. Extraordinary performance in the face of non-routine problems greatly depends on the expansion and enrichment of one's habitual domain and the application of new or unusual operators to restructure the habitual domain.

3 Three Principles for Expanding Habitual Domains

Newfound intelligence in problem solving and decision making depends on the effective expansion of one's habitual domain, including breaking his pre-programmed behavior patterns, generating and applying new operators to enrich reachable domains, and activating fresh ideas and bright solutions in actual domains. Because of the rigidity of how people think and behave, expanding habitual domains is not an easy, routine task. The difficulty makes innovation scarce and valuable. The difficulty calls for more scholastic effort dedicated to the understanding of habitual domains and their expansion laws. Based on two decades of research and practice, Yu [1990, 1991, 1995] proposes eight basic tools and nine general principles as the guidance for expanding and enriching habitual domains. These methods and principles are all important and complement each other to form an unbreakable approach for deep knowledge. The following lists three of them and their associated explanations. In the next section, we shall provide a theoretic foundation to justify their effectiveness.

The projecting principle[1]: This principle of expanding habitual domains is applied when one considers the functions, problems, and outlooks of someone else in a different position. In an organizational structure, it is especially useful to consider these matters from the point of view of someone in a higher position, as well as someone in a lower one. For example, a production manager can expand his habitual domain by considering the president's problems of marketing, finance, personnel management, public relations, and so on. Similarly, he can expand his habitual domain by considering the situation of a production operator. In general, by projecting his thinking to other important functions of the company, a middle manager can expand his habitual domain to include most of the major concerns of the organization; indeed, such an expansion makes him suitable and eligible for promotion to more challenging responsibilities. As another example, an OR worker in an organization can—and should—expand his habitual domain by considering the concerns, problems and outlooks of a variety of its executives, as well as its other staff members. If he does so, he will have a habitual domain large enough to envision all of the important dimensions of the organization's problems, and thus be able to bring appropriately broad analyses and solution proposals to bear on them.

Furthermore, since such considerations will encompass wide classes of problems (for example, in a manufacturing company, those of production, marketing, finance, and human factors) they will bring to the analyst's habitual domain a wide spectrum of intellectual structures, many of which can prompt new theoretical research, as well as the op-

[1] This principle is listed as one of the eight basic methods in Yu [1990].

portunities of bringing findings to bear on the practical problems of the organization. In summary, expanding their habitual domains through projection is especially important for OR workers, both in and out of hierarchical organizations.

The Deep-and-Down Principle: In our daily lives we are continuously presented with routine--and occasionally urgent--events and problems, and our attention is thus forced to shift to them as they come to our attention. Against this background it is easy to imagine that some good ideas stored in our potential domains may not catch our attention, and thus may not contribute to enriching our actual domains.

To overcome this neglect, and to prepare our habitual domains for better responses to non-trivial problems, we need to let our minds be peaceful and open--that is, go deep and down--to allow ideas in the potential to catch our attention, and thus move to the actual domain. It is a commonly reported experience that good ideas often surface during the common routines of life, such as taking a shower and a walk, or during times set aside for quiet thought.

The other inference that should be drawn from this principle is that it is useful to assume a low and humble posture, thus making it easy for other people to offer ideas and suggestions. Such an open environment stimulates new and creative ideas. Lao Tze, a Chinese philosopher, put it this way: "Why is the ocean the king of all rivers? Because the ocean assumes a lower position" and thus allows all the water to flow into it easily.

The Alternating Principle: Just as a door that is always closed or always open loses its functions as a door, so an assumption that is always either imposed or omitted loses its value. To counteract the common tendency always to make the same assumptions as part of our habitual domains, this method urges that they be changed from time to time to see what new ideas can emerge.

A linear program typically has a single criterion and a fixed set of feasible solutions defined by linear inequalities. But the single criterion can be replaced by multiple criteria (MC), and an MC-simplex method can be developed. Similarly, the fixed sets of feasible solutions can be replaced by a flexible set defined by the multiple constraints (also denoted by MC) of several levels of inequalities. These two changes lead to the development of an MC^2-simplex method (Yu [1985, pp.219-263]).

Solving a linear program finds an optimal solution to a given system. By removing the assumption that the system is given a priori, one can begin to develop preferred system designs and their associated contingency plans (Lee, Shi and Yu [1990]).

The alternating principle is also useful in daily life, where it can suggest new ideas or help to solve complex problems. Business now recognizes that complaints from workers and customers, formerly brushed off as painful nuisances, are actually bearers of important new information that can usefully be pondered and acted on. Persons would do well to appreciate the same point. Such complaints can not only become the bases for solving problems important to the complainers, they can also suggest new and improved forms of service.

4 A Theoretical Foundation: Mental Focus Control

Suppose that there are m mental items $I_1, I_2, \ldots, I_m$ in the PD, which can be potentially recalled from one's memory system. Let $\mu_t(E, I_i)$ denote the average connectivity or mental association from E to I_i. Given an initial mental item E, assume that each potential items I_i (i = 1, 2, ..., m) has a positive probability, no matter how small, to be recalled from the memory. Since their connectivity to E are different, their possibilities of being recalled or activated are different. To describe this feature of activation, we define the notion of an activation probability as follows:

Definition 1. Given an initial item E, the propensity for a mental item I to be activated is defined as an activation probability. Let $P_t(I|E)$ denote the probability of activating I given E at time t.

Note that $P_t(I|E)$ is the conditional probability that I is recalled given the appearance of E. $P_t(I|E)$ follows the general rules of a probability function and represents the uncertainty of the activation of mental items under a certain condition. Since the larger $\mu_t(E, I)$ is, the more likely I will be recalled, we assume that $P_t(I|E)$ is positively proportional to $\mu_t(E, I)$. Thus, we may define activation probability $P_t(I_i|E)$ to be the normalized average connectivity as follows:

$$P_t(I_i|E) = \frac{\mu_t(E, I_i)}{\sum_{k=1}^{m} \mu_t(E, I_k)} \quad (i = 1, 2, \ldots, \mathrm{m}) \tag{1}$$

According to neural science (for instance see Fischback [1992]), we may postulate that, given E, the recalling of an item I_i depends on its average connectivity $\mu_t(E, I_i)$ with E, and its threshold $\theta(I_i)$ for activation. The threshold is determined by various factors. First, the threshold θ value is partially determined by one's biophysical and psychological state. This is evidenced by the fact that bio-rhythms has effect on our daily thinking performance and that we often have strange dreams at night or generate great ideas during a break. As brain scientists have shown, one's memory of any object exponentially decays over time. Thus, we usually have extremely weak memory trace on long past events such that our experience with the events is almost untraceable at a conscious level of psychological state. However, it may be recallable if our psychological state makes the threshold value θ extremely low, such as during a deep sleep. Similarly, great ideas often have very weak associations with other memory items. They can be reachable from other memory items only if we are given a proper initial stimulus and the threshold value is very low.

The second factor that can affect threshold is attention allocation. When facing a problem, one cannot pay attention to all relevant information at one time. He usually focuses his attention on some arousing or promising directions, which may be guided by external advice or internal belief. If something is given attention, the noises that prevent its associated ideas being recalled will be reduced and the thresholds for those ideas will be decreased. The more attention is focused on a memory item, the more deeply one thinks about it, the lower its threshold value, and the more likely it is recalled.

Therefore, it is reasonable to assume that a threshold value of item I given E, $\theta(I)$, is a function of one's psychological state and his attention allocated on I.

Definition 2. A distribution of threshold values at time t across all mental items I_1, $I_2, \ldots, I_m$ is called an *attention spectrum* and is denoted by Θ_t:

$$\Theta_t = \{\theta_t(I) \mid I = I_1, I_2, \ldots, I_m\}. \tag{2}$$

Let

$$\alpha_t(I|E) = \frac{\theta_t(I)}{\sum_{k=1}^{m} \mu_t(E, I_k)} \quad I = I_1, I_2, \ldots, I_m \tag{3}$$

and

$$\alpha_t = \{\alpha_t(I|E) \mid I = I_1, I_2, \ldots, I_m\}. \tag{4}$$

$\alpha_t(I|E)$ is thus the *normalized attention spectrum*.

In terms of activation probabilities and attention spectra, I may be activated from E if and only if

$$P_t(I|E) > \alpha_t(I|E) \tag{5}$$

Therefore, the mental items in the following set:

$$A_t(E) = \{I \mid P_t(I|E) > \alpha_t(I|E), I = I_1, I_2, \ldots, I_m\} \tag{6}$$

may be activated from E. $A_t(E)$ is a part of the reachable domain PD. Which mental item(s) in $A_t(E)$ is actually activated is determined by a certain random mechanism and/or a certain decision mechanism. More research on identifying these mechanisms needs to be done.

By comparing (6) with the definition of an operator, we find that A_t acts as an operator on E. Given an activation probability distribution, each attention spectrum determines an operator. On the other hand, given any operator O and an activation probability distribution, there exists an attention spectrum $\alpha_t(I|E)$ such that

$$O(E) = \{I \mid P_t(I|E) > \alpha_t(I|E),\ I = I_1, I_2, \ldots, I_m\}. \tag{7}$$

Therefore, an attention spectrum and an operator are one-to-one correspondent. Finding an operator is equivalent to finding an attention spectrum.

Equations (5) and (6) are graphically illustrated in Fig. 3. Here the horizontal axis shows the potential domain, the vertical axis represents values of activation probability distributions and attention spectra, and the bold-faced intervals on horizontal axis represent the set of items that may be activated.

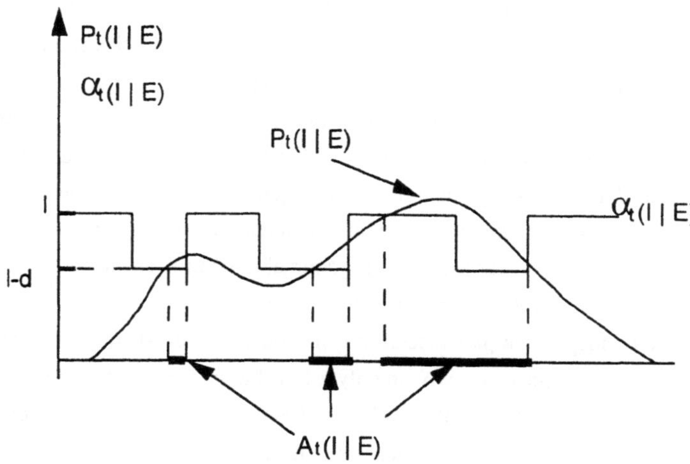

Fig. 3 Activation Probability

As shown by Fig. 3, we assume that an attention spectrum $\alpha_t(I|E)$ takes a form of saw-teeth. It has a horizontal line reflecting the overall level (l) of attention determined by the level of one's charge structure (which is a reflection of his psychological state) in terms of Yu [1990, 1995]; it has saw-teeth, whose depth (d) reflects how deeply a person thinks, whose locations (c) reflect the focus of his attention. Depending on the arrangement of items, a spectrum may have one or multiple saw-teeth. Notice that, the locations of saw-teeth in an attention spectrum point to the mental items that are to be recalled. They indicate a subject field, a topic, a methodology, a direction, or some specific mental items, to which one pays attention.

When E is obvious in the context, the attention spectrum will be simply written as α_t or $\alpha_t(l,c,d)$. When an item I is emphasized, the attention spectrum will be written as $\alpha_t(I)$ or $\alpha_t(I|l,c,d)$. Generally, an attention spectrum can be written as

$$\alpha_t(I,E|l_t,c_t,d_t)$$

which specifies initial stimulus E, recalling item I, and the settings of parameters l, c, and d over time.

The recall of a mental item depends on both the activation probability distribution and the attention spectrum. The activation probability distribution has hardware characteristics and is determined by the physical property of a brain. Of course, one can change associations between mental items and so an activation probability distribution by learning. However, during recalling, this hardware characteristic can be regarded as fixed.

On the other hand, an attention spectrum has software characteristics and is adjustable and controllable. For a given activation probability distribution, one may have a different performance on problem solving and decision making by controlling his psychological state and paying attention to different focuses with different depth and width. For example, by lowering l and adjusting c and d, one may generate some great ideas that would have been submerged in the deep memory at a different setting of l, c, and d. Thus, an attention spectrum can be regarded as a programmable control on mental performance with an adjustable configuration of parameters of l, c, and d.

Are l, c, and d independent parameters? We may observe that when one thinks deeply, he can think about and focus narrowly on only a few things. On the other hand, when he thinks superficially, he can think about and focus broadly on many things. In terms of an attention spectrum, this phenomenon suggests that the width and the depth of the saw-teeth in an attention spectrum are negatively related. Without much supportive evidence, we offer the following speculation:

Hypothesis: The total area of the saw-teeth in an attention spectrum measures one's total attention energy. It is a constant at a time (but may vary from time to time). Therefore, the depth and the width of the saw-teeth are negatively proportional.

This hypothesis can be explained by assuming that attention is a kind of electric signal. The frequency spectrum of an electric signal does possess the property spelled out in the hypothesis (Usher 1984). Of course, to verify the assumption, more research needs to be done.

When activation probability distribution $P_t(I|E)$ is fixed, $A_t(E)$ becomes a functional of $\alpha_t(l,c,d)$, according to (6), which in turn is a function of three parameters *l, c,* and *d.* As we have already discussed, expanding a habitual domain is equivalent to adjusting operators from time to time or situation to situation instead of applying a preexisting rigid operator always. Since an attention spectrum is one-to-one correspondent to an operator, expanding a habitual domain is thus equivalent to adjusting an attention spectrum by changing the settings of parameters *l, c,* and *d,* and breaking the habitual way of thinking, responding, and behaving--the old, rigid attention spectrum. Section 3 lists three general principles for expanding habitual domains. Now let us use the notion of attention spectra to understand how they are working.

(1) The alternating principle basically states a method of adjusting focuses on alternative assumptions. In terms of an attention spectrum, this principle urges us to adjust parameter *c* (the location of the saw-teeth) of an attention spectrum. Alternating assumptions includes alternating views for observation, alternating beliefs about reality, alternating promises of external advice, and alternating expectations for a process. It results in the generation of new lines of thinking and the change of attention allocation. In scientific research, one tends to make rigid assumptions unconsciously. These assumptions then guide one through a fixed line of reasoning. Consequently, he always sees only one face of reality and is unable to appreciate the existence of other faces and the beauty of the whole. He may run out of fresh ideas when he walks along the same line again and again and ends up making crumbles.

(2) The projecting principle states a way of adjusting the location and the width of attention focuses. Of course, the change of the width of the saw-teeth in an attention spectrum will automatically lead to the change of the location and the depth of the saw-teeth according to the hypothesis. By taking other people's positions and thinking from their viewpoints, one tends to see more and becomes more broadly-minded. We are told that a self-centered person is usually narrowly-minded. This may imply a fact that such a person habitually does not apply the projecting principle. He does not consider other people's interests, benefits, functions, concerns, and problems. He sees everything from his own standpoints. Under some circumstances, he tends to be stubborn, near-sighted, narrowly-minded, or even selfish.

(3) The deep-and-down principle states a way of shifting downward the overall level of charge structures and consequently that of an attention spectrum. By using the deep-

and-down principle, we deepen our psychological states in order to be able to pay attention to a large scope of ideas. This principle also allows external advice to flow easily into our mind, which can change the location of the saw-teeth in an attention specturm in order to expand our mind. Note that without changing the focuses and the depth of thinking, lowering parameter l can still help one enlarge $A_t(E)$ and activate more fresh ideas and solutions from his potential domain.

The above three principles are not exclusive. There are other methods of adjusting l, c, and d and so changing an attention spectrum and alternating operators. Besides, the effectiveness of any one principle is different under different situations. In self-controlled situations such as thinking, a thinker may not know which focus could be more fruitful. It is effective to control psychological states and the depth of thinking. In non self-controlled situations such as teaching, negotiating, and persuading, appropriate external contexts or hints can be effective to control one's focus of attention and keep his thinking on a certain track. However, his psychological state and the depth of thinking may be less controllable.

5 Conclusion

We have introduced the basic concepts of habitual domains and illustrated the importance of their expansion and enrichment. To empower ourselves we could use mental focus control, allowing us to find and locate better ideas more readily. We have also described three principles for us to break undesirable habits and expand and enrich our habitual domains, and proposed a mental focus control model to heuristically justify why the proposed principles are effective.

Many research problems are open for exploration. For instance, when a problem domain (such as finding a good job, research and development, marketing a special product, managing a newly merged company, ... etc.) is fixed, how do we diagnose the problems, identify the habitual domain of the decision maker and help him to expand and enrich his habitual domain so that he could become confident and competent to make good decisions? This kind of problem can be solved by competence set analysis. Some preliminary results are reported in Yu and Zhang [1989, 1990], Li and Yu [1994], and Shi and Yu [1996]. However, much more work needs to be done, including the related computer decision support systems. Because the problems are based on the "set covering process", instead of the more traditional "numerical ordering." Many challenges and opportunities remain to be explored by bright researchers. The reader should have no problem visualizing that people who master both the concepts of the "set covering process (habitual domain studies)" and traditional numerical ordering will have a much larger domain of expertise and competence than those who limit themselves to numerical ordering.

Lao-Tzu, a great philosopher, once said that depending on their intellectual development people can be classified into three categories. Upon hearing a new theory, the top intellectuals will actively learn it and use it; the middle intellectuals will do nothing; the lower intellectuals will laugh at and ridicule it. Lao-Tzu's observation is substantiated by Barker [1988]. The phenomenon of habitual domains is a piece of "very fertile land" which needs many people to cultivate. We invite all those who could appreciate themselves and other people as priceless living systems of unlimited potentials to work together to empower and benefit ourselves and our societies.

6 References

Alinsky, S. D. (1972). *Rules for Radicals*, Vintage Books, New York.

Barker, J. A. (1988). *Discovering the Future: the Business of Paradigms*, I.L.I Press, St. Paul, MN.

Chan, S. J. and **Yu, P. L.** (1985). "Stable Habitual Domains: Existence and Implications," *Journal of Mathematical Analysis and Applications* 110, 469-482.

Churchman, C.W. (1979). *The Systems Approach and Its Enemies*, Basic Books, New York.

The Diagram Group (1982). *The Brain: A User's Manual*, Berkeley Books, New York.

Fischbach, G.D. (1992). *Mind and Brain*, Scientific American. 48-57.

Lee, Y. R., **Shi, Y.** and **Yu, P. L.** (1990). "Linear Optimal Designs and Optimal Contingency Plans," *Management Science* 36, 1106-1119.

Li, H. L. and **Yu, P. L.** (1994). "Optimal Competence Set Expansion Using Deduction Graph," *Journal of Optimization Theory and Applications* 80, 75-91.

Mackenzie, K.D. (1988). "The Process approach to Organizational Design," *Human Systems Management* 8, 31-43.

Shi, D. S. and **Yu, P. L.** (1996). "Optimal Expansion and Design of Competence Sets with Asymmetric Acquiring Costs," *Journal of Optimization Theory and Applications* 88, 643-658.

Usher, M. I. (1984). *Information Theory for Information Techniques*, Macmillan, London.

Yu, P. L. (1985). *Multiple Criteria Decision-Making: Concepts*, Techniques and Extensions, Plenum, New York, N. Y.

Yu, P. L. (1990). *Forming Winning Strategies, An Integrated Theory of Habitual Domains*, Springer-Verlag, Berlin, Heidelberg, New York, London, Paris, Tokyo.

Yu, P. L. (1991). "Habitual Domains," *Operations Research* 39, 869-876.

Yu, P. L. (1995). *Habitual Domains: Freeing Yourself from the Limits on Your Life*, Highwater Editions, Kansas City.

Yu, P. L. and **Zhang D.** (1989). "Competence Set Analysis for Effective Decision Making," *Control - Theory and Advanced Technology* 5, 523-547.

Yu, P. L. and **Zhang D.** (1990). "A Foundation for Competence Set Analysis," *Mathematical Social Sciences* 20, 251-299.

Exploration of Multi-objective Problems in Artificial Intelligence

Hirotaka Nakayama

Department of Applied Mathematics, Konan University, Kobe 658, JAPAN

Abstract: One of main themes in artificial intelligence is to simulate brains of human beings. This can not be almost realized by only a model of brain, because brains of human beings have a great number of functions and perform them in complex manners. Therefore, techniques in artificial intelligence are usually discussed for those functions separately. For example, decision making is a major function of brain. Since it seems not to be practical to automatize decision making by artifical intelligence, this function has been studied from a viewpoint of decision support by a system based on collaboration of machines (computers) and human beings. Interactive multi-objective programming techniques are good examples in this field.

On the other hand, intelligence such as inference and pattern recognition has been developed to be automatized. In particular, machine learning is gaining much popularity for pattern recognition. Usually, it utilizes some model (of brain), e.g., artificial neural networks. Then, we encounter trade-off between the richness of representation and simplicity of model as well as in usual modeling.The performance of machine learning depends on the balancing on this trade-off.

This paper shows several kinds of multi-objective problems which appear in artificial intelligence along with some trials for solving them.

Key Words: artificial intelligence, machine learning, multi-objective problems

1 Introduction

So far, artificial intelligence (AI) has aimed to automatize several kinds of intelligent activities of human beings by machines (or, computers). For example, expert systems for medical and engineering diagnosis have been developed to some extent along this line. However, it is not so easy to extract "rules" from experts and maintain or revise the rule base. In order to overcome this difficulty, many expert systems are recently used by combination with another technique of artificial intelligence, say, machine learning such as artificial neural networks.

On the other hand, for decision making, which is one of the most major intelligent activities of human beings, its automatization is very difficult: That is because decision making is originally a problem of value judgment of decision makers, and hence may depend on decision makers themselves in many cases. Moreover, the environment of decision making in many practical probelms is uncertain and dynamically varied. In addition, the value judgment of decision makers is not always consistent during the decision process. This is very natural because the ability of discrimination of human beings is limited, and moreover decision makers get additional information one after another during the process. Since it is difficult to extract the value judgment of decision makers in a complete form in advance, it has been observed that decision support systems should be "open" to the value judgment of decision makers in many practical problems. Many interactive multi-objective programming techniques have been developed along this line. Although artificial intelligence can play an important role as a part of decision support systems, it is not sufficient by itself in many decision support systems.

It should be noted, however, that although artificial intelligence is not sufficient by itself in many decision support systems, it can play an important role as "a part" of decision support systems. For example, the number of alternatives may be reduced by using some pattern classification techniques. Decision makers can realize a new alternative by using some inference engine. Like this, appropriate utilization of artificial intelligence can reduce the burden of decision makers, and even create a new alternative. In the following, we discuss pattern classification as one of artificial intelligence, and explore multi-objective problems in it.

2 Modeling in AI for Pattern Classification

Pattern classification is one of the most major themes in artificial intelligence (AI), which is widely applied to medical and engineering diagnosis, portfolio selection, pattern recognition including image/sound data processing and so on. Historically, statistical methods such as discrimination analysis are famous in this field. Statistical methods, however, require some assumption on data distribution, e.g. the normal distribution. In addition, the number of data is needed to be sufficiently large in order that they are statistic.

Another artificial intelligence technique, artificial neural network, is gaining much popularity as machine learning for pattern classification. Above all, the back propagation algorithm is applied to various kinds of pattern classification problems. It has been observed that the method shows an excellent generalization ability by "appropriate" structuring and learning. Here, by structuring we mean to decide the structure of the artificial neural network, and by learning to follow the training data.

Usually, it is difficult to decide the structure of artificial neural network such as the number of hidden layers and hidden units in advance. In modeling brains of human beings as well as many other modeling, there is a conflict between the richness of representation and the simplicity of model. Namely, although we need the freedom of model much enough to represent well the given data, a too much freedom causes the so called over-learning.

For example, linear functions, one of the simplest mathematical model, can not represent complicated nonlinear behaviors. No matter how faithfully the model fit the complicated nonlinear curve, however, it is not necessarily a "good" model. In other words, even if the model represent well the given data, it can not necessarily the unknown data. This is due to the fact that data usually have noise, and over-learning follows even noise. In order to avoid this difficulty, two major approaches are applied: One of them is to limit the freedom of artificial neural networks, and the other one is to stop the learning at some "appropriate" degree.

In artificial neural networks, the structure of network reflects the richness of representation: It is said that the fewer hidden units provide in general the more generalization ability. However, artificial neural networks with too few hidden units can not follow the training data well, namely can not learn so well. Here, in mahine learning such as artificial neural networks, we have a conflict between learning and generalization. It is important to construct artificial neural networks by making this trade-off appropriately.

To this end, several approaches, e.g., AIC(Akaike's Information Criterion; Akaike 1973), GPE(Generalized Prediction Error; Moody 1992) and NIC(Network Information Criterion; Murata *et al.* 1994) have been proposed so far. These methods suggest unified criteria taking into account these two conflicting objectives. It is of course hard to make each criterion separately in an explicit form which measures over-learning (or, over-freedom of model) and generalization. In addition, even though these two criteria can be established, as many researchers who are engaged in multiple criteria decision making (MCDM) may easily see, it is not so easy to construct a universal unified criterion (i.e., a scalarization function) making trade-offs effectively for all cases. In fact, to be effective, the stated criteria impose some statistical assumptions.

Another approach is the heuristic one. For example, CV(cross validation) test decides values of parameters under consideration by examining the performance of artificial neural networks for training sets and validation sets selected with some proportion from a given data set. This corresponds to drawing the Pareto curve on the criteria space in MCDM. Recall that the best way to see the trade-off between two criteria is to draw the Pareto curve. From this reason, CV test is applied effectively in many cases.

Furthermore, another trial to build a "good" network is to prune unnecessary units starting from a network with redundant units. Since relatively large networks can learn well, i.e., can decrease the training error well, we add a term for reducing the scale of network as a penalty. To this end, the sum of squarred (or absolute value of) connection weights is usually used. This technique is called the weight decay. On the other hand, there is a method which removes unnecesary units examinning the correlation among units. Since the final goal in machine learninig is to maximize the generalization, these methods attain the goal in an asymptotic way on the trade-off curve between the generalization and the freedom of model from the side of over-freedom.

There are also methods for constructing an appropriate structure of artificial neural networks from the side of "under-freedom". Since artificial neural networks with too few hidden units can neither follow the given data so well nor provide a good generalization. Cascaded correlation networks generate hidden units gradually by taking into account of correlation between the output of hidden units and the error of output of the network. Also, MSM (multi-surface method), which will be described below, can be regarded as the one in this category.

Finally, we mention a method which makes the discriminant hypersurface as smooth as possible in order to increase the generalization ability. As is readily seen, pattern classification is solved by finding the discriminant hypersurface. Artificial neural networks for pattern classification can be considered as a function providing the discriminant hypersurface. Roughly speaking, it can be regarded that making the discriminant hypersurface as smooth as possible corresponds to making artificial neural networks as simple as possible. It is intuitively clear that sufficiently smooth discriminant hypersurfaces are not affected by noise so severely. To this end, artificial neural networks using radial base functions apply a penalty term for the smoothness. On the other hand, pattern classification techniques using mathematical programming apply several techniques for making the discriminant surface as smooth as possible. This will be stated in the next section.

3 Multi-objective Programming for Pattern Classification

Among several machine learninig techniques for pattern classification, we focus on methods based on mathematical programming in this section. In particular, Multisurface Method (MSM), which was developed by Mangasarian (1968), is attractive because it is very simple and can provide an exact discrimination hypersuface even for highly nonlinear problems without any assumptions on the data distribution.

The method finds a piecewise linear discrimination hypersurface by solving linear programming iteratively. One of most prominent features of MSM is that 1) there is no parameters to be decided in advance, 2) relatively short computing time is required, and 3) the resulting solution is guaranteed to be the global optimal. Mangasarian *et al.* (1992) reported its good performance in application to medical diagnosis problems.

However, we observed that the method tends to produce too complicated discrimination surfaces, which cause a poor generalization ability because it is affected by noise. In order to overcome this difficulty, we proposed a multi-objective optimization approach (Nakayama *et al.* 1994). On the other hand, Benett-Mangasarian proposed a method called Robust Linear Programming

Discrimination (RLPD) whose idea is based on the goal programming. We shall make a brief survey of these methods in the following.

3.1 Multisurface Method (MSM)

Suppose that given data in a set X of n-dimensional Euclidean space belong to one of two categories $\mathcal{A}$ and $\mathcal{B}$. Let A be a matrix whose row vectors denote points of the category $\mathcal{A}$. Similarly, let B be a matrix whose row vectors denote points of the category $\mathcal{B}$. For simplicity of notation, we denote the set of points of $\mathcal{A}$ by A. The set of points of $\mathcal{B}$ is denoted by B similarly. MSM suggested by Mangasarian (1968) finds a piecewise linear discrimination surface separating two sets A and B by solving linear programming problems iteratively:

Step 1 . Solve the following linear programming problem at k-th iteration (set $k = 1$ at the beginning):

$$\text{(I)} \qquad \begin{aligned} &\text{Maximize} \quad \phi_i(A,B) = \alpha - \beta \\ &\text{subject to} \end{aligned}$$

$$\begin{aligned} A\boldsymbol{u} &\geq \alpha\mathbf{1} \\ B\boldsymbol{u} &\leq \beta\mathbf{1} \\ -\mathbf{1} &\leq \boldsymbol{u} \leq \mathbf{1} \\ \boldsymbol{p}_i^T\boldsymbol{u} &\geq \frac{1}{2}\left(\frac{1}{2} + \boldsymbol{p}_i^T\boldsymbol{p}_i\right) \end{aligned} \tag{1}$$

where $\boldsymbol{p}_i$ is given by one of

$$\begin{aligned} \boldsymbol{p}_1^T &= (\frac{1}{\sqrt{2}}, 0, \ldots, 0) \\ \boldsymbol{p}_2^T &= (-\frac{1}{\sqrt{2}}, 0, \ldots, 0) \\ &\vdots \\ \boldsymbol{p}_{2n}^T &= (0, \ldots, 0, -\frac{1}{\sqrt{2}}) \end{aligned}$$

Here, the constraints (1) is introduced to avoid a trivial solution $\boldsymbol{u} = 0,\ \alpha = 0,\ \beta = 0$. The objective function reflects the distance between two parallel hyperplanes $g(\boldsymbol{u}) := \boldsymbol{x}^T\boldsymbol{u} = \alpha$ and $g(\boldsymbol{u}) := \boldsymbol{x}^T\boldsymbol{u} = \beta$. After soving LP problem (I) for each i such that $1 \leq i \leq 2n$, we take a hyperplane which classify correctly as many given data as possible. Let the solution be $\boldsymbol{u}^*, \alpha^*, \beta^*$, and let the corresponding value of objective function be $\phi^*(A,B)$.

If $\phi^*(A,B) > 0$, then we have a complete separating hyperplane $g(\boldsymbol{u}^*) = (\alpha^* + \beta^*)/2$. Set $\tilde{A}^k = \{\boldsymbol{x} \in X |\ g(\boldsymbol{u}^*) \geq (\alpha^* + \beta^*)/2\}$ and $\tilde{B}^k = \{\boldsymbol{x} \in$

$X|\ g(\boldsymbol{u}^*) < (\alpha^* + \beta^*)/2\}$. $\tilde{A}^k$ and $\tilde{B}^k$ represent subregions of the category $\mathcal{A}$ and $\mathcal{B}$ in X, respectively, which is decided at this stage. Go to Step 3.

Otherwise, go to Step 2.

Step 2 . First, remove the points such that $\boldsymbol{x}^T\boldsymbol{u}^* > \beta^*$ from the set A. Let A^k denote the set of removed points. Take the separating hyperplne as $g(\boldsymbol{u}^*) = (\beta^* + \tilde{\beta})/2$ where $\tilde{\beta} = \text{Min}\ \{\boldsymbol{x}^T\boldsymbol{u}^*|\ \boldsymbol{x} \in A^k\}$. Let $\tilde{A}^k = \{\boldsymbol{x} \in X|\ g(\boldsymbol{u}^*) > (\beta^* + \tilde{\beta})/2\}$. The set $\tilde{A}^k$ denotes a subregion of the category $\mathcal{A}$ in X which is decided at this stage. Rewrite $X\backslash\tilde{A}^k$ by X and $A\backslash A^k$ by A.

Next, remove the points such that $\boldsymbol{x}^T\boldsymbol{u}^* < \alpha^*$ from the set B. Let B^k denote the set of removed points. Take the separating hyperplne as $g(\boldsymbol{u}^*) = (\alpha^* + \tilde{\alpha})/2$ where $\tilde{\alpha} = \text{Min}\ \{\boldsymbol{x}^T\boldsymbol{u}^*|\ \boldsymbol{x} \in B^k\}$. Let $\tilde{B}^k = \{\boldsymbol{x} \in X|\ g(\boldsymbol{u}^*) < (\alpha^* + \tilde{\alpha})/2\}$. The set $\tilde{B}^k$ denotes a subregion of the category $\mathcal{B}$ in X which is decided at this stage. Rewrite $X\backslash\tilde{B}^k$ by X and $B\backslash B^k$ by B.

Set $k = k + 1$ and go to Step 1.

Step 3. Construct a piecewise linear separating hypersurface for A and B by adopting the relevant parts of the hyperplanes obtained above.

Remark At the final p-th stage, we have the region of $\mathcal{A}$ in X as $A^1 \cup A^2 \cup \ldots \cup A^p$ and that of $\mathcal{B}$ in X as $B^1 \cup B^2 \cup \ldots \cup B^p$. Given a new point, its classification is easily made. Namely, since the new point is either one of these subregions in X, we can classify it by checking which subregion it belongs to in the order of $1, 2, \ldots, p$.

3.2 Reformulation as Multi-objective Programming

3.2.1 Multi-objective Programming Approach

It has been observed that MSM yields sometimes too complex discrimination hypersurfaces. One of reasons for this phenomenon may be that the discrimination surface is constructed by two parallel hyperplanes solving a LP problem at each iteration.

In order to overcome the difficulty of MSM, we can consider a hyperplane which has the set B on its one side and classifies correctly as many points of A as possible. This is performed by solving the following LP problem:

$$
\text{(II)} \qquad \begin{array}{l} \text{Maximize} \quad (A\boldsymbol{u} - \beta\mathbf{1}) \\ \text{subject to} \\ \qquad B\boldsymbol{u} - \beta\mathbf{1} \leq \mathbf{0} \\ \qquad -\mathbf{1} \leq \boldsymbol{u} \leq \mathbf{1} \end{array}
$$

The problem (II) is a multi-objective optimization problem. We can take a scalarization of linear sum of the objective functions. For this scalarization, we have

$$\text{(II')} \quad \begin{array}{l} \text{Maximize} \quad \sum_{i=1}^{m}(y_i^+ - y_i^-) \\ \text{subject to} \end{array}$$

$$\begin{aligned} A\boldsymbol{u} - \beta\mathbf{1} &= \boldsymbol{y}^+ - \boldsymbol{y}^- \\ B\boldsymbol{u} - \beta\mathbf{1} &\leq \mathbf{0} \\ -\mathbf{1} \leq \boldsymbol{u} &\leq \mathbf{1} \\ \boldsymbol{y}^+, \ \boldsymbol{y}^- &\geq \mathbf{0} \end{aligned}$$

More generally, considering the symetric role of A and B, we solve another LP problem in addition to (II'):

$$\text{(III)} \quad \begin{array}{l} \text{Minimize} \quad (B\boldsymbol{u} - \alpha\mathbf{1}) \\ \text{subject to} \end{array}$$

$$\begin{aligned} A\boldsymbol{u} - \alpha\mathbf{1} &\geq \mathbf{0} \\ -\mathbf{1} \leq \boldsymbol{u} &\leq \mathbf{1} \end{aligned}$$

from which

$$\text{(III')} \quad \text{Minimize} \quad \sum_{i=1}^{m}(z_i^+ - z_i^-)$$

$$\text{subject to}$$

$$\begin{aligned} A\boldsymbol{u} - \alpha\mathbf{1} &\geq \mathbf{0} \\ B\boldsymbol{u} - \alpha\mathbf{1} &= \boldsymbol{z}^+ - \boldsymbol{z}^- \\ -\mathbf{1} \leq \boldsymbol{u} &\leq \mathbf{1} \\ \boldsymbol{z}^+, \ \boldsymbol{z}^- &\geq \mathbf{0} \end{aligned}$$

In oder to avoid a tivial solution $\boldsymbol{u} = 0,\ \boldsymbol{y} = 0,\ \boldsymbol{z} = 0,\ \alpha = 0,\ \beta = 0$, we add the constraint (1) to each LP problem above.

At the beginning, set k=1. Let A^k be the set of points of A such that $\boldsymbol{x}^T\boldsymbol{u}^* > \beta^*$ for the solution to (II'), and let n_A denote the number of elements of A^k. Similarly, let B^k be the set of points of B such that $\boldsymbol{x}^T\boldsymbol{u}^* < \alpha^*$ for the solution to (III'), and let n_B denote the number of elements of B^k.

If $n_A \geq n_B$, then we take a separating hyperplane as $g(\boldsymbol{u}^*) = (\beta^* + \tilde{\beta})/2$ where β^* is the solution to (II') and $\tilde{\beta}$=Min$\{\boldsymbol{x}^T\boldsymbol{u}^* |\ \boldsymbol{x} \in A^k\}$. Let $\tilde{A}^k = \{\boldsymbol{x} \in X|\ g(\boldsymbol{u}^*) > (\beta^* + \tilde{\beta})/2\}$. The set $\tilde{A}^k$ denotes a subregion of the category $\mathcal{A}$ in X which is decided at this stage. Rewrite $X\backslash\tilde{A}^k$ by X and $A\backslash A^k$ by A.

If $n_A < n_B$, then we take a separating hyperplane as $g(\boldsymbol{u}^*) = (\alpha^* + \tilde{\alpha})/2$ where α^* is the solution to (III') and $\tilde{\alpha}$=Max$\{\boldsymbol{x}^T\boldsymbol{u}^* |\ \boldsymbol{x} \in B^k\}$. Let $\tilde{B}^k = \{\boldsymbol{x} \in X|\ g(\boldsymbol{u}^*) < (\alpha^* + \tilde{\alpha})/2\}$. The set $\tilde{B}^k$ denotes a subregion of the category $\mathcal{B}$ in X which is decided at this stage. Rewrite $X\backslash\tilde{B}^k$ by X and $B\backslash B^k$ by B.

Setting $k = k$+1, solve the problem (II') and (III') until we have $A\backslash\tilde{A}^k = \emptyset$ and $B\backslash\tilde{B}^k = \emptyset$.

Remark At the final p-th stage, we have the region of $\mathcal{A}$ in X as

$A^1 \cup A^2 \cup \ldots \cup A^p$ and that of $\mathcal{B}$ in X as $B^1 \cup B^2 \cup \ldots \cup B^p$. Given a new point, its classification is easily made. Namely, since the new point is either one of these subregion in X, we can classify it by checking which subregion it belongs to in the order of $1, 2, \ldots, p$.

3.2.2 Goal Programming Approach

Unlike the method by multi-objective optimization above, we can also consider another method obtaining a hyperplane which has one of the sets on its one side and minimizes the sum of distances between misclassified data of another set and the hyperplane at each iteration. This is formulated as a goal programming, and was originally given by Benett-Mangasarian (1992), who named it the robust linear programming discrimination (RLPD). The algorithm is summarized as follows:

In order that the separating hyperplane minimizes the sum of distances between misclassified data in the class $\mathcal{A}$ and the hyperplane, we can solve the linear programming problem

$$\text{(IV)} \quad \begin{array}{rcl} \multicolumn{3}{l}{\text{Minimize} \quad \sum_{j=1}^{k} y_j} \\ \multicolumn{3}{l}{\text{subject to}} \\ -A\boldsymbol{u} + \beta\mathbf{1} & \leq & \boldsymbol{y} \\ B\boldsymbol{u} - \beta\mathbf{1} & \leq & \mathbf{0} \\ -\mathbf{1} \leq \boldsymbol{u} & \leq & \mathbf{1} \\ \boldsymbol{y} & \geq & \mathbf{0} \end{array}$$

Similarly, for the class $\mathcal{B}$, we have

$$\text{(V)} \quad \begin{array}{rcl} \multicolumn{3}{l}{\text{Minimize} \quad \sum_{i=1}^{m} z_i} \\ \multicolumn{3}{l}{\text{subject to}} \\ A\boldsymbol{u} - \alpha\mathbf{1} & \geq & \mathbf{0} \\ B\boldsymbol{u} - \alpha\mathbf{1} & \leq & \boldsymbol{z} \\ -\mathbf{1} \leq \boldsymbol{u} & \leq & \mathbf{1} \\ \boldsymbol{z} & \geq & \mathbf{0} \end{array}$$

In order to avoid a trivial solution $\boldsymbol{u} = 0,\ \boldsymbol{y} = 0,\ \boldsymbol{z} = 0,\ \alpha = 0,\ \beta = 0$, we add the constraint (1) to each LP problem above. We can make the same procedure as the multi-objective optimization stated above taking (IV) in place of (II') and (V) in place of (III').

3.2.3 Classification of New Data

A point $\boldsymbol{x} \in R^n$ is classified into $\mathcal{A}$ or $\mathcal{B}$ as follows: The judgment is made by hyperplanes obtained as solutions to corresponding LP problems. Suppose that the given new point is not classified definitely for the first k-1 hyperplanes, and that it is classified definitely by the k-th hyperplane (i.e., it is in the subset removed by the k-th hyperplane).

- If the k-th hyperplane is obtained by solving (II') or (IV), then the point is judged to be in $\mathcal{A}$.
- If the k-th hyperplane is obtained by solving (III') or (V), then the point is judged to be in $\mathcal{B}$.

3.2.4 Examples

In the above, we reformulated MSM as multi-objective programming problems. There, a simple linear sum of objective functions are applied for scalarization. However, other scalarization methods well known in MCDM are of course possible to apply.

The performance of stated methods above can be seen in Figs. 1-3. The black part in figures shows the region of cross points. It is readily seen that MSM using multi-objective programming provides simpler discrimination functions than the original MSM. It has been observed that MSM using goal programming also shows a similar behaivor (Nakayama-Kagaku 1996). However, they still provide sometimes too complicated discrimination hypersurfaces. This is because that they try to find an exact discrimination surface which classifies all data correctly.

In order to avoid the over-learning, we suggested methods which do not require the perfect classification for given data, but allow unclassified data (Nakayama-Kagaku 1996). This idea can be performed by applying fuzzy mathematical programming techniques which produce "gray zones". The effectiveness of the modified methods can be proved by several applications.

In this approach, however, we observed that it is important how to decide the degree of fuzziness, i.e., the membership function of fuzzified constraints. As was already shown in Nakayama-Tanino (1994) and Nakayama (1996), there is a close relationship between the aspiration level approach and fuzzy mathematical programming. Taking this into account, it is readily seen that the aspiration level approach can be applied for deciding the degree of fuzziness. This will be an interesting research subject in the future.

4 Concluding Remarks

In this paper, we discussed multi-objective problems in artificial intelligence, in particular, in machine learning. So far, main topics in MCDM have been methods for supporting decision makers. However, the burden of decision makers can be decreased by utilizing artificial intelligence techniques. In addition, artificial intelligence itself has multi-objective problems in it as stated in this paper. We can apply several ideas and techniques which have been developed in MCDM to these new fields. Further researches for intelligent decision support systems are expected in the future.

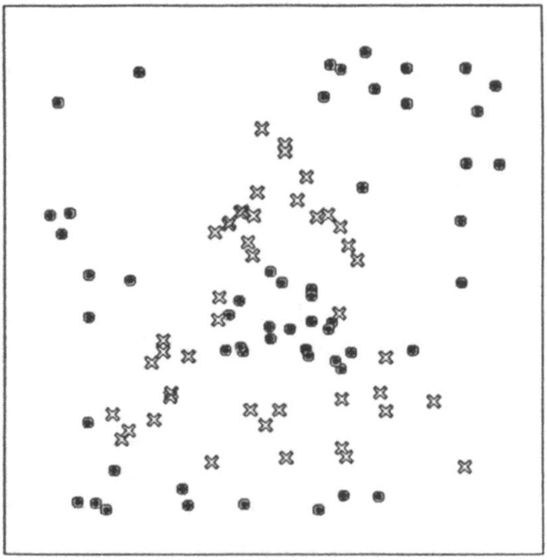

Fig. 1. Data with Two Categories

Fig. 2. MSM

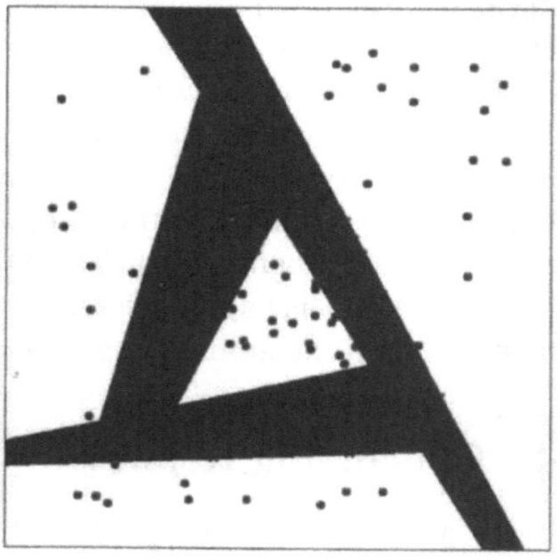

Fig. 3. MSM using MOLP

References

[1] Akaike, H. (1973), Information Theory and an Extension of the Maximum Likelihood Principle, in B.N. Petrov and F. Csaki (eds.) *2nd International Symposium on Information Theory*, pp.267-281, Tsahkadsov/Armenia

[2] Benett, K. P. and O. L. Mangasarian (1992), Robust Linear Programming Discrimination of Two Linearly Inseparable Sets, *Optimization Methods and Software*, Vol. 1, pp. 23-34.

[3] Mangasarian, O. L. (1968), Multisurface Method of Pattern Separation, *IEEE Trans. on Information Theory* Vol.IT-14, No. 6, pp. 801-807.

[4] Moody, J. E. (1992), The Effective Number of Parameters: An Analysis of Generalization and Regularization in Nonlinear Learning Systems, in J.E. Moody, S.J. Hanson and R.P. Lippmann (eds.) *Advances in Neural Information Processing Systems*, Vol.4, pp.847-854, San Mateo/USA, Morgan Kaufmann, 1992

[5] Murata, N., S. Yoshizawa and S. Amari (1996), Network Information Criterion-Determining the Number of Overtraining and Cross-Validation, *IEEE Trans*NN, (to appear)

[6] Nakayama, H. (1996), Aspiration Level Approach to Interactive Multi-objective Programming and its Applications, in P.M. Pardalos *et al.* eds., *Advances in Multicriteria Analysis*, pp. 147-173, Kluwer Academic Publishers

[7] Nakayama, H. and T. Tanino (1994), *Theory and Applications of Multi-objective Programming*, Japanese Society of Instrument and Control Engineering, (in Japanese)

[8] Nakayama, H. and K. Nasu (1994), A Pattern Classifier using Linear Programming and Multi-objective Linear Programming, presented at XI-th International Conf. on Multiple Criteria Decision Making, Coimbra/Portugal

[9] Nakayama, H. and N. Kagaku (1996), Pattern Classification by Linear Goal Programming and its Applications, presented at II-nd International Conf. on Multi-objective Programming and Goal Programming, Malaga/Spain

[10] Nakayama, H., N. Kagaku and M. Yoshida (1996), Learning for Pattern Classification, Proc. of IV-th International Conf. of EUFIT, Aachen/Germany

PART 3:

APPLICATIONS

DESIGN DECOMPOSITION IN QUALITY FUNCTION DEPLOYMENT

Kwang-Jae Kim[1], Herbert Moskowitz[2], and Jong-Seok Shin[1]

[1] Department of Industrial and Manufacturing Engineering, The Pennsylvania State University, University Park, PA 16802, USA
[2] Krannert Graduate School of Management, Purdue University, West Lafayette, IN 47907, USA

Abstract. Quality function deployment (QFD) is a cross-functional planning tool which ensures that the voice of the customer is systematically deployed throughout the product planning and design stages. One of the impediments to the adoption of QFD as a product design aid has been the large size of the design matrix, called a house of quality (HOQ) chart. This paper presents a formal approach to reducing the size of a HOQ chart using the concept of design decomposition combined with multiattribute value theory. The proposed approach aims to decompose a HOQ chart into smaller subproblems which can be solved efficiently and independently. The smaller size of the subproblems would make the QFD process simpler, easier, and cheaper. The decomposition algorithm proposed in this paper can also be applied to forming machine cells and part families in flexible manufacturing systems when the design incidence matrix is nonbinary with no diagonal structure and the column entities are correlated.

Keywords. Quality Function Deployment (QFD), House of Quality (HOQ), Decomposition, Multiattribute Value Theory

1 Introduction

Today, many companies are facing rapid changes stimulated by technological innovations and changing customer demands. These companies realize that the effort to develop new products faster that customers want and continue to purchase is crucial for their survival. Quality Function Deployment (QFD) provides a specific approach for translating the "voice of the customer" through the various stages of product planning, engineering, and manufacturing into a final product. The overall objective of QFD is to reduce the length of the product development cycle, while simultaneously improving product quality and delivering the product at a lower cost; a broader objective of QFD is to increase market share by gaining competitive advantage [Sullivan 1986].

QFD was originally developed and implemented in Japan at the Kobe Shipyards of Mitsubishi Heavy Industries, Ltd. in 1972. During the 1970's, Toyota and its suppliers developed QFD further in order to address design problems associated with automobile manufacturing (body rust). During the 1980's many U.S.-based companies began employing QFD. It is believed that there are now over 100 companies using QFD in the U.S. [Griffin and Hauser 1992], including the Budd Corporation [Morrell 1987], the Kelsey Hayes Corporation [Gipprich 1987], Motorola [Bosserman 1992], Digital Equipment Corporation, Hewlett Packard, Xerox, AT&T, ITT, NASA, Eastman Kodak, Goodyear, Proctor and Gamble, Polaroid, NCR, Ford, General Motors [Griffin 1991; Shipley 1992], and United States housing industry [Armacost, Componation, Mullens, and Swart 1994]. Many successes across a broad range of industries have been reported at the USA QFD symposia, which have been held annually since 1989 [ASI 1989].

The basic concept of QFD is to translate the desires of the customer (potential purchaser of the product) into product design or engineering characteristics, and subsequently into parts' characteristics, process plans, and production requirements associated with its manufacture. Ideally, each translation uses a chart, called "houses of quality (HOQ)," which relates the variables associated with one "design phase" (e.g., customer attributes) to the variables associated with the subsequent "design phase" (e.g., engineering characteristics). The structure of an HOQ depends on the objective, stage, and scope of the QFD project. However, there are a set of standard components of an HOQ, including customer attributes (CAs, "what to do") and their relative importance rating, engineering characteristics (ECs, "how to do it"), relationship matrix between CAs and ECs, correlation matrix among ECs, and marketing and technical benchmarking data (customer and technical competitive analysis). The schematic of an HOQ is shown in Figure 1.

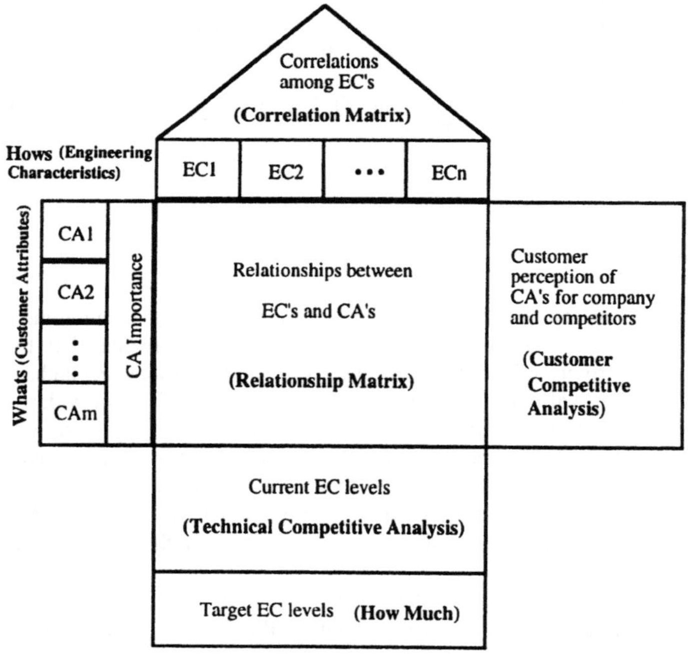

Figure 1. Schematic of a House of Quality Chart

Notwithstanding the rapid growth of the QFD literature, QFD has had limited success as a product design aid. One of the impediments to the adoption of QFD has been the difficulty in interpreting and incorporating the complicated interactions between CAs and ECs (relationship matrix in an HOQ) and correlation among the ECs (correlation matrix in the roof of an HOQ) in design analysis phase. A design problem for a simple product can have a large number of CAs and ECs. The complexity of the QFD process increases with the size of the HOQ chart, i.e., the number of CAs and ECs included in the HOQ. If an HOQ contains 20 CAs and 30 ECs (which is not an unreasonable size), there are more than 1,000 relationships to be analyzed if every cell was to be addressed. This implies the need for a huge amount of time, effort, and cognitive burden devoted to filling out the HOQ, as well as the great amount of energy involved in the analysis.

The HOQ in the QFD study of a relatively simple product (an indoor connector for the cable television market), performed by Raychem Corporation, had 28 CAs and 52 ECs. The QFD team reported that the HOQ was too large to deal with, and they cautioned future teams against creating large HOQs [Liner 1994]. The QFD teams of the U.S. Navy [DeVries and Barton 1995] and Siemens Industrial Automation [Hunter and Landingham 1994], a manufacturer of industrial control equipment, had a similar problem. The Siemens' HOQ chart contained 40 CAs and 103 ECs. It was practically infeasible to analyze all the relationships involved within the given time and resources. In order to reduce the HOQ chart to a manageable size, the QFD team deleted 21 CAs and 63 ECs that were deemed less significant, which might be a risky approach if not done very carefully. Considering the "hows" (e.g., ECs in phase I) of a design phase become the "whats" of the next phase and are expanded into a more detailed level, the size of an HOQ can have an even bigger impact in the downstream phases.

One way to reduce the complexity of a large scale design problem is to invoke the concept of decomposition. As a result of decomposition, the size of a design unit becomes smaller and thus the complexity of the design process is reduced. The computational time usually increases at a higher rate, e.g., exponentially, as a function of problem size [Kusiak and Wang 1993]. Therefore solving a number of smaller subproblems generated by decomposing a larger problem would be computationally more efficient. The challenge is to decompose the whole design problem into subproblems of acceptable size so that the subproblems can be easily solved independently, and solutions of the subproblems can collectively generate the overall design.

Clustering techniques have been used in forming part families and machine groups for cellular manufacturing. More recently, the use of decomposition in a concurrent engineering design environment has been examined by many authors. However, previous research results may not be directly applied to decomposing an HOQ in QFD. There exists a common limitation in the sense that most decomposition techniques were designed for binary incidence matrices, which represent the relationship between rows (e.g., tasks, machines) and columns (e.g., design parameters, components) in a binary fashion. The binary scheme is not proper to represent an HOQ in general. Also there has been no method developed for the case where column entities are correlated, which is the case in an HOQ.

The purpose of this paper is to develop a formal approach to reducing the size of an HOQ chart using the concept of design decomposition combined with multiattribute value theory. The proposed approach aims to decompose an HOQ into smaller subproblems which can be solved efficiently and independently. The correlation among the ECs as well as the strength of the relationship between CAs and ECs are considered in the decomposition process. The smaller size of the subproblems would make the QFD process simpler, easier, and cheaper.

Section 2 briefly reviews the related literature in the areas of group technology and flexible manufacturing systems. In Section 3, the proposed approach to decomposing an HOQ is presented. A numerical example is provided in Section 4 to illustrate the proposed approach. Finally, conclusions and future research are discussed in Section 5.

2 Review of Related Literature

There are many algorithms for solving the problem of grouping machines and components in group technology and flexible manufacturing systems. The algorithms can be classified with respect to the characteristics of the design incidence matrices that they were developed for: (1) binary or nonbinary matrices, and (2) diagonal or nondiagonal structure matrices.

Binary and Nonbinary Matrices

A matrix is called a binary matrix if each element takes a value of either 1 or 0, indicating there is or is not a relationship between the corresponding row and column, respectively. In the machine-component matrix for forming part families and machine cells in group technology, "1" in the cell (i,j) means that machine i is able to manufacture component j.

A matrix is called a nonbinary matrix if the elements are allowed to have a range of values, instead of just 1's or 0's. In such a case, the value of the cell (i,j) indicates the strength of the association between the entities in row i and column j. For example, in the machine-component matrix, the entry in (i,j) would indicate the degree of appropriateness of using machine i in manufacturing component j, with a higher value representing a better match. Therefore the values in column j can express the alternative process routings or the priority of using the machines to manufacture component j.

Diagonal and Nondiagonal Matrices

An incidence matrix is called a diagonal structure matrix (or diagonal matrix) if its rows and columns can be arranged in such a way that the matrix separates into mutually exclusive submatrices, with no non-zero entries outside the submatrices. Figure 2-(a) shows an example of a diagonal matrix which consists of three submatrices. Note that all the entries in a submatrix (shaded region) do not have to be non-zero. However, all the entries outside the submatrices must be zero. An incidence matrix which does not have the diagonal structure property is called a nondiagonal matrix, an example of which is shown in Figure 2-(b).

Existing Algorithms

Most of the grouping algorithms developed so far focus on decomposing binary matrices. Various approaches have been proposed to decompose an incidence matrix which is binary and diagonal, such as single linkage cluster analysis [McAuley 1972], bond energy algorithm [McCormick, Schweitzer and White 1972], rank order clustering [King 1980], and cluster identification algorithm [Iri 1968; Kusiak and Chow 1987], to name a few.

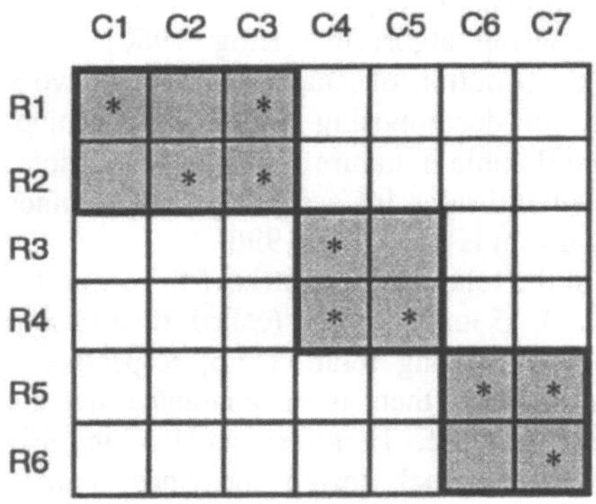

(a) Diagonal Matrix

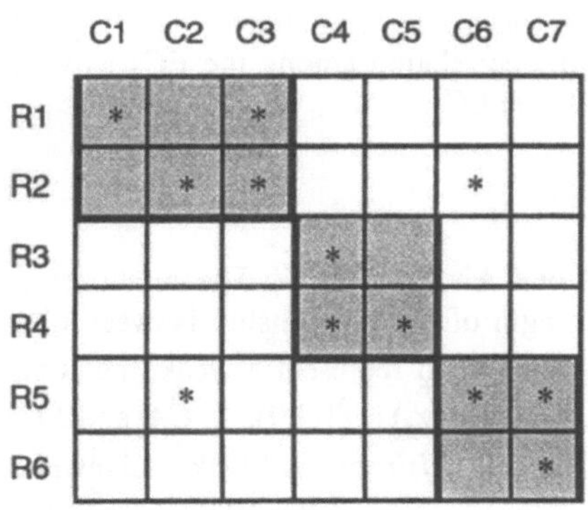

(b) Nondiagonal Matrix

Note: * represents a non-zero entry.

Figure 2. Examples of Diagonal and Nondiagonal Incidence Matrices

The study on decomposing a binary and nondiagonal design matrix has received relatively less attention, mainly due to the complexity associated with the nonseparable property of the matrix. Kumar, Kusiak and Vannelli [1986] developed an analytic (Algorithm I) and a heuristic algorithm (Algorithm II) using an optimal k-decomposition model in network theory. Kusiak and Wang [1993] used a branch-and-bound algorithm to efficiently identify the overlapping rows and columns, which cause the matrix to be nondiagonal. After removing the overlapping rows and columns, the matrix may be decomposed as if it were a diagonal structure matrix.

The decomposition of a nonbinary matrix has scarcely been discussed in the literature. The only grouping methodology to date which formally deals with a nonbinary matrix can be found in Zhang and Wang [1992]. The authors applied the single linkage clustering algorithm [McAuley 1972] for a nonbinary and diagonal matrix, and the rank order clustering algorithm [King 1980] for a nonbinary and nondiagonal matrix. The contribution of Zhang and Wang's work is that the authors provide an approach for decomposing nonbinary machine-component matrices, which are much more flexible in nature than binary matrices. However, the authors did not provide systematic rules for assigning the machines and components into groups when the algorithm is done [Wu 1996].

In typical QFD applications a cell (i,j) in the relationship matrix (i-th row and j-th column) of an HOQ is assigned 1, 3, 5 or 1, 3, 9 (called relationship coefficients) to represent a weak, medium, and strong relationship, respectively, between the i-th CA and the j-th EC. In addition, there is no guarantee that the relationship matrix would have the diagonal structure. Therefore, an HOQ in QFD can be considered to be a nonbinary nondiagonal design incidence matrix. Moreover, the ECs (columns) in an HOQ are correlated with one another, as manifested by the entries in the roof part. This is a unique feature of the HOQ, which has not been considered in the machine-component matrix literature.

The next section presents a new approach which can be applied to decomposing an HOQ into subproblems. The HOQ is viewed as a nonbinary nondiagonal matrix. The proposed approach considers the correlation among the ECs as well as the relationship coefficients between CAs and ECs.

3 Model Formulation

Consider an HOQ chart which consists of m CAs and n ECs. The relationship coefficient, denoted as f_{ij}, represents the strength of the relationship between CA_i and EC_j. f_{ij} is assigned α_1, α_2, α_3 (α_i>0, i=1,2,3) to represent a weak, medium, and strong relationship, respectively; e.g., (α_1, α_2, α_3) = (1,3,9). If CA_i and EC_j have no relationship, f_{ij} is set at zero, and the cell (i,j) remains blank. Likewise, the correlation coefficient, denoted as $g_{j_1j_2}$, in the roof part of the HOQ is assigned β_1, β_2, β_3 (β_j>0, j=1,2,3) to represent a weak, medium, and strong correlation, respectively, between EC_{j_1} and EC_{j_2}. (The coefficients may have signs (plus or minus) depending on the nature of the relationship (or correlation), the signs are disregarded in our model and just the strengths are considered.) An illustrative HOQ consisting of 5 CAs and 6 ECs, with the relationship and correlation coefficients recorded, is given in Figure 3-(a).

Suppose it is desired to decompose the HOQ into K subproblems, or sub-HOQs. In the k-th sub-HOQ, r_k is defined as the number of rows, and c_k as the number of columns, k=1, ..., K. (Hereafter, a sub-HOQ will be referred to as a group, and denoted as G_k, k=1, ..., K.) It is assumed that the value of K as well as the size of the groups (r_k and c_k) are determined by the user.

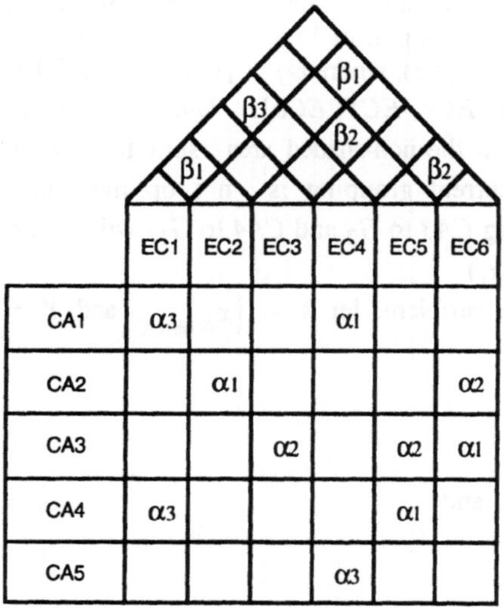

(a) An Illustrative HOQ with 5 CAs and 6 ECs

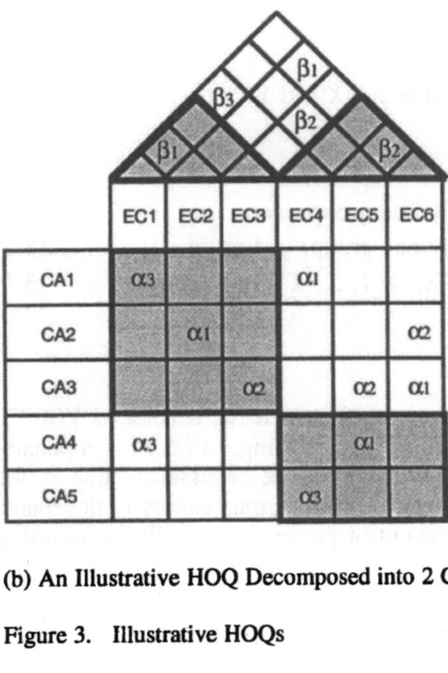

(b) An Illustrative HOQ Decomposed into 2 Groups of Sizes (3, 3) and (2, 3)

Figure 3. Illustrative HOQs

The problem is to decompose the given HOQ into K groups of sizes (r_1, c_1), (r_2, c_2), ..., (r_k, c_k) in order to minimize the sum of the coefficient entries (i.e., f_{ij} and $g_{j_1j_2}$) which do not reside in the resulting groups among all possible groupings. Such entries are referred to as "out-of-group" entries. The sum of the out-of-group entries depends on the assignment of CAs and ECs to groups, and is minimized at the optimal grouping. For example, in Figure 3-(b), the HOQ is decomposed into 2 groups of sizes (3,3) and (2,3), where $G_1 = \{CA1, CA2, CA3, EC1, EC2, EC3\}$ and $G_2 = \{CA4, CA5, EC4, EC5, EC6\}$. Under the current decomposition, the out-of-group entries in the non-shaded area sum to $(2\alpha_1 + 2\alpha_2 + \alpha_3 + \beta_1 + \beta_2 + \beta_3)$. The current grouping is not optimal since interchanging $CA3$ and $CA4$ (i.e., moving $CA3$ to G_2 and $CA4$ to G_1) will reduce the sum by α_3 (= $+\alpha_2 - \alpha_2 - \alpha_1 - \alpha_3 + \alpha_1$).

In order to mathematically model the problem, let $X = \left[x_{ik}\right]_{m\times K}$ and $Y = \left[y_{jk}\right]_{n\times K}$, where

$$x_{ik} = \begin{cases} 0 & \text{if } CA_i \notin G_k, \\ 1 & \text{if } CA_i \in G_k, \text{ and} \end{cases}$$

$$y_{jk} = \begin{cases} 0 & \text{if } EC_j \notin G_k, \\ 1 & \text{if } EC_j \in G_k. \end{cases}$$

Then the decomposition problem can be formulated as an optimization problem as follows:
Find the grouping X and Y which

Minimizes Overall Dissatisfaction for X and Y (3.1)

subject to

CA_i is included in one and only one group, i=1, ..., m, (3.2)
EC_j is included in one and only one group, j=1, ..., n, (3.3)
G_k contains r_k CAs and c_k ECs, k=1, ..., K. (3.4)

Objective Function

The objective is to minimize the overall dissatisfaction level, denoted as $V(X,Y)$, due to the out-of-group entries as a result of the grouping. $V(X,Y)$ is a penalty function which associates a real number with the degree of dissatisfaction in the grouping result. In essence, the existence of the out-of-group entries is the source of dissatisfaction. If the penalty due to the out-of-group entries in the relationship matrix and the penalty in the correlation matrix have the additive independence property, $V(X,Y)$ can be expressed as [Keeney and Raiffa 1976]:

$$V(X,Y) = w_R V_R(X,Y) + w_C V_C(X,Y), \tag{3.5}$$

where $V_R(X,Y)$ and $V_C(X,Y)$ denote the dissatisfaction level associated with the relationship and the correlation matrix, respectively. They are scaled in such a way that $0 \le V_R(X,Y) \le 1$ and $0 \le V_C(X,Y) \le 1$, with 0 being the best and 1 the worst. w_R and w_C represent the relative weight assigned to each criterion for an optimal grouping, namely, minimizing out-of-group entries in the relationship and the correlation matrix, respectively. They are scaled in such a way that $0 \le w_R \le 1$, $0 \le w_C \le 1$, and $w_R + w_C = 1$.

Assuming that the sum of the out-of-group entries in the relationship matrix affects $V_R(X,Y)$ in a linear fashion, $V_R(X,Y)$ can be expressed as:

$$V_R(X,Y) = \frac{\sum_{i=1}^{m}\sum_{j=1}^{n} f_{ij}(1 - \sum_{k=1}^{K} x_{ik} y_{jk}) - R_{min}}{R_{max} - R_{min}}, \tag{3.6}$$

where R_{min} (R_{max}) represents the lower (upper) bound on the sum of the out-of-group entries in the relationship matrix. Likewise,

$$V_C(X,Y) = \frac{\sum_{j_1=1}^{n-1}\sum_{j_2=j_1+1}^{n} g_{j_1 j_2}(1 - \sum_{k=1}^{K} y_{j_1 k} y_{j_2 k}) - C_{min}}{C_{max} - C_{min}}, \tag{3.7}$$

where C_{min} (C_{max}) represents the lower (upper) bound on the sum of the out-of-group entries in the correlation matrix. (The derivation of R_{min}, R_{max}, C_{min}, C_{max} is given in the Appendix.)

<u>Constraints</u>

As mentioned in (3.2)-(3.3), each CA and EC should be included in one and only one group throughout the grouping process. This condition can be represented as

$$\sum_{k=1}^{K} x_{ik} = 1, \quad i=1, \ldots, m, \tag{3.8}$$

$$\sum_{k=1}^{K} y_{jk} = 1, \quad j=1, \ldots, n. \tag{3.9}$$

In addition, the group size has to be maintained as specified by the user:

$$\sum_{i=1}^{m} x_{ik} = r_k, \quad k=1, \ldots, K, \tag{3.10}$$

$$\sum_{j=1}^{n} y_{jk} = c_k, \quad k=1, ..., K. \tag{3.11}$$

<u>Model Formulation</u>

Combining the objective function and the constraints described above, the model for optimal decomposition of an HOQ can now be formulated as follows:

$$\text{Minimize } V(X,Y) = w_R V_R(X,Y) + w_C V_C(X,Y) \tag{3.12}$$

subject to

$$(3.8) - (3.11) \tag{3.13}$$
$$x_{ik} = 0 \text{ or } 1, \quad i=1, ..., m; k=1, ..., K, \tag{3.14}$$
$$y_{jk} = 0 \text{ or } 1, \quad j=1, ..., n; k=1, ..., K. \tag{3.15}$$

$V_R(X,Y)$ and $V_C(X,Y)$ in (3.12) are defined in (3.6) and (3.7), respectively. Additional constraints may be introduced to the above formulation in order to enforce the user's preemptive requirements on grouping. For example, if it were required that EC_{j_1} and EC_{j_2} be always included in the same group, then the following equation can be added as a constraint:

$$y_{j_1 k} - y_{j_2 k} = 0, \quad k=1, ..., K.$$

The formulation given in (3.12) - (3.15) is a 0-1 quadratic programming problem with linear constraints. Once the optimal solution, $X^* = \left[x_{ik}^*\right]_{m \times K}$ and $Y^* = \left[y_{jk}^*\right]_{n \times K}$, is found, the HOQ can be decomposed into K sub-HOQs by assigning individual CAs and ECs to proper groups as dictated by X^* and Y^*. The k-th group G_k is composed of CA_i's and EC_j's such that $x_{ik}=1$ and $y_{jk}=1$, $k=1$, ..., K.

4 Illustrative Example

To illustrate the proposed approach, we consider the HOQ developed for the ERAM (Engine Room Arrangement Modeling) project with some embellishments. The ERAM project is part of the mid-term sealift ship technology development program of the Navy. The objective of the program is to reduce ship construction time by 40% and initial acquisition cost by 15-21%. QFD has been chosen as an implementation tool in order to achieve the objective [DeVries and Barton 1995]. A part of the ERAM HOQ is shown in Figure 4, which consists of 10 CAs and 18 ECs.

When the HOQ was first developed and used, the design team experienced much difficulty in reaching a consensus in the analysis phase, mainly due to its complexity. In order to alleviate the difficulty, the design team decided to decompose the HOQ. It was considered that groups (i.e., sub-HOQs) with 6 ECs would be reasonably easy to deal with. Thus the number of groups to be formed (K) was set at 3. Since there was no reason for one group to have significantly more CAs than the other groups, the group sizes were set at $(r_k, c_k) = (3, 6)$, $(3, 6)$, and $(4, 6)$, k=1,2,3. R_{min} and R_{max}, which are used to scale $V_R(X, Y)$ in (3.6), were computed to be 0 and 230. C_{min} and C_{max}, which are used to scale $V_C(X, Y)$ in (3.7), were computed to be 0 and 164. (A weak, medium, strong relationship (correlation) is assigned 1, 3, 9, respectively, for the computational purpose. See Appendix for the computational procedure.)

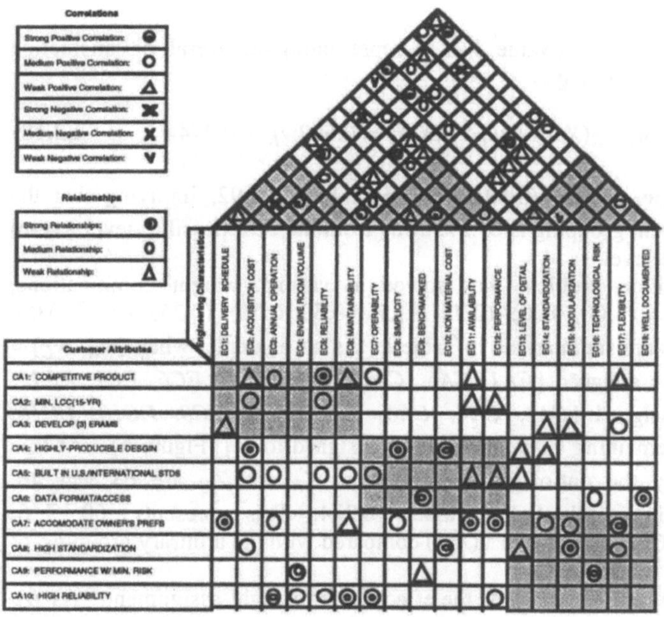

Figure 4. The ERAM HOQ

The original HOQ, given in Figure 4, has 54 non-zero entries in the relationship matrix and 40 non-zero entries in the correlation matrix. The non-zero entries, occupying 30% (=54/180) of the relationship matrix cells and 26% (=40/153) of the correlation matrix cells, are scattered all over the matrices in a "random" manner. For the sake of illustration, three groups are formed in an arbitrary way; the first 3 CAs and 6 ECs constitute G_1, the next 3 CAs and 6 ECs constitute G_2, and the remaining CAs and ECs constitute G_3 (as indicated by the thick lines in Figure 4). Then the sum of the out-of-group entries, denoted as Z_R

and Z_C, equals to 140 and 127 for the relationship and the correlation matrix, respectively. The level of dissatisfaction of this arbitrary grouping is computed as

$$V_R(X,Y) = \frac{\sum_{i=1}^{10}\sum_{j=1}^{18} f_{ij}(1-\sum_{k=1}^{3} x_{ik}y_{jk}) - R_{min}}{R_{max} - R_{min}} = \frac{140-0}{230-0} = 0.609, \qquad (3.16)$$

$$V_C(X,Y) = \frac{\sum_{j_1=1}^{17}\sum_{j_2=j_1+1}^{18} g_{j_1 j_2}(1-\sum_{k=1}^{3} y_{j_1 k}y_{j_2 k}) - C_{min}}{C_{max} - C_{min}} = \frac{127-0}{164-0} = 0.774. \quad (3.17)$$

Thus, the objective function value, $V(X,Y)$, representing the overall dissatisfaction level at the current grouping is

$$V(X,Y) = w_R V_R(X,Y) + w_C V_C(X,Y) = 0.609 w_R + 0.774 w_C. \qquad (3.18)$$

For example, at (w_R, w_C) = (0.50, 0.50), $V(X,Y)$ is 0.692, implying that the quality of the current grouping is 69.2% away from the best possible level toward the worst possible level.

The decomposition problem was solved at various weight combinations, namely, (w_R, w_C) = (1.00, 0.00), (0.75, 0.25), (0.50, 0.50), (0.25, 0.75), (0.00, 1.00). The results are summarized in Table 1. As an example, when (w_R, w_C) = (0.50, 0.50), G_1 is assigned {*CA1, CA5, CA10, EC3, EC5, EC6, EC7, EC11, EC12*}, G_2 is assigned {*CA2, CA4, CA6, EC2, EC8, EC9, EC10, EC16, EC18*}, and the remaining CAs and ECs are assigned to G_3 (Figure 5). At this optimal grouping, the out-of-group entry sums Z_R and Z_C are 63 and 45, respectively, resulting in the $V(X,Y)$ value of 0.274. This represents a 60.4% (= (0.692-0.274) / 0.692) decrease in $V(X,Y)$ compared with the arbitrary grouping of the original HOQ in Figure 4.

The results given in Table 1 show the effect of the weight assignment. As w_R increases, Z_R decreases and thus V_R also decreases. Similarly, as w_C increases, Z_C and V_C decrease (in fact, do not increase). It is notable that the optimal solutions from (w_R, w_C) = (0.50, 0.50) and (0.25, 0.75) are identical. This solution could have been obtained as the optimal solution for the (w_R, w_C) = (0.00, 1.00) case since the Z_C value of 45 is the minimum attainable value in this example. It is evident that there exist many alternative optimal groupings for (w_R, w_C) = (0.00, 1.00) because, starting from the current optimal grouping, swapping CAs in different groups will not affect the Z_C value.

The fuzzy rank order clustering (ROC) algorithm [Zhang and Wang 1992], which is the only grouping methodology proposed so far for a nonbinary nondiagonal matrix, has also been applied to decompose the ERAM HOQ. The result is given as the last row of Table 1, and is also depicted in Figure 6. The ROC algorithm rearranges rows and columns based just upon the relationship

Table 1. Summary of Decomposition Results

Method	Weights $\begin{pmatrix} W_R \\ W_C \end{pmatrix}$	Decomposition: Group1	Decomposition: Group2	Decomposition: Group3	$\begin{pmatrix} Z_R \\ Z_C \end{pmatrix}^*$	Penalty Function: $\begin{pmatrix} V_R \\ V_C \end{pmatrix}^{**}$	Penalty Function: $V(X, Y)^{***}$
Arbitraty Grouping (Figure 4)	N/A	*CA1,CA2,CA3* *EC1,EC2,EC3,EC4,EC5,EC6*	*CA4,CA5,CA6* *EC7,EC8,EC9,EC10,EC11,EC12*	*CA7,CA8,CA9,CA10* *EC13,EC14,EC15,EC16,EC17,EC18*	140 127	0.609 0.774	(0.692)†
Proposed Algorithm	1.00 0.00	*CA1,CA5,CA10* *EC3,EC5,EC6,EC7,EC13,EC14*	*CA4,CA6,CA9* *EC2,EC4,EC8,EC9,EC16,EC18*	*CA2,CA3,CA7,CA8* *EC1,EC10,EC11,EC12,EC15,EC17*	41 79	0.178 0.482	0.178
	0.75 0.25	*CA1,CA2,CA10* *EC3,EC5,EC6,EC7,EC13,EC14*	*CA4,CA5,CA6* *EC2,EC8,EC9,EC10,EC16,EC18*	*CA3,CA7,CA8,CA9* *EC1,EC4,EC11,EC12,EC15,EC17*	57 64	0.248 0.390	0.283
	0.50 0.50	*CA1,CA5,CA10* *EC3,EC5,EC6,EC7,EC11,EC12*	*CA2,CA4,CA6* *EC2,EC8,EC9,EC10,EC16,EC18*	*CA3,CA7,CA8,CA9* *EC1,EC4,EC13,EC14,EC15,EC17*	63 45	0.274 0.274	0.274
	0.25 0.75	*CA1,CA5,CA10* *EC3,EC5,EC6,EC7,EC11,EC12*	*CA2,CA4,CA6* *EC2,EC8,EC9,EC10,EC16,EC18*	*CA3,CA7,CA8,CA9* *EC1,EC4,EC13,EC14,EC15,EC17*	63 45	0.274 0.274	0.274
	0.00 1.00	*CA1,CA2,CA3* *EC3,EC5,EC6,EC7,EC11,EC12*	*CA4,CA5,CA6* *EC2,EC8,EC9,EC10,EC16,EC18*	*CA7,CA8,CA9,CA10* *EC1,EC4,EC13,EC14,EC15,EC17*	103 45	0.448 0.274	0.274
ROC Algorithm	N/A	*CA1,CA5,CA10* *EC2,EC3,EC5,EC6,EC7,EC12*	*CA2,CA4,CA7* *EC1,EC4,EC8,EC10,EC11,EC17*	*CA3,CA6,CA8,CA9* *EC9,EC13,EC14,EC15,EC16,EC18*	68 108	0.296 0.659	(0.478)†

* Z_R (Z_C) denotes the sum of the out-of-group entries in the relationship (correlation) matrix.

** $V_R = (Z_R - R_{min})/(R_{max} - R_{min})$, where $R_{min} = 0$, $R_{max} = 230$.

$V_C = (Z_C - C_{min})/(C_{max} - C_{min})$, where $C_{min} = 0$, $C_{max} = 164$.

*** $V(X, Y) = w_R V_R + w_C V_C$

† The $V(X, Y)$ values are computed at $(w_R, w_C) = (0.5, 0.5)$

matrix part. Consequently, it achieved a fairly good result in Z_R (=68), while the Z_C value (=108) was very poor. At $(w_R, w_C) = (0.50, 0.50)$, the overall value $V(X,Y)$ (=0.477) represents 74.1% higher dissatisfaction level compared with the optimal value from the proposed approach (=0.274).

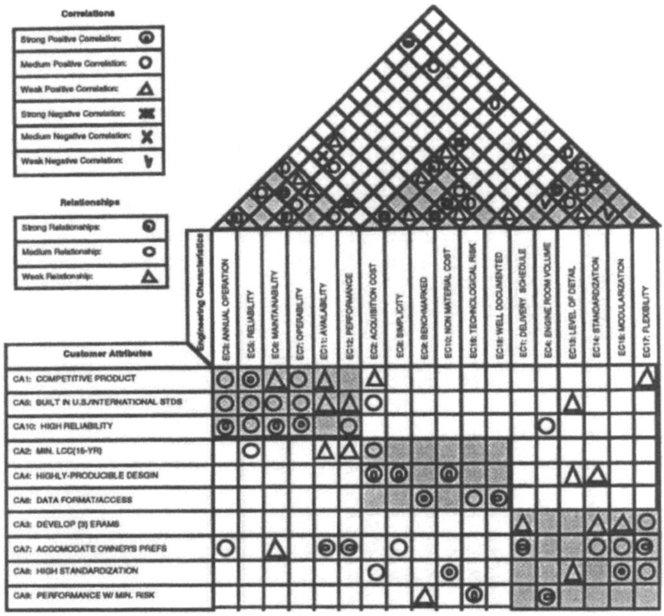

Figure 5. Decomposition Result for $(w_R, w_C) = (0.5, 0.5)$

The groups formed by the decomposition algorithm can be considered as component design modules which collectively constitute the whole design problem. In Figure 5, G_1 comprises such ECs as annual operation (*EC3*), reliability (*EC5*), maintainability (*EC6*), operability (*EC7*), availability (*EC11*), and performance (*EC12*). It seems fairly clear that G_1 represents the operation/maintenance aspect of the product once it is put to use, and thus might be called an *operation* module. G_2 includes such ECs as acquisition cost (*EC2*), simplicity (*EC8*), benchmarked (*EC9*), non-material cost (*EC10*), technological risk (*EC16*), and well-documented (*EC18*), and might be called a *development cost* module. The last group, G_3, includes delivery schedule (*EC1*), engine room volume (*EC4*), level of detail (*EC13*), standardization (*EC14*), modularization (*EC15*), flexibility (*EC17*). This group might be called a *designability* module.

While such a labeling would not be always feasible, there exists a high chance of forming cohesive and meaningful groups if the original HOQ contains consistent information between CAs and ECs and among the ECs. The decomposition thus permits the design team to cross-check the consistency of the information provided in the HOQ. Lack of cohesive grouping may indicate omission of important relationships and correlations or inclusion of irrelevant ones in the HOQ.

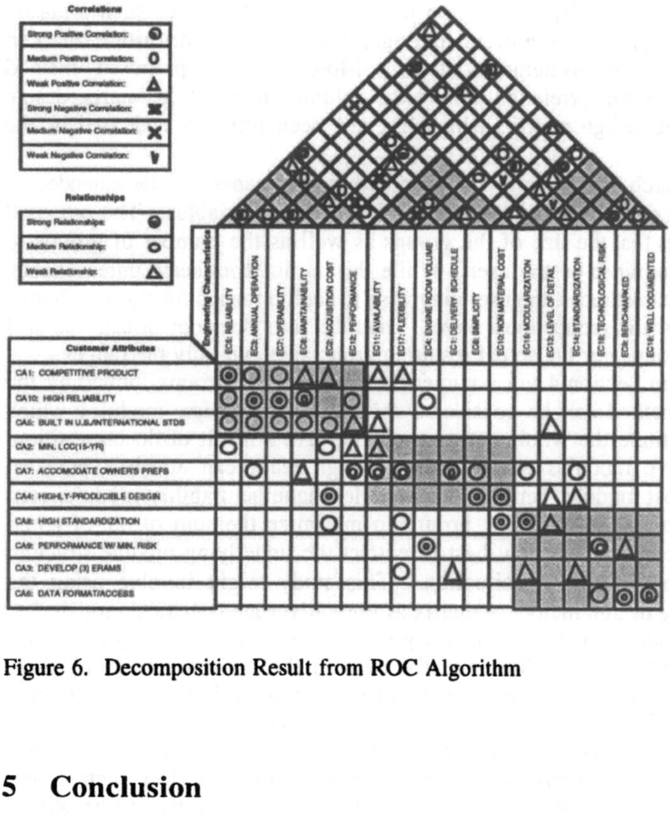

Figure 6. Decomposition Result from ROC Algorithm

5 Conclusion

This paper has developed a formal approach to reducing the size of an HOQ chart using the concept of design decomposition combined with multiattribute value theory. The proposed approach aims to decompose an HOQ chart into smaller subproblems which can be solved efficiently and independently. The correlation among the ECs as well as the strength of the relationship between CAs and ECs are considered in the decomposition process. The smaller size of the subproblems would make the QFD process simpler, easier, and cheaper.

This research should have an immediate impact on QFD practice. One of the impediments to the adoption of QFD has been the difficulty in interpreting and incorporating the complicated interactions between CAs and ECs and correlation among the ECs in the design analysis phase. By decomposing a large HOQ into smaller sub-HOQs of manageable size, the design team cannot only enhance the concurrency of the design activities, but also reduce a significant amount of time, effort, and cognitive burden required for analysis. This would help to obviate the objections to the adoption of QFD as a product design aid and improve the efficiency of QFD process.

From the methodological viewpoint, the contribution of the research is the development of a systematic method to decompose a nonbinary nondiagonal design incidence matrix. The research on decomposition of a nonbinary nondiagonal matrix has received virtually no attention in spite of the flexibility such a matrix can provide (e.g., alternative process routings, machine usage priorities, etc.) in flexible manufacturing systems context. Moreover, the proposed method explicitly considers the correlation among the column entities (e.g., parts, design parameters) of the design matrix, which has not been properly addressed in the literature.

As future research, the basic model proposed in this paper can be extended to incorporate the size of the groups as part of its decision variables. The proposed approach assumes that the size of the groups as well as the number of groups to be formed are determined by the user. While this is not considered unreasonable, under certain circumstances, the user may wish to set the ranges for the number of rows and columns the groups must have. Thus the user does not directly control the number of groups nor the actual resulting group size, but only the ranges.

As mentioned in Section 4 via a numerical example, the groups formed by the decomposition method can be considered as component design modules which collectively constitute the whole design problem. As a result of decomposition, obtaining cohesive modules to which the design team can attach a practical meaning (e.g., cost module, performance module, manufacturability module, etc.) would be as important as obtaining groups to minimize the sum of out-of-group entries. Then the question is how best to extract the underlying module structure that may exist in the problem situation. This study might involve using the eigenvalues of the design incidence matrix as in multivariate statistical models.

If the given incidence matrix is a nondiagonal matrix, it cannot be seperated into mutually exclusive groups. Consequently, even in the optimal grouping, there must be some non-zero out-of-group entries, for example, as in Figure 5. In such a case, the collection of the subproblem solutions would not perfectly generate (i.e., would be worse than) the optimal solution of the whole design problem. The difference in the resulting designs may depend on many factors such as the size of the matrix, the sparsity/density of the matrix, and the number of groups to be formed. It would be worthwhile to examine under what situations the decomposition scheme works best in the sense of faithfully restoring the overall optimal design.

Appendix. Derivation of R_{max}, R_{min}, C_{max}, C_{min}.

R_{max} (R_{min}) represents the upper (lower) bound on the sum of the out-of-group entries in the relationship matrix. Similarly, C_{max} (C_{min}) represents the upper (lower) bound on the sum of the out-of-group entries in the correlation matrix.

For a given HOQ consisting of m CAs and n ECs, the sizes of the groups to be formed are specified as (r_k, c_k), where r_k is the number of rows and c_k is the number of columns in G_k (group k), $k = 1, \ldots, K$. R_{max}, R_{min}, C_{max}, C_{min} can be determined using the number of weak, medium, strong relationships and correlations existing in the HOQ with minimal computational effort. The following subsections describe the procedure, and also demonstrate it using the numerical example given in Section 4.

In the following discussion, let $T_{\alpha 1}$, $T_{\alpha 2}$, $T_{\alpha 3}$ denote the total number of weak, medium, strong relationships in the relationship matrix, respectively. Similarly, in the correlation matrix, $T_{\beta 1}$, $T_{\beta 2}$, $T_{\beta 3}$ represents the total number of weak, medium, strong correlations (either positive or negative), respectively.

(1) R_{max}

The sum of the out-of-group entries would be maximized if the out-of-group cells have as many relationships (i.e., non-zero entries) as possible, with priority given to stronger relationships. Therefore, R_{max} can be defined as the optimal objective function value of the following optimization problem:

$$\begin{aligned}
&\text{Maximize } R_{max} = \alpha_1 N_{11} + \alpha_2 N_{12} + \alpha_3 N_{13} \\
&\text{s.t.} \quad N_{11} + N_{12} + N_{13} \leq \left(mn - \sum_{k=1}^{K} r_k c_k \right) \\
&\qquad N_{11} \leq T_{\alpha 1} \\
&\qquad N_{12} \leq T_{\alpha 2} \\
&\qquad N_{13} \leq T_{\alpha 3} \\
&\qquad N_{11}, N_{12}, N_{13} \geq 0 \text{ and integer,}
\end{aligned} \tag{A.1}$$

where N_{11} = Total number of weak relationships placed outside the groups,
N_{12} = Total number of medium relationships placed outside the groups,
N_{13} = Total number of strong relationships placed outside the groups.

For the example problem given in Section 4, the HOQ has 10 CAs and 18 ECs, or $(m, n) = (10, 18)$. It is desired that three groups be formed of sizes (3, 6), (3, 6), and (4, 6). The total number of cells within the three groups is $\sum_{k=1}^{3} r_k c_k = 3\times6 + 3\times6 + 4\times6 = 60$, and thus the total number of cells outside the groups is $mn - \sum_{k=1}^{3} r_k c_k = 10\times18 - 60 = 120$. In the HOQ shown in Figure 4, $T_{\alpha 1} = 17$, $T_{\alpha 2} = 20$, and $T_{\alpha 3} = 17$. For the illustration purpose, 1-3-9 scale is employed, i.e., $(\alpha_1, \alpha_2, \alpha_3) = (1, 3, 9)$. Then, the formulation given in (A.1) can be re-written for the example problem as

$$
\begin{aligned}
\text{Maximize } & R_{max} = 1 \cdot N_{11} + 3 \cdot N_{12} + 9 \cdot N_{13} \\
\text{s.t. } & N_{11} + N_{12} + N_{13} \leq 120 \\
& N_{11} \leq 17 \\
& N_{12} \leq 20 \\
& N_{13} \leq 17 \\
& N_{11}, N_{12}, N_{13} \geq 0 \text{ and integer.}
\end{aligned}
$$

R_{max} is obtained as 230 at the optimal solution $N_{11}^* = 17$, $N_{12}^* = 20$, and $N_{13}^* = 17$.

(2) R_{min}

On the contrary to the R_{max} case, the sum of the out-of-group entries would be minimized if the cells in the groups have as many relationships (i.e., non-zero entries) as possible, with priority given to stronger relationships. Therefore, R_{min} can be defined as the optimal objective function value of the following optimization problem.

$$
\begin{aligned}
\text{Minimize } & R_{min} = \alpha_1(T_{\alpha 1} - N_{21}) + \alpha_2(T_{\alpha 2} - N_{22}) + \alpha_3(T_{\alpha 3} - N_{23}) \\
\text{s.t. } & N_{21} + N_{22} + N_{23} \leq \left(\sum_{k=1}^{K} r_k c_k \right) \\
& N_{21} \leq T_{\alpha 1} \\
& N_{22} \leq T_{\alpha 2} \\
& N_{23} \leq T_{\alpha 3} \\
& N_{21}, N_{22}, N_{23} \geq 0 \text{ and integer,}
\end{aligned}
\tag{A.2}
$$

where N_{21} = Total number of weak relationships placed inside the groups,
N_{22} = Total number of medium relationships placed inside the groups,
N_{23} = Total number of strong relationships placed inside the groups.

The formulation given in (A.2) can be re-written for the example problem as:

$$
\begin{aligned}
\text{Minimize } & R_{min} = 1 \cdot (T_{\alpha 1} - N_{21}) + 3 \cdot (T_{\alpha 2} - N_{22}) + 9 \cdot (T_{\alpha 3} - N_{23}) \\
\text{s.t. } & N_{21} + N_{22} + N_{23} \leq 60 \\
& N_{21} \leq 17 \\
& N_{22} \leq 20 \\
& N_{23} \leq 17 \\
& N_{21}, N_{22}, N_{23} \geq 0 \text{ and integer.}
\end{aligned}
$$

R_{min} is obtained as 0 at the optimal solution $N_{21}^* = 17$, $N_{22}^* = 20$, and $N_{23}^* = 17$.

(3) C_{max}

As in the R_{max} case, the sum of the out-of-group entries would be maximized if the out-of-group cells have as many correlations (i.e., non-zero entries) as possible, with priority given to stronger correlations. Therefore, C_{max} can be defined as the optimal objective function value of the following optimization problem:

$$
\begin{aligned}
&\text{Maximize } C_{max} = \beta_1 N_{31} + \beta_2 N_{32} + \beta_3 N_{33} \\
&\text{s.t.} \quad N_{31} + N_{32} + N_{33} \le \left({}_nC_2 - \left(\sum_{k=1}^{K} {}_{c_k}C_2 \right) \right) \\
&\qquad N_{31} \le T_{\beta 1} \\
&\qquad N_{32} \le T_{\beta 2} \\
&\qquad N_{33} \le T_{\beta 3} \\
&\qquad N_{31}, N_{32}, N_{33} \ge 0 \text{ and integer,}
\end{aligned}
\tag{A.3}
$$

where N_{31} = Total number of weak correlations placed outside the groups,
N_{32} = Total number of medium correlations placed outside the groups,
N_{33} = Total number of strong correlations placed outside the groups.

For the example problem given in Section 4, the total number of cells within the three groups is ${}_6C_2 + {}_6C_2 + {}_6C_2 = 45$, and thus the total number of cells outside the groups is ${}_{18}C_2 - ({}_6C_2 + {}_6C_2 + {}_6C_2) = 108$. In the HOQ shown in Figure 4, $T_{\beta 1} = 14$, $T_{\beta 2} = 14$, $T_{\beta 3} = 12$. For the correlation part, 1-3-9 scale is also employed, i.e., $(\beta_1, \beta_2, \beta_3) = (1, 3, 9)$. (In this computation, we do not differentiate a positive and a negative correlation because we are concerned about the strength of the correlation which is left out of the groups rather than its sign.) Then, the formulation given in (A.3) can be re-written for the example problem as:

$$
\begin{aligned}
&\text{Maximize } C_{max} = 1 \cdot N_{31} + 3 \cdot N_{32} + 9 \cdot N_{33} \\
&\text{s.t.} \quad N_{31} + N_{32} + N_{33} \le 108 \\
&\qquad N_{31} \le 14 \\
&\qquad N_{32} \le 14 \\
&\qquad N_{33} \le 12 \\
&\qquad N_{31}, N_{32}, N_{33} \ge 0 \text{ and integer.}
\end{aligned}
$$

C_{max} is obtained as 164 at the optimal solution $N_{31}^* = 14$, $N_{32}^* = 14$, $N_{33}^* = 12$.

(4) C_{min}

The sum of the out-of-group entries would be minimized if the cells in the groups have as many correlations (i.e., non-zero entries) as possible, with priority given to stronger correlations. Therefore, C_{min} can be defined as the optimal objective function value of the following optimization problem:

$$\begin{aligned}
\text{Minimize } \; & C_{min} = \beta_1(T_{\beta 1} - N_{41}) + \beta_2(T_{\beta 2} - N_{42}) + \beta_3(T_{\beta 3} - N_{43}) \\
\text{s.t.} \quad & N_{41} + N_{42} + N_{43} \leq \left(\sum_{k=1}^{K} {}_{c_k}C_2 \right) \\
& N_{41} \leq T_{\beta 1} \\
& N_{42} \leq T_{\beta 2} \\
& N_{43} \leq T_{\beta 3} \\
& N_{41}, N_{42}, N_{43} \geq 0 \text{ and integer,} \qquad \text{(A.4)}
\end{aligned}$$

where N_{41} = Total number of weak positive correlations placed inside the groups,
N_{42} = Total number of medium positive correlations placed inside the groups,
N_{43} = Total number of strong positive correlations placed inside the groups.

The formulation given in (A.4) can be re-written for the example problem as:

$$\begin{aligned}
\text{Minimize } \; & C_{min} = 1 \cdot (T_{\beta 1} - N_{41}) + 3 \cdot (T_{\beta 2} - N_{42}) + 9 \cdot (T_{\beta 3} - N_{43}) \\
\text{s.t.} \quad & N_{41} + N_{42} + N_{43} \leq 45 \\
& N_{41} \leq 14 \\
& N_{42} \leq 14 \\
& N_{43} \leq 12 \\
& N_{41}, N_{42}, N_{43} \geq 0 \text{ and integer}
\end{aligned}$$

C_{min} is obtained as 0 at the optimal solution $N_{41}^* = 14$, $N_{42}^* = 14$, $N_{43}^* = 12$.

References

Armacost, R., Componation, P., Mullens, M. and Swart, W. (1994), "An AHP Framework for Prioritizing Customer Requirements in QFD: An Industrialized Housing Application," *IIE Transactions* 26 (4), 72-79.

ASI (1989 - 1994), *Transactions from the Symposium on Quality Function Deployment*, American Supplier Institute/GOAL/QPC.

Bosserman, S. (1992), "Quality Function Deployment: The Competitive Advantage," Private Trunked Systems Division, Motorola.

DeVries, R. and Barton, M. (1995), Personal Communication.

Gipprich, J. (1987), "QFD Case Study - Kelsey Hayes Corporation," *Quality Function Deployment: A Collection of Presentations and QFD Case Studies*, SectionVIII, ASI Inc., Dearborn, Michigan.

Griffin, A. (1991), "Evaluating Development Processes: QFD as an Example," Report 91-121, Marketing Science Institute, Cambridge, Massachusetts.

Griffin, A. and Hauser, J. (1992), "Voice of The Customer," Report 92-106, Marketing Science Institute, Cambridge, Massachusetts.

Hunter, M. and Van Landingham, R. (1994), "Listening to the Customer Using QFD," *Quality Progress* 27, 55-59.

Iri, M. (1968), "On the Synthesis of Loop and Cutset Matrices and the Related Problems," in K. Kondo (ed.) *RAAG Memories* 4 (A-XIII), 376-410.

Keeney, R. and Raiffa, H. (1976), *Decisions with Multiple Objectives*, John Wiley and Sons, New York.

King, J. R. (1980), "Machine-Component Grouping in Production Flow Analysis: An Approach Using a Rank Order Clustering Algorithm," *International Journal of Production Research* 18 (2), 213-232.

Kumar, K., Kusiak, A. and Vannelli, A. (1986), "Grouping Parts and Components in Flexible Manufacturing System," *European Journal of Operational Research* 24, 387-397.

Kusiak, A. and Chow, W. (1987), "An Efficient Cluster Identification Algorithm," *IEEE Transactions on System, Man, and Cybernetics* SMC-17 (4), 696-699.

Kusiak, A. and Wang, J. (1993), "Decomposition of the Design Process," *Journal of Mechanical Design* 115, 687-695.

Liner, M. (1994), "QFD Comes to Raychem Corporation: The Study of a Pilot Project," in Sammy G. Sinha (ed.), *Successful Implementation of Concurrent Engineering Products and Processes*, Van Nostrand Reinhold, New York.

McAuley, J. (1972), "Machine Grouping for Efficient Production," *Production Engineer* 51, 153-57.

McCormick, W. T., Schweitzer, P. J. and White T. E. (1972), "Problem Decomposition and Data Recognition by a Clustering Technique," *Operations Research* 20, 993-1009.

Morrell, N. (1987), "Quality Case Studies - Budd Corporation," *Quality Function Deployment: A Collection of Presentations and QFD Case Studies*, SectionVII, ASI Inc., Dearborn, Michigan.

Shipley, T. (1992), "Quality Function Deployment: Translating Customer Needs into Product Specifications," Working Paper, Department of Industrial Engineering and Management Systems, University of Central Florida, Orlando, Florida.

Sullivan, L.P. (1986), "Quality Function Deployment," *Quality Progress* 19, 39-50.

Wu, K. (1996), "Decomposition of the House of Quality", Master Thesis, The Pennsylvania State University, University Park, Pennsylvania.

Zhang, C. and Wang, H. (1992), "Concurrent Formation of Part Families and Machine Cells Based on the Fuzzy Set Theory," *Journal of Manufacturing Systems* 11 (1), 61-67.

A STUDY OF EVALUATIONS OF MUTUAL FUND INVESTMENT STRATEGIES*

PATRICK L. BROCKETT[1], W.W. COOPER[1], KU-HYUK KWON[2] and TIMOTHY W. RUEFLI[1]

[1]College of Business Administration, The University of Texas at Austin, Austin, Texas, USA 78712
[2]Department of Business Administration, College of Business Administration & Economics, The Yonsei University, 134 Shinchon-Dong,Seodaemun-Gu, Seoul, KOREA, 120-749

Evaluations of publicly announced investment strategies of open-end mutual funds were secured from a questionnaire sent to brokers and editors of financial newsletters. *Ex Ante* evaluations secured in this manner are compared with *ex post* performance over the period 1984-1988 and rated via data from Morningstar, Inc., and *Business Week* on mutual fund performances over this same period. The present study is designed so that separations can be effected between risk and return evaluations and performance over all pertinent *ex ante* and *ex post* pairings. *Ex ante* evaluations of risk and return are found to be positively correlated--as posited in the finance, decision theory and economics literatures--but their *ex post* pairings are negatively related. Somewhat surprisingly, *ex ante* to *ex post* evaluations of risk are positively correlated while *ex ante* to *ex post* evaluations of return are negatively correlated. The source of the *ex ante* to *ex post* negative relations is therefore in the return rather than the risk evaluations. This casts additional light on the Bowman Paradox (and related topics) which have been extensively studied in the strategic management literature. It also adds to these paradoxes and suggests programs for further research in a manner that is discussed in the addendum to this paper.

Key words: Strategy, Risk, Return, Mutual Funds

1. INTRODUCTION

The pioneering study by Bowman [1980] reawakened interest in risk and return relations in the strategic management literature. We do not examine this literature here because we have elsewhere reviewed it in detail[1] and because, for the most part, these studies have been confined to *ex post* data. Discussions of the strategies which subjects used to direct their *ex ante* evaluations of risks and returns have either been omitted or else have been only indirectly inferred from *ex post* data. In addition, with few exceptions, this literature does not attempt to ascertain the meanings that might have been assigned by subjects to terms like "risk" and/or the "returns" with which they have been concerned. Even fewer of these studies have attempted to ascertain how the subjects implemented their definitions en route to arriving at evaluations of prospective strategies. Thus, this literature may best be regarded as bearing only indirect relations to the present study which is concerned not only with the meanings assigned to terms like "risk" and "return" but also with how these terms are used in arriving at risk and return evaluations of proposed strategies as well as how they are measured and used, on an *ex ante* basis en route to seeing how these evaluations match with *ex post* performance.

In a sense, one part of this study--i.e., the part concerned with the meanings assigned to these terms--renews contact with work, in the 1960s and 1970s such as Mao [1970], Byrne [1971], and others who tried to determine possible meanings that practitioners might accord to terms such as "risk." Mao [1970], for example, interviewed executives (numbers were not reported) to ascertain the meanings they assigned to "investment risk" in a capital budgeting context. He found that most executives thought of risk as "the prospect of not meeting a target rate of return." Hence, in the minds of these executives, risk was related to "downside deviations from the target rate of return," rather than to the symmetric deviations that are inherent in commonly used risk measures like variances and standard deviations or in the regression measure referred to as "beta" in the literature on capital markets. See Lakonishok *et al.* [1993]. See also the reports on interviews with U.S. and Israeli executives in March and Shapira [1987] as well as the studies by Crum, Laughhunn, and Payne [1981], McWhinney [1967], and Petty and Scott [1981] where the risk orientations reported are also pointed toward downside deviations. Furthermore, as reported in Mao [1970], when large portions of company resources are involved in these capital budgeting investment decisions, the meaning assigned to risk by these executives was found to focus on the "danger of insolvency."[2]

Other approaches to this topic have also been essayed. The monumental study of U.S. and Canadian company executives by MacCrimmon and Wehrung [1986], for example, employed a very sophisticated decision theoretic approach which may be summarized as follows. Using von Neumann-Morgenstern [1947] utility-theoretic axioms, a "behavioral questionnaire" was employed as a research instrument to ascertain how 509 top-level U.S. and Canadian managers would

---

[1] See, e.g., Brockett *et al.* [1993]. See also Charnes, *et al.* [1993] and Kwon [1991].

[2] Which is to say that the "level" at which an undesirable event occurs needs to be considered along with (or complementary to) its chance of occurrence. Cf. the discussion of "risks," "levels," or "exposure" and "coverage" (when a multiplicity of risks are to be involved) as discussed in the opening section of Brockett *et al.* [1993].

respond to business and personal situations by reference to decisions that these managers reported they would make in the presence of different risk-reward possibilities. MacCrimmon and Wehrung did ask participants to state what they meant by "risk"--or, more precisely, what they meant by a "risky" situation--but these attempts played only a minor role in their study. More generally, MacCrimmon and Wehrung relied on decision theoretic methods to deduce risk-preference attitudes from decisions that the participants reported they would make when confronted with different combinations of risk-reward possibilities. See, e.g., their discussion of the "in-basket/out-basket" experiment in their questionnaire.

Turning from these attempts to ascertain meanings that practitioners might assign to terms like risk and return, we next note that risk and strategy are, for the most part, *ex ante* concepts. This carries with it the possibility that the strategies that might be followed and the results realized may differ, and the same is true for risks. This brings us to the literature that has been devoted to forecasts and related attempts at *ex ante* evaluations of the strategies that might be employed and the risks that might be taken in the pursuit of managerial objectives.

There is a long history of studies in the literature of finance going back at least to Cowles [1933] on the quality of analyst forecasts of stock prices and extending to forecasts of company earnings and subsequent performance as in Cragg and Malkiel [1968]. Most of these studies have been directed to evaluating the accuracy of the forecasts, as measured relative to subsequent performance of the forecasted values. It is only very recently that attempts have been made to use other criteria such as the profitability that might be obtained by adhering to the forecasts, as is done in Leitch and Tanner [1991]. Implicitly this use of profit rather than forecast realizations as a benchmark raises the possibility that the published research might have employed criteria which differed from that employed by forecasters or users of such forecasts. However, Leitch and Tanner seem rather to be concerned with providing an *ex ante* theory of their own devising (consistent with economics) to explain why such forecasts continue to be made and used despite what appears to be a long record of spotty and unsatisfactory performance. In any case, we do not discuss this literature in more detail since no effort appears to have been made to ascertain what criteria are employed by forecasters or users of forecasts.

It is, of course, possible to use such forecasts for other purposes, as is done by Bromiley [1991] who used data from IBES (Institutional Brokers Estimate System) to study risk in relation to organization behavior such as the slack accumulations that are posited by Cyert and March [1963] as a management response to uncertainty. For these purposes Bromiley used the forecasts of security analysts with respect to expected returns (income streams) of manufacturing enterprises and the variance of these forecasts as a measure of risk (or uncertainty). This is probably about all that could be done in the way of securing an *ex ante* measure of risk with these data. Nevertheless, it is necessary to note that this variance is a measure of variations between <u>forecasters</u> rather than variations between forecasts and actuals or other criteria that might have been employed by makers and users of these forecasts.

Bromiley [1991] is not the only example of an attempt to relate *ex ante* evaluations to organization (slack) performance.[3] Singh [1986] utilized results from a questionnaire survey as reported in Appendix A of Khandwalla [1977, 1981] with returns from 112 Canadian and 61 U.S. firms[4] in each of which 3 or more respondents provided information on decentralization, risk taking, environmental turbulence, mass output orientation, and competitive pressure during the 3 year period 1973-1975. Singh then guides his analysis in accordance with the following quotation:[5] "The decision rule that the organization uses is to maximize the cumulative probability of obtaining an outcome in terms of performance that exceeds the minimum acceptable [e.g., a target rate of return]." According to Singh this provides access to the prospect theory of Kahneman and Tversky [1979] in which risks and returns may be negatively related because returns below minimum acceptable levels will induce increased risk taking behavior. See also Dubin and Savage [1963].

Using archival data on the rates of return reported for these firms, Singh is unable to assess (1) their probability of occurrence, as his formulations require, and/or (2) whether the returns actually realized correspond to levels that would be regarded as satisfactory. Instead, he turns attention to the correlations obtained between returns, as reported, and, using "absorbed" and "unabsorbed slack,"[6] as in Bourgeois [1981] and Bourgeois and Singh [1983], he attempts to relate his findings to items like degrees of decentralization as responses to different degrees of risk.

Trying to obtain various measures of risk in this manner--e.g., by reference to slack accumulation--is indirect, at best, and differs in other ways from the direct approach to risk and the return results which we report. Another important difference is that Singh does not separate risk and return characterizations and evaluations such as we have effected in Charnes, *et al.* 1993 and as we shall report in the present study. Finally, neither Singh nor Khandwalla had access to publicly announced strategies to which their firms were publicly committed--as is the case for the mutual funds covered in the present study. Thus we cannot know with any assurance whether the archival data used by Singh [1986] reported results which were obtained from conformance to the strategies reported for these firms by Khandwalla [1977, 1981].

---

[3]Sitkin and Pablo [1993] provide a detailed and up-to-date review of the literature which treats risk in relation to its organization theory implications.

[4]As discussed on p.571 of Singh [1986], lack of archival data and other considerations caused the number included in his study to be reduced to 64 medium and large U.S. and Canadian corporations.

[5]Singh [1986: 565] believes that this formulation is consistent with the position taken by most organization theorists. We might also note that it is also consistent with the more general chance constrained programming formulations described in Brockett, *et al.* [1993] where the target is a moving one given by the objective of maximizing the probability of equaling or exceeding the S&P 500 rate of return subject to a constraint on the allowable risks that can be taken.

[6]These "financially" oriented measures are of limited usefulness for the study of slack in other contexts. See, for instance, the discussion of the extended field experiment used to study slack accumulations for vehicle maintenance activities in the U.S. Air Force that is discussed in Charnes, Clarke, and Cooper [1989] where such financial statement data are neither directly pertinent nor available.

2. PURPOSE

There is a paucity of published literature which attempts to evaluate strategies in a manner that makes it possible to compare *ex ante* strategy evaluations with the *ex post* results that followed. Apart from a few studies, like those of Jemison [1987], there are even fewer attempts to ascertain how evaluators proceed to effect their evaluations and then examine how these relate to subsequent outcomes by reference to measures designed to reflect these orientations. The purpose of the present study is to begin to help fill these gaps in the literature by using results from a survey of brokers and editors of financial newsletters which we can relate to widely used criteria for performance evaluation by, e.g., organizations specializing in such evaluations.

Our focus will be on mutual funds where the requisites for effecting these kinds of evaluations of strategies and subsequent performance are available. Our interest, however, is not in individual funds which might offer good or poor investment opportunities. It centers, rather, on broad classifications of announced mutual fund strategies as obtained from Morningstar, Inc. of Chicago. Our use of Morningstar services also provided access to performance data on 830 mutual funds over the period 1984-1988 which we could study in detail as reported in Brockett *et al.* [1993]. It also provided access to a subset of 383 of these companies with *ex post* risk and return ratings over this same period as published by *Business Week*.

Among the class of persons concerned with evaluating mutual fund strategies we selected brokers and editors of financial newsletters and further restricted our choices to persons or organizations specializing in evaluating the behavior of such funds. This provided access to expertise. Using brokers and editors of financial newsletters also provided access to persons who are accustomed to articulating their evaluations and explaining or justifying them to others and, it should be noted, these evaluations enter into the behavior of many others such as subscribers to these newsletters and clients of these brokers. Finally, we confined attention to open-end mutual funds since this avoids possible difficulties such as might occur with the inclusion of closed-end funds where the share values, and hence returns, can differ from the portfolio values.

3. CRITERIA FOR PERFORMANCE EVALUATION

The sections that follow will describe the way the present study was made and how the results were validated. Before entering into the detail required to do this we can qualitatively summarize some of the major findings as follows: "Chance of loss," by a wide margin, is the definition of risk favored by respondents. It is nearly "twice as favored" as the second choice, "large fluctuations in value of portfolio." This is of interest because it is the latter, less favored definition, which lends itself to the use of the variance and variance related measures that are usually used in the studies reported in the strategic management and finance literatures.

The term "loss" may be extended to "opportunity losses" with the supposedly riskless T-Bill rate providing a very natural (and widely used) measure of this kind of loss which is often used by practitioners. We can therefore extend the downside orientation of "chance of loss" to include such opportunity losses in a way that also comprehends negative returns. Using this expanded definition makes it possible to obtain a measure of risk by use of a chance constraint on returns below the T-Bill rate as a measure of risk. This provides contact with the results reported in Brockett *et al.* [1993]. It also provides contact with the ratings published in *Business Week* where risk is measured by reference to the number of periods and the amounts by which each fund's returns fell below the T-Bill rates.

Confining attention to open-end mutual funds, as already noted, makes it clear that return on the portfolio is the appropriate measure of return.[7] More precisely, the measure to be used is net return, or yield, which is equal to gross yield less the expense ratio of a fund. This measure which has long been used in evaluating mutual fund performance--cf., Friend, *et al.* [1970], Blume and Friend[8] [1977] or Baumol, *et al.* [1989]--is clearly the appropriate measure in the case of open-end mutual funds where an investor may liquidate his investment for a pro-rata share of the fund's portfolio on any day. Furthermore, there is no divergence between customers and owners in such funds since they are one and the same, employees are few, and even the supposed possibility of conflict between management and shareholders has now been shown to be virtually non-existent in the study by Baumol, *et al.* [1989].[9]

Selection of net return as a fund objective, however, does not end the matter of choice. There is also the issue of how performance is to be evaluated. We need to select a target for evaluating performance in terms of this objective if we are to maintain contact with the findings we cited earlier in a literature extending from Mao [1970] to March and Shapira [1987]. The target we select is the return for the S&P 500 index which many funds (and others) also use.[10] This choice also provides access to the *Business Week* ratings where ability to equal or exceed the return on the S&P 500 in each year over the period 1984-1988 was used to rate each fund's performance.

Proceeding in the manner we have just indicated enables us to evaluate risk and return separately. This, in turn, enables us to continue with the line of development initiated in Brockett *et al.* [1993] where this separation was effected

---

[7]Attention is confined to equity funds--i.e., funds investing in equities rather than bonds--and we might also note that for all practical purposes total assets and total portfolio values are about equal. See Weiss [1990] for ratings of bond (=fixed income) funds.

[8]This study by Blume and Friend is of special interest for its bearing on the "Bowman paradox" by reference to its early finding of negative relations between risk and return which, more recently, have been extended to by Fama and French [1992] to securities in a wide range of industries--with consequent concern for persons committed to the "efficient markets" hypothesis.

[9]Chapters 1 and 2 of this book provide an excellent overview of the history and current state of this huge and growing industry--now well in excess of three 18
trillion dollars in asset values.

[10]The methods for calculating return and even the terminology have been pretty well standardized. See, e.g., "Choosing a Mutual Fund, What the Prospectus Tells You" on p. A-16 of the Management Digest in *Newsweek*, Dec. 1991: "You can quickly read the explanation of how the fund calculates...return. These calculations are governed by SEC regulations and the language has been largely standardized. [The prospectus will contain a table] that will show the fund's total return performance over a period of time. Usually this performance will be compared to an index such as the S&P 500...."

via a mathematical model using a chance constrained programming formulation in which (1) return was expressed as an "objective" in terms of maximizing the probability of exceeding the return on the S&P 500 while (2) risk was formulated as a chance constraint which limited the chance of falling below the T-Bill rate of return. Hence it is of interest to see whether the negative relation between risk and return which was identified in Brockett *et al.* [1993] as being "more in the return than in the risk performances" as found in market data continues to hold in the present study which is based on broker and newsletter editor evaluations as well as market data.

Using responses to the first set of questions in the questionnaire, which is presented in the Appendix, it was found that the *ex ante* relations between risk and return were positive. We interpret this to be consistent with the finance, economics and decision theory literatures--which have an *ex ante* (before the fact) orientation wherein relatively high risks must be accompanied by expectations of relatively high returns, if these alternatives are to be selected. This is to be contrasted with the Bowman paradox and related findings in the strategic management literature where interest centers on the negative relations that have been found to hold empirically between risk and return based exclusively on *ex post* data.

Here we are able to confirm the findings in Brockett *et al.* [1993] of a negative relation between risk and return and are also able to carry this further. Based on the responses from our panel of practitioners we find that they do very well in evaluating risk, where their *ex ante* evaluations of fund strategies relate positively to *ex post* results. The trouble, i.e., the source of the negative relations between risk and return, seems to lie rather in the return evaluations where the *ex ante* to *ex post* relation is negative, as is also true for the relation between *ex ante* risk and *ex post* return.

This kind of separation could be important in many practical contexts by showing where evaluation abilities might best be improved. Here, however, we simply note the somewhat surprising nature of this result. It is surprising, first, because risk is a more recondite concept and thus is supposedly more difficult to evaluate. Second, it is return rather than risk which is "the name of the game," since this is the objective to be sought, whereas risk enters only to constrain the choices allowed in pursuit of this objective!

4. STUDY DESIGN

We now turn to the evidence on which these conclusions are based. In assembling this evidence, our study used the questionnaire in the Appendix to solicit responses from brokers and editors of financial newsletters. As can be seen, the cover page of the questionnaire consists of a letter[11] which is followed on page 1 by a listing of possible ratings directed to securing evaluations of relative risks and returns associated with the strategies listed in the stub. These ratings range

[11] See Dillman [1983] p. 362 for a discussion of approaches to good questionnaire design. See also Fink [1995].

from "very low" to "very high." Page 2 of the questionnaire is then directed toward ascertaining how closely any one of several different meanings correspond to a respondent's use of the terms "risk" and "return." A final section of the questionnaire allows respondents to set forth their own usage in cases where none of the definitions provided in the questionnaire were sufficiently satisfactory. See Tables A-1 and A-2 in Appendix for comments received in response to this final section---all of which were in addition to supplying the requested ratings.

The thus structured questionnaire was used to minimize time and inconvenience for respondents, with the final section providing a further opportunity to help us ascertain whether we had overlooked important alternative definitions of risk or return. The requested rankings refer to what are called "objectives" but might also be called "strategies" as defined in Brockett, *et al.* [1993].[12] Witness, for example, the following quotation from the May 1, 1990, Prospectus of National Liquid Reserves, Inc. (a money market fund):

> The Fund's investment objectives are maximum current income and preservation of capital. The Fund seeks to achieve its objectives with (a strategy to guide its) portfolio selections (as follows) ..., etc.

To cast this in chance constrained programming terms, we might say that the fund's objective is to maximize current income---as given by, say, maximizing its expected value or by maximizing the probability of doing better than an index fund. To complete this characterization we might say that the maximization is to be undertaken for each portfolio under the constraints imposed by its strategies, including risk allowances (e.g., on preservation of capital or chance of loss) as formulated in chance constraints.

These topics are developed in detail with accompanying definitions in Brockett, *et al.* [1993] for analytical and modeling purposes. Here we use the terms "objectives" and "strategies" interchangeably, as is common in the industry and as was done in the questionnaire. Thus, referring to the questionnaire we see that the risks associated with each strategy were to be rated from very low to very high, after which the returns associated with these same strategies were to be similarly rated. There are questions of bias to be dealt with, of course, and we will now begin to attend to these after first noting that the categories "very low" and "very high" were included mainly to provide us with further guidance. We will report on these expanded categories as necessary but, for the most part, we collapse the categories to Low, Moderate, and High for simplicity of presentation and discussion.[13]

To eliminate or reduce the possibility of "position bias," which can arise from the order in which questions are presented [Becker, 1954; Cantril, 1944; Carpenter and Blackwood, 1979; Pike, Dalgleish, and Jackson, 1985], we followed the

---

[12]See also the dictionary edited by Cooper and Ijiri [1983] where "strategy" is defined as "A plan of action used to guide or control other plans of action."

[13]A full and very detailed series of tests and examinations are reported in Kwon [1991] with results which are consistent with what is reported in the present paper.

recommendation of Campbell and Mohr [1950] and Grant [1948] and used a Latin square design in which the order of the questions on page 2 were systematically rotated through their possible permutations. We also pilot-tested the questionnaire to ensure that respondents would understand our statement of the strategies as presented on page 1 in the questionnaires. These strategy categories, we might note, had *a priori* plausibility for ease of understanding in their favor because they were secured from *Business Week* and Morningstar Inc., which are widely known and used references---at least by respondents like ours. The pilot test was conducted by means of mailings with accompanying telephone inquiries to 18 brokers in the Austin, Texas, area who possessed special expertise in mutual funds. Returns were received from 17 of the 18 brokers and telephone follow-up calls made it clear that these persons had no difficulty with the terminology being used. In fact, several were able to identify the sources as *Business Week* and/or Morningstar while others expressed a preference for other usages.

5. JUXTAPOSITION BIAS

We also used our pilot test to ascertain whether biases might arise from juxtapositioning risk and return ratings as is done on page 1 of the questionnaire.[14] One third of the 18 broker respondents were sent only the questions related to risk, another third were sent the questions relating to return and a final third were sent the full questionnaire and asked to supply both risk and return ratings. This part of the pilot test was intended to provide an opportunity to check for a possible tendency of the third group to have juxtaposition affect their risk and return ratings. Thus it might be expected that differences in the ratings of risk and return might be smaller for the third group compared to the ratings that might be expected when the risk and return questions are separated. Table 1 shows the contrary results from our field study with, on average, differences between risk and return ratings smaller for (A)-(B), the separated ratings of risk and return, than for (C)-(D), the ratings from the questionnaire requesting both.

[14]We are indebted A. Lewin for advice on these and other tests used for this questionnaire.

TABLE 1: Average Risk-Return Evaluations of Strategies in Pilot Study (1)

Type of Questionnaire: STRATEGIES	Risk (A)[a]	Return (B)[b]	Risk-Return (C)[c]	(D)[d]	Differences (A)-(B)	(C)-(D)
Asset Allocation	4.00	3.20	3.80	3.80	0.80	0.00
Aggressive Growth	4.50	4.20	4.30	3.30	0.30	1.00
Balanced	2.70	2.80	2.70	2.80	-0.10	-0.10
Equity Income	2.30	3.20	2.70	3.00	-0.90	-0.30
Growth	3.20	3.40	3.20	3.50	-0.20	-0.30
Growth-Income	2.20	3.20	2.30	2.80	-1.00	-0.50
Income	1.80	2.40	2.30	2.70	-0.60	-0.40
International	3.20	3.20	3.80	3.70	0.00	0.10
Option Income	3.00	3.00	2.80	2.20	0.00	0.60
Specialty	4.30	3.20	3.80	2.80	1.10	1.00
Small Company	3.80	3.60	4.00	2.70	0.20	1.30
Financial	4.00	2.60	3.50	3.00	1.40	0.50
Health	4.20	3.60	3.50	2.60	0.60	0.90
Precious Metals	4.70	2.20	4.50	2.20	2.50	2.30
Natural Resources	3.80	2.60	3.80	2.50	1.20	1.30
Technology	4.30	3.60	4.50	3.00	0.70	1.50
Utilities	2.80	2.40	2.00	3.00	0.40	-1.00
Average					0.376	0.465
Variance					0.778	0.767

[a]Risk Ratings in Risk-Only Questionnaire. [b]Return Ratings in Return-Only Questionnaire. [c]Risk Ratings in Risk-Return Questionnaire. [d]Return Ratings in Risk-Return Questionnaire.

Our sample was fairly small and several cells contained only one or two responses so that wide confidence intervals were involved in the statistical analyses. Nevertheless, we were tentatively able to conclude that no strong evidence of juxtaposition bias was secured from this experiment. For further confirmation, we also analyzed the data as in Table 2 where the average ratings from these returns are also recorded for risk and return separately and compared to the averages for these same categories from respondents who received the juxtaposed questions. As can be seen, there is as much as a one-point difference in only one case and this occurs in the return rather than the risk category--where it might be more reasonably expected to occur--in the case of funds investing in stocks of health and health-care companies. Finally, we use the variances listed in Tables 1 and 2 to obtain values of $F=1.01399$ and 1.86577, respectively, and note that neither achieves significance at the customary level of 5%. Thus in terms of differences in variability, as well as in terms of differences in average ratings, we concluded that there is no strong evidence of juxtaposition bias.

TABLE 2: Average Risk-Return Evaluations of Strategies in Pilot Study

Type of Questionnaire: STRATEGIES	Risk (A)[a]	Return (B)[b]	Risk-Return (C)[c]	(D)[d]	Differences (A)-(C)	(B)-(D)
Asset Allocation	4.00	3.20	3.80	3.80	0.20	-0.60
Aggressive Growth	4.50	4.20	4.30	3.30	0.20	0.90
Balanced	2.70	2.80	2.70	2.80	0.00	0.00
Equity Income	2.30	3.20	2.70	3.00	-0.40	0.20
Growth	3.20	3.40	3.20	3.50	0.00	-0.10
Growth-Income	2.20	3.20	2.30	2.80	-0.10	0.40
Income	1.80	2.40	2.30	2.70	-0.50	-0.30
International	3.20	3.20	3.80	3.70	-0.60	-0.50
Option Income	3.00	3.00	2.80	2.20	0.20	0.80
Specialty	4.30	3.20	3.80	2.80	0.50	0.40
Small Company	3.80	3.60	4.00	2.70	-0.20	0.90
Financial	4.00	2.60	3.50	3.00	0.50	-0.40
Health	4.20	3.60	3.50	2.60	0.70	1.00
Precious Metals	4.70	2.20	4.50	2.20	0.20	0.00
Natural Resources	3.80	2.60	3.80	2.50	0.00	0.10
Technology	4.30	3.60	4.50	3.00	-0.20	0.60
Utilities	2.80	2.40	2.00	3.00	0.80	-0.60
Average					0.077	0.165
Variance					0.159	0.297

[a]Risk Ratings in Risk-Only Questionnaire. [b]Return Ratings in Return-Only Questionnaire. [c]Risk Ratings in Risk-Return Questionnaire. [d]Return Ratings in Risk-Return Questionnaire.

There is still other evidence which can be brought to bear on the possibility of juxtaposition bias, as we shall see, and we will also be able to use Tables 1 and 2 in other ways. Notice that emphasis is on ratings of *ex ante* expectations rather than on the *ex post* results on mutual fund behavior that are reported in Brockett, *et al.* [1993]. We did not ask respondents to indicate actions they might take on these *ex ante* ratings so that, for instance, the high-risk and low-return average ratings accorded to the "Precious Metals" investing strategy does not mean that these brokers would never recommend such a mutual fund.

6. UNIVERSE FRAME AND RESPONSE RATE

Having satisfied ourselves with respect to possible juxtaposition biases, we mailed the entire questionnaire as shown in the Appendix to 85 editors of financial newsletters specializing in mutual funds. The names and addresses were obtained from various sources,[15] with the objective of a mailing to the entire "universe" of

[15]These sources consisted of the mutual fund section of the Dick Davis Digest, Hudson's Newsletter Directory, Gale Directory of Publications, and The Individual Investor's Guide to Investment Publications during the period 1988-1989.

such editors,[16] at least as far as we could identify it---although the term "frame" might be a better choice (Cochran, 1961:4) because of the practical impossibility of identifying all members of this universe. Eleven of these 85 mailings were returned because of post office inability to locate respondents from the addresses used. This reduced the list to 74 potential respondents from which we obtained 17 returns in a first round, with only 13 being usable, after which we undertook a second mailing to the remaining 57 non-respondents (including those who sent unusable returns) and obtained 16 returns, all being usable in a second, and final, round.

The response rate from the brokers $(17/18)100 \doteq 95\%$ is very high. The response rate from the editors of financial newsletters is $(29/74)100 \doteq 40\%$ after allowing for the 11 returns which were considered by the post office to be undeliverable. This is low compared to the 52% questionnaire response rate from newspaper business editors reported in Albaum, Kozmetsky and Peterson [1984] but it is high relative to the median values of 19% reported for industrial surveys and 30% for consumer surveys.[17] Although higher than the latter pair of return rates, it is lower than the rate reported for newspaper editors but does not seem to be very far out of line even in the latter case and so we deem our response rate to be somewhat better than minimally satisfactory.

7. RESPONSE BIAS

Even with a seemingly satisfactory response rate, further attention is required because there is a possibility of non-response bias from the financial newsletter editors which could have dictated additional attempts to secure returns from the some 60% of the potential respondents who failed to return our questionnaire. Non-response bias is a topic which has long been of concern in survey design including sociological and social psychology surveys, opinion polling and marketing. Hence we turned to these literatures for help. As suggested by Armstrong and Overton [1977], Pace [1939], Ferber [1949], and Zimmer [1956], we could test for such bias if we regarded the second round respondents as more like non-respondents. Having maintained separate records for each of the two rounds, we thus could test to see if there were statistically significant differences between the responses in the first and second rounds. For this purpose, we applied a t-test to each question. As is well known, such repeated applications of t-tests should result in some rejections even when the null hypothesis of "no-difference" is true. Hence, we also conducted χ^2 tests as a further check even though the validity of this test may also be questioned.[18]

---

[16]From the total of 500 investment-related newsletters estimated by Hulbert [1989], our selection of 85 was restricted to editors of financial newsletters dealing with mutual funds at least to the extent of devoting special sections or special issues of their letters to this topic.

[17]Also see the references cited in Albaum, Kozmetsky and Peterson [1984] for further discussion.

[18]Cochran [1954], for example, recommends against using the χ^2 test when more than 20 percent of the cells have a frequency less than 5 and any cell has a frequency less than 1.

TABLE 3: Results of t-Test and χ^2-Tests for Return Ratings (1st Round vs. 2nd Round)

STRATEGIES	t-Tests		χ^2 Tests	
	t-Values	Prob.(t>|α|)	χ^2-Values	Prob.(χ^2>|α|)
Asset Allocation	-0.915	0.368	1.197	0.550
Aggressive Growth	-0.316	0.754	0.108	0.743
Balance	-1.222	0.233	2.354	0.308
Equity Income	-0.176	0.861	0.108	0.948
Growth	-0.670	0.509	0.480	0.488
Growth-Income	-0.673	0.507	0.862	0.650
Income	-0.356	0.725	0.532	0.766
International	1.520	0.141	3.073	0.215
Option Income	-0.318	0.753	0.767	0.857
Specialty	0.066	0.948	0.475	0.789
Small Company	-1.609	0.120	4.624	0.099
Financial	0.044	0.965	0.392	0.942
Health	0.409	0.689	2.730	0.435
Precious Metals	-0.737	0.469	0.622	0.733
Natural Resources	0.094	0.926	2.380	0.304
Technology	0.891	0.382	3.595	0.166
Utilities	0.409	0.686	7.045	0.134

However, even in the presence of the high risk of error arising from our repeated use of these tests, Table 3 shows that the null hypothesis of no difference between the two groups of respondents cannot be rejected in any category. This table, which refers to the differences in risk ratings between these two groups of respondents, shows that all t and χ^2 values have very high probabilities of occurrence on the null hypothesis so that the hypothesis of no difference between the two groups cannot be rejected without a very high risk of committing a Type I error.

Similar results are apparent from Table 4 where the probabilities associated with the t and χ^2 values are recorded for differences in return ratings of these different strategies. As can be seen, only "Precious Metals" reaches significance at the 0.05 level and this could have resulted because of relatively large discrepancies in risk and return ratings---e.g., as recorded in the returns from the brokers listed in Table 1.

As a still further check on non-response bias, we examined for possible inconsistencies in ranking on both risk and return ratings between the first and second round respondents. For instance, the mutual fund strategy with highest average rating was accorded a rank of one, followed by a rank of two for the next highest average rating, and so on. Using rank correlations between the two rounds

TABLE 4: Results of t-Test and χ^2-Tests for Return Ratings (1st Round vs. 2nd Round)

STRATEGIES	t-Tests		χ^2 Tests	
	t-Values	Prob.(t>|α|)	χ^2 Values	Prob.(χ^2>|α|)
Asset Allocation	-0.839	0.410	1.867	0.393
Aggressive Growth	0.310	0.759	1.448	0.694
Balance	-0.765	0.454	0.675	0.411
Equity Income	0.895	0.382	6.070	0.108
Growth	1.150	0.261	1.417	0.492
Growth-Income	-0.762	0.453	0.862	0.650
Income	-1.368	0.184	2.392	0.302
International	0.139	0.891	5.385	0.146
Option Income	-0.588	0.562	5.359	0.252
Specialty	-0.651	0.522	3.42	0.331
Small Company	0.401	0.692	2.269	0.686
Financial	-0.730	0.473	3.458	0.326
Health	-0.373	0.712	0.251	0.969
Precious Metals	-2.210	0.037*	8.325	0.040*
Natural Resources	-1.377	0.181	2.557	0.465
Technology	-0.425	0.674	5.094	0.278
Utilities	-1.141	0.267	4.822	0.306

*statistically significant at 0.05 level

resulted in a value of 0.87 for Kendall's τ_β for risk ratings and a value of 0.69 for return ratings, with both being statistically significant at levels of 0.0001 and 0.001, respectively, and this high positive and statistically significant correlation fails to support evidence of differences between the two rounds such as could be expected in the presence of non-response bias.

Because respondents were selected from editors of financial newsletters specializing in mutual fund evaluations it seems reasonable to suppose that they are not subject to biases arising from differences in occupational background [Franzen and Lazarsfeld, 1945], socioeconomic status, [Bebbington, 1970], or lack of familiarity with the topic [Edgerton, Britt and Norman, 1947; Suchman and McCandless, 1940]. However, we undertook such further checks as were possible by reference to data in sources such as *The Individual Investor's Guide to Investment Publications* which suggested that, if anything, our respondents were persons with longer experience and their letters had larger circulations than the average of such financial newsletters. These "above average" findings help to confirm that our respondents can be regarded as knowledgeable experts (as desired for our purposes) by reference both to length of experience and the recognition associated with their relatively large circulations.

TABLE 5: Results of t-Tests and χ^2 -Tests for Risk Ratings (Brokers vs. Editors)

Strategies	t-Tests		χ^2 Tests	
	t-Values	Prob.(t > |a|)	χ^2 Values	Prob.(χ^2 > |a|)
Asset Allocation	6.690	0.001*	27.143	0.001*
Aggressive Growth	-0.719	0.482	2.414	0.29
Balanced	1.056	0.287	1.187	0.552
Equity Income	-1.136	0.267	1.364	0.506
Growth	-2.302	0.033*	5.802	0.055
Growth-Income	-3.191	0.005*	9.031	0.029*
Income	0.840	0.411	0.818	0.664
International	-2.269	0.0388	8.056	0.049*
Option Income	-2.287	0.031*	4.852	0.183
Specialty	-0.719	0.478	1.391	0.499
Small Company	-1.509	0.143	3.235	0.198
Financial	0.573	0.571	2.818	0.420
Health	-0.093	0.927	0.610	0.894
Precious Metals	-0.303	0.764	4.418	0.110
Natural Resources	0.718	0.479	2.528	0.282
Technology	-0.783	0.441	5.766	0.056
Utilities	-0.187	0.853	1.134	0.889

*Statistically significant at 0.05 level

TABLE 6: Results of t-Tests and χ^2 -Tests for Risk Ratings (Brokers vs. Editors)

Strategies	t-Tests		χ^2 Tests	
	t-Values	Prob.(t > |a|)	χ^2 Values	Prob.(χ^2 > |a|)
Asset Allocation	1.816	0.096	18.734	0.001*
Aggressive Growth	-1.053	0.311	3.205	0.361
Balanced	-0.517	0.612	0.342	0.559
Equity Income	1.017	0.322	1.387	0.709
Growth	-0.626	0.542	2.964	0.397
Growth-Income	-0.340	0.739	1.253	0.534
Income	0.907	0.377	2.942	0.401
International	-0.949	0.358	2.417	0.490
Option Income	-0.08	0.937	5.0998	0.277
Specialty	-1.940	0.068	6.278	0.179
Small Company	-2.354	0.032*	8.038	0.090
Financial	-0.970	0.347	1.155	0.764
Health	-1.729	0.109	7.401	0.116
Precious Metals	-3.111	0.005*	8.908	0.063
Natural Resources	-2.010	0.061	5.440	0.245
Technology	-0.977	0.339	5.378	0.251
Utilities	-0.521	0.608	1.825	0.768

*Statistically significant at 0.05 level

With these tests completed for our financial newsletter editors, we turned to our broker responses to see if they might reasonably be combined with those of our newsletter editors. The results of t-tests and χ^2 calculations comparing averages for the ratings from the two groups are shown in Tables 5 and 6. As can be seen, significant differences in risk ratings for the various strategies appear in 5 cases for the t-values and 3 cases for the χ^2 values in Table 5. Turning to Table 6, which deals with the return ratings, only 2 of the t-values are seen to attain significance and only one χ^2 value has this property.

As noted earlier, repeated uses of t and χ^2 values for significance is very likely to lead to rejection of true hypotheses and an accompanying commission of Type I errors under the play of chance alone. These significance results are therefore subject to serious qualification, but we shall nevertheless allow for their possible validity by conducting separate analyses of these two groups (brokers and newsletter editors) when it appears important for us to take the possibility of differences in their responses into account.

8. INTER-RATER RELIABILITY AND INTER-RATER AGREEMENT

Having explicated our design we next turn to how the results we reported above are validated. We start with an evaluation of inter-rater reliability and agreement where, following Tinsley and Weiss [1975], we distinguish between the two as follows.

> *Inter-rater agreement* [italics added] represents the extent to which the different judges tend to make exactly the same judgments about the rated subject. When judgments are made on a numerical scale, Inter-rater agreement means that the judges assigned exactly the same values when rating the same person [Tinsley and Weiss, 1975: 359].

In contrast,

> *Inter-rater reliability* [italics added] represents the degree to which the ratings of different judges are proportional when expressed as deviations from their means. In practice, this means that the relationship of one rated individual to other rated individuals is the same although the absolute numbers used to express this relationship may differ from judge to judge. Inter-rater reliability usually is reported in terms of correctional or analysis of variance indexes [Tinsley and Weiss, 1975: 359].

In summary, inter-rater agreement refers to different judges supplying the same ratings while inter-rater reliability refers to ratings that conform to the same ordering even when the values supplied by the judges differ numerically.

It is important that we take both possibilities into account since, when combining responses from brokers and editors of financial newsletters, we may need to allow for differences in motivation as well as differences in experience and expertise. For instance, as Hulbert [1989] observes "Brokers are compensated on a commission basis, i.e., frequencies of transactions and number of shares transacted. In contrast, newsletter editors are directly paid by the value of their advice, i.e., the size of subscribers and profitability of subscribers. Consequently, while incentives of brokers [may] sometimes conflict with incentives of investors, incentives for newsletter editors in general coincide with incentives of investors and their performance data are objectively evaluated and discussed in the public domain." For further discussion on this topic, see Hulbert [1989].

9. Inter-Rater Reliability

In their review article, Tinsley and Weiss [1975] suggest the use of Snedecor's formula of Inter-rater reliability [Snedecor, 1940: 205; Ebel, 1951: 410] when ordinal data are involved - *viz.*,

$$S = \frac{MS_s - MS_e}{MS_s + (k-1)MS_e}$$

where MS_S = mean square of ratings within judges, MS_e = mean square of ratings between judges, and k = the number of raters or judges.

The results from applying this formula to our data are shown in Table 7. As can be seen, coefficients of inter-rater reliability for risk evaluations of mutual fund strategies are high (0.613 for brokers, 0.571 for newsletter editors, and 0.554 for the total sample) and statistically significant. Inter-rater reliability coefficients for return evaluations of mutual fund strategies are lower, but nonetheless statistically significant (0.107 for brokers, 0.286 for newsletter editors, and 0.211 for the total). Thus, we conclude that the orderings supplied in the returns for the two groups are consistent.

TABLE 7: Inter-rater Reliability of Individual Ratings

	Brokers	Editors	Totals
Ratings of Risks for Strategies	0.613**	0.571**	0.554**
Rankings of Different Definitions of Risks	0.383**	0.312**	0.297**
Ratings of Returns for Strategies	0.107*	0.286**	0.211**
Rankings of Different Definitions of Returns	0.341*	0.284**	0.302**

**statistically significant at 0.0001 level
*statistically significant at 0.005 level

The analyses for Table 7 were based on individual ratings. For many purposes, our interest will center rather on the average ratings for which we use the following formula, from [Ebel, 1951: 411],

$$R_A = \frac{MS_s - MS_e}{MS_s},$$

where the symbols have the same meaning as in formula (1).

TABLE 8 Inter-rater Reliability of Average Ratings

	Brokers	Editors	Totals
Ratings of Risks for Strategies	0.950**	0.975**	0.981**
Rankings of Different Definitions of Risks	0.882**	0.929**	0.946**
Ratings of Returns for Strategies	0.569**	0.921**	0.915**
Rankings of Different Definitions of Returns	0.838**	0.920**	0.945**

**statistically significant at 0.0001 level

As Table 8 shows, inter-rater reliability coefficients for average ratings are statistically significant and much higher than those exhibited in Table 7, as might be expected because the latter refers to results obtained from individual ratings. Also brought out is the fact that reliability coefficients for newsletter editors are higher than those for stock brokers in every category of Table 8 but not in Table 7. In both cases, however, our ratings and rankings easily pass the customary significance tests for reliability.

10. Inter-Rater Agreement

To examine agreement in ratings of risk and return and rankings of the definitions of risks and returns, Kendall's coefficient of concordance (W) was used. As Table 9 shows, general agreement is exhibited in the responses. In particular, agreement for the risk ratings assigned to mutual fund strategies is high and highly significant while agreement on the ranking of definitions is lower (but still highly significant). Apparently different definitions of risk are compatible with similar ratings of risks assigned to various mutual fund strategies by both groups of experts. This carries over to the results shown in the final column on the right of Table 9 where both groups are combined.

TABLE 9: Kendall's Coefficients of Concordance (W)

	Brokers	Editors	Totals
Ratings of Risks for Strategies	0.6708 (0.00001)	0.6215 (0.00001)	0.5921 (0.00001)
Rankings of Different Definitions of Risks	0.4197 (0.0003)	0.3122 (0.00001)	0.2862 (0.00001)
Ratings of Returns for Strategies	0.2439 (0.0015)	0.3960 (0.00001)	0.2914 (0.00001)
Rankings of Different Definitions of Returns	0.3512 (0.0036)	0.3400 (0.00001)	0.3331 (0.00001)

( ) Significance Level

11. RESULTS

Definitions of Risk and Return

The results which are all summarized in the final section of this paper are here treated one a time to show how these results (and their interpretations) were arrived at. We begin with the concept of risk.

Concepts of Risk

In some respects, our study is similar to Mao [1970] although we are oriented toward mutual funds and our respondents are persons who are accustomed to evaluating *ex ante* statements of strategies and the risks and returns with which they might be identified and then explicitly reporting the results of their evaluations to others. We also differ from Mao as well as others, like March and Shapira [1987], by attempting to provide measures or at least rankings for the definitions of return and risk in a manner which enabled us to take advantage of the fact that the questionnaire was sent to persons who are professionally occupied with these topics. To capture the responses to our questionnaire, we used a "Likert type" ordinal scale to allow qualitative evaluations in the form of rankings ranging from "Very Low" to "Very High," to be converted to numerical values, as explained in the questionnaire.

TABLE 10: Average Ratings of Risk Concepts

Risk Definitions	Average Ratings
Chance of Loss	1.80
Large Fluctuations in Value of Portfolio	2.50
Chance of Unfavorable Events	3.30
Chance of Making Less Than a Target Rate of Return	3.90
Chance of Bankruptcy	4.20
Chance of Missing More Favorable Opportunities	4.30

Table 10 shows the average values accorded to the ratings for risk by all 46 respondents, where the ordering is from 1.8, the "highest" average, to 4.3, the "lowest" average. These results indicate that, "chance of loss," such as might be associated with a chance constraint, is rated first and "large fluctuations in value of portfolio," such as might be associated with a variance measure of risk is second, while, finally, "chance of missing more favorable opportunities" (an opportunity cost definition) is last. Even though many respondents appear to be familiar with sophisticated concepts and measures of risk, such as variances and the betas used in CAPM (Capital Asset Pricing Models)---see, e.g., Ch. V in Elton and Gruber [1987]---brokers and newsletter editors seem to regard risk as a "downside phenomena," as witness "Chance of Loss" being in the first position with further enforcement from the 3 categories following "Large Fluctuations in Value of Portfolio."

The results reported in Table 10 are consistent with other survey findings, e.g., Mao [1970], Shapira [1986], and MacCrimmon and Wehrung [1986]. Wilcoxon rank sum tests, as recorded in Table 11, were used to ascertain whether pairwise comparisons of average rating differences were statistically significant.

TABLE 11: Wilcoxon Rank Sum Tests for Ratings of Risk Definitions

Comparisons of Risk Definitions	Z Values	Prob. > |Z|
Chance of Loss Vs. Large Fluctuations in Value of Portfolio	2.36	0.0185**
Large Fluctuations in Value of Portfolio Vs. Chance of Unfavorable events	3.10	0.0019**
Chance of Unfavorable Events Vs. Chance of Making Less Than a Target Rate of Return	-1.86	0.0630
Chance of Making Less Than a Target Rate of Return Vs. Chance of Bankruptcy	1.22	0.2224
Chance of Bankruptcy Vs. Chance of Missing More Favorable Opportunities	0.34	0.7354
Chance of Making Less Than a Target Rate of Return Vs. Chance of Missing More Favorable Opportunities	0.94	0.3497
Chance of Unfavorable Events Vs. Chance of Bankruptcy	2.49	0.0129*

** Statistically Significant at Least at 0.01 Level * Statistically Significant at Least at 0.05 Level

As Table 11 shows, significant differences occurred between the average ratings for "chance of loss" and "large fluctuations in value of portfolio" and between the latter and "chance of unfavorable events," <u>but not between the other pairs</u>. Hence we may, if we wish, combine the other categories with each other and with Chance of Loss on grounds of lack of statistical significance and consistency of meaning, but we cannot similarly combine them with Large Fluctuations in Value of Portfolio.[19]

Returning to the individual categories, "Chance of Loss" might therefore seem the most appropriate concept of risk to use in evaluating these strategies. On the other hand, limitations imposed by the structured nature of the questionnaire may have produced limitations on alternate possibilities. See the additional responses reported in the Appendix. Finally, allowance should be made for other possibilities such as the possibility that risk is a multi-dimensional concept so that combinations of the above might also need to be considered. See, e.g., Gooding [1978] for supporting material which bears directly on the behavior of institutional investors such as mutual funds.

Concepts of Return

Turning from risk to return, we perform a similar analysis. Table 12 shows the average ratings for return characteristics of mutual fund strategies with "large rate of return on assets" being rated first and "large dividends" rated last. This not only supports the use of return on asset (ROA) as in Brockett, *et al.* [1993], it also supports a literature that goes back at least to the studies of Friend, Blume, and Crockett [1970]. Perhaps of more interest from the standpoint of long-standing controversies in the finance literature[20] is the fact that large dividends was ranked

[19]"Chance of loss" is, of course, the definition of risk commonly used in the insurance industry. Responses favoring "chance of unfavorable events" might be combined logically with ``chance of loss" but we do not do this since respondents might have other things in mind such as the occurrence of recession, etc.

[20]E.g., the controversies surrounding the Modigliani-Miller hypothesis. See the summary and discussion in Gordon [1993].

last and, moreover, its "average" value of 5.0 means that it was ranked last by all respondents.

TABLE 12: Average Ratings of Return Concepts

Return Definitions	Average Ratings
Large Rate of Return on Assets	2.10
Large Ending Value of Portfolio	2.50
Large Profit	2.70
High Chance of Exceeding a Target Rate of Return	3.40
Large Capital Gains	3.50
Large Dividends	5.00

This result is further confirmed in Table 13 where the Wilcoxon rank sum tests accord statistical significance to every alternative to which large dividends are compared in the last 5 rows. We might, however, again note the need for qualification along lines like those noted in our discussions of the rankings for risk in Tables 9 and 10. This extends to qualification for what may be overly sharp interpretations of some of the pairings presented in Table 13. For instance, a large value of ROA may be preferred to a high chance of exceeding a target return without implying that the two are mutually exclusive. Furthermore, this result as reported in Table 13 might have been reversed if we allowed for moving targets such as "performing better than the market average."[21] The fact that a high chance of exceeding a target return ranks higher (see the negative value of Z) than a high ending value of the portfolio might lend further credence to such remarks.

TABLE 13: Wilcoxon Rank Sum Tests for Ratings of Return Definitions

COMPARISONS OF RETURN DEFINITIONS	Z Values	Prob. > |Z|
Large Rate of Return on Assets Vs. High Chance of Exceeding a Target Rate of Return	3.30	0.001**
Large Ending Value of Portfolio Vs. High Chance of Exceeding a Target Rate of Return	-2.34	0.02**
Large Ending Value of Portfolio Vs. Large Capital Gains	-2.39	0.003**
Large Profit Vs. Large Capital Gains	2.51	0.01**
Large Rate of Return on Assets Vs. Large Dividends	6.19	0.0001**
Large Ending Value of Portfolio Vs. Large Dividends	5.54	0.0001**
Large Profit Vs. Large Dividends	5.43	0.0001**
High Chance of Exceeding a Target Rate of Return Vs.	4.21	0.0001**
Large Capital Gains Vs. Large Dividends	4.37	0.0001**

Note: Only statistically significant results are reported. ** Statistically significant at least at 0.05 level

Risk and Return Characteristics of Mutual Funds

We now turn from rankings of definitions of risk and return to the ratings accorded by respondents to the strategies shown in Table 14. As already noted, respondents were asked to rate these strategies in a way that would reflect their classifications

[21]E.g., as represented by the S&P 500 in Brockett, *et al.* [1993] which reports on a study based on actual performance for these same funds that were also under way at the time this questionnaire study was initiated.

on an order ranging from very low to very high. Using a value of 5 for very high and continuing numerically to a value of 1 for very low, we obtain the averages for these ratings that are shown alongside the strategies to which they apply with 4.63, the highest, being assigned to funds investing in "Precious Metals" and 1.95, the lowest, being assigned to funds investing for "Income."

TABLE 14: Classification of Average Ratings of Risk for 17 Mutual Fund Strategies

High Risk		Moderate Risk		Low Risk	
Small Company	(4.13)	Growth	(3.48)	Income	(1-95)
Specialty	(4.18)	Option Income	(3.5)	Utilities	(2.45)
Aggressive Growth	(4.53)	Financial	(3.65)	Balanced	(2.53)
Technology	(4.53)	Natural Resources	(3.73)	Equity Income	(2.65)
Precious Metals	(4.63)	Health	(3.85)	Growth Income	(2.73)
		International	(3.9)	Asset Allocation	(3.08)

( ) Average Ratings

Next we utilize these numerical results to reduce the risk classes from 5 to 3. This does not cause much trouble in the reclassifications from very high to high risk or very low to low risk, but there is some ambiguity in deciding which numerical limits define the range of moderate risk. The robustness of our conclusions was checked, as reported in Kwon [1991], by shifting the break points over a relatively wide range without affecting any of our major conclusions. Hence, for simplicity and succinctness, we here present results from using the numerical value 3.0 as the break point. However, to avoid being unduly restrictive we relax this somewhat, as indicated in Table 14, with Asset Allocation being assigned to the Low rather than the Moderate risk class.

Statistical tests were conducted with respect to the investment strategies noted at the cutting points between these risk classifications. The results are noted in Table 15. Asset Allocation vs. Growth achieved statistical significance but

TABLE 15: t-Tests and Wilcoxon Rank Sum Tests

Strategies	t-Tests		Wilcoxon-Tests	
	t-Values	Prob.(t > |a|)	z Values	Prob.(z > |a|)
Asset Allocation [3.08] vs. Growth [3.48]	-2.768	0.0071*	-2.638	0.0083*
International [3.90] vs. Small Company [4.13]	-1.525	0.1314	-1.437	0.1506
Specialty [4.18] vs. Aggressive Growth [4.53]	-2.85	0.0056*	-2.735	0.0062*

( ) Average Ratings * Statistically Significant at Least at 0.05 Level

International vs. Small Company did not. However, reallocating Small Company and Specialty to the Moderate risk class does not affect our results with respect to the risk-return relations reported below. Hence, we continue with the classifications exhibited in Table 14.[22]

[22]We also conducted analyses using the full range of 5 classes from Very Low to Very High with similar results as reported in Kwon [1991].

Ex Ante Risk-Return Relations of Mutual Funds

We next proceed to classify the averages of the ratings on return in a fashion similar to Table 16. The Moderate Return class is here more clearly clustered and distinguished from the other two classes and so we did not perform comparable significance and reclassification tests. As implied in the sentence of the questionnaire introducing the requested ratings, respondents were asked to rate risks and returns associated with these strategies in a typical year. The orientation was to be directed toward "an expected value," at least in a loose sense, rather than a particular or recent experience. We cannot guarantee that this was what all respondents had in mind but we assume that on the average this is true.

TABLE 16: Classification of Average Ratings of Return for 17 Mutual Fund Strategies

High Return		Moderate Return		Low Return	
Specialty	(3.43)	Financial	(3.03)	Income	(2.36)
Technology	(3.53)	Asset Allocation	(3.05)	Option Income	(2.57)
Health	(3.58)	Growth Income	(3.05)	Utilities	(2.82)
Growth	(3.59)			Precious Metals	(2.84)
International	(3.67)			Balanced	(2.87)
Small Company	(3.74)			Natural Resources	(2.95)
Aggressive Growth	(4.00)			Equity income	(2.95)

( ) Average Ratios

Since the average values shown in Table 14 and Table 16 are based on an ordering, we employed a rank correlation, assigning 1 to the highest rank average and 17 to the lowest. Spearman's rank correlation coefficient between the resulting *ex ante* risk rankings for mutual fund strategies as given in Table 14 and the expected return rankings obtained from Table 16 was found to be 0.55 and statistically significant at the 0.05 level.[23]

Table 17 further confirms this positive relation between *ex ante* risk and *ex ante* return with a value of $\chi^2 = 308.692$ and significance at 0.0001, and a value of Somers' D = 0.680 which has an asymptotic standard error estimate of 0.042.

TABLE 17: *Ex Ante* Risk Vs. *Ex Ante* Return

Return (*ex ante*)	High	0	175	65
	Moderate	67	2	0
	Low	54	10	10
		Low	Moderate	High

Risk (*ex ante*)

Note *ex ante* return categories are based on ratings of total respondents

[23]As noted in Table 5, significant differences in risk ratings for strategies (5 cases for t values and 3 cases for χ^2 values) appeared between brokers and editors of financial newsletters. Spearman's rank correlation between risk and return rankings evaluated by editors only is found to be positive (0.75) and statistically significant at 0.05 level, while Spearman's rank correlation between risk and return rankings evaluated by brokers only is found to be positive (0.29) but statistically not significant.

As we noted earlier, this result is consistent with the precepts of decision theory, economics, and finance where *ex ante* orientations apply to risk-return choices and "rationality" is assumed (e.g., as in risk aversion). Note, however, that our approach bypasses difficulties in the literature dealing with this research, such as research on the capital asset pricing model of the finance where, *faute de mieux*, "instrumental variables" and like uses of *ex post* data are employed as surrogates for the *ex ante* risk evaluations and a positive linear relationship between risk and return is assumed. Generally, such instrumental variables use regressions on *ex post* data in earlier periods as proxies for *ex ante* characterizations of risk in later periods. See Ch. 13 in Elton and Gruber[1987]. This is not needed here since we have both *ex ante* ratings by our experts based on explicitly stated strategies as well as data on actual performance to compare *ex ante* ratings to *ex post* (actual) behaviors.

Ex Ante Risk VS. *Ex Post* Return

Being in position to examine relations between *ex ante* and *ex post* risks and returns we can now expand on studies of topics like the Bowman paradox in the strategic management literature--which have heretofore been confined mainly to the use of *ex post* data without ensuring that they emanate from the *ex ante* strategy choices (with accompanying *ex ante* risks) that might have given rise to the observed realizations.

Using return ratings for performances of each of the open-end mutual funds over the period 1984-1988, as reported in Laderman [1989],[24] we can match these *ex post* ratings of return with the ratings of *ex ante* risks in Table 14 and obtain the results exhibited in Table 18.

TABLE 18: *Ex Ante* Risk Vs. *Ex Post* Return

Return (*ex post*)	High	64	36	4
	Moderate	53	73	10
	Low	4	78	61
		Low	Moderate	High
		Risk (*ex ante*)		

Note *ex ante* return categories are based on ratings of total respondents

The frequencies shown in the cells of this table exhibit a tendency for low *ex post* return to be associated with high *ex ante* risk categories, and *vice versa*, over this period[25] and this appears to extend the Bowman paradox to this *ex ante*-to-*ex post* relation between risk and return. This is formally confirmed with a value of Kendall's τ_β = -0.5223 which is significant at 0.0001 and also by a value of

[24]See the discussion in Brockett, *et al.* [1993], which relates these *Business Week* ratings to other ratings.

[25]This was the most recently available data for the period 1989-1990 when the survey underlying this study was initiated.

-0.507 for Somers' D with an asymptotic standard error estimate of -.033 and, finally, a χ^2 test is consistent with strong interdependence with a value of 137.731 which is significant at 0.0001.[26]

Ex Ante Risk VS. *Ex Post* Risk

We now carry the above analysis a stage further which we can do because our questionnaire design enables us to separately identify performances in risk and return evaluations. Thus, turning to Table 19 we see that a positive relation is exhibited between the *ex ante* risk characterizations supplied by our respondents and the risk ratings effected from *ex post* data as reported in *Business Week*-Morningstar. See Laderman [1989]. This positive relation is formally confirmed via a value of τ_β = 0.56321, which is significant at 0.0001 and a value of Somers' D = 0.545 with an asymptotic standard error of 0.033 and the χ^2 test value is also consistent with strong interdependence with a value of $\chi^2 = 167.22$ which is significant at 0.0001.[27] Evidently our *ex ante* risk characterizations are consistent with these risk characterizations obtained *ex post facto* whereas the relation between *ex ante* risk and *ex post* return is not and, indeed, the latter pair appears to be negatively related.

TABLE 19: *Ex Ante* Risk Vs. *Ex Post* Risk

Risk *(ex post)*	High	2	63	59
	Moderate	39	86	11
	Low	80	38	5
		Low	Moderate	High
		Risk *(ex ante)*		

Note *ex ante* return categories are based on ratings of total respondents

Ex Ante Return VS. *Ex Post* Return

Finally, we turn to Table 20 which provides a similar portrayal for the *ex ante* ratings of return from our respondents and the *ex post* ratings of return for these same strategies obtained from *Business Week*. The frequencies shown in the cells of this Table exhibit a tendency for low *ex ante* return to be associated with high *ex post* return categories and *vice versa*. This is formally confirmed with a value of Kendall's τ_β = -0.3794 which is significant at a level of 0.0001 and also by the value of -0.342 for Somers' D with an asymptotic standard error estimate of -.039 and, finally, a χ^2 test is consistent with strong interdependence with $\chi^2 = 83.709$ which is significant at 0.0001.[28]

---

[26]The *ex ante* risk categorizations used here are based on the ratings by both brokers and editors of financial newsletters. However, similar results are secured with analyses for brokers only and editors only.

[27]Again, similar results are secured for brokers only and editors only. See Kwon [1991]

[28]Similar results are obtained for brokers only and editors only.

TABLE 20: *Ex Ante* Return Vs. *Ex Post* Return

	High	125	76	39
Return	Moderate	4	40	25
(ex post)	Low	14	20	40
		Low	Moderate	High

Return *(ex ante)*

Note *ex ante* return categories are based on ratings of total respondents

12. SUMMARY AND CONCLUSION

We have now added quantitative measures and supporting evidence to our earlier (qualitative) summary of the main conclusions to be drawn from our survey. Reiterating a point made much earlier by Mao [1970] and Byrne [1971] we find that risk is generally viewed as a downside phenomenon by our respondents. Variance, or even the semi-variance measure introduced by Markowitz [1959], does not reflect this view of risk as well as an orientation which locates this behavior in the tails of statistical distributions that are supposed to represent the chance of loss and gain. That is, it is the tails rather than deviations from the center which best reflect this downside orientation and measures need to be developed for this purpose as is done in Brockett, *et al.* [1993].[29]

It is also useful to separate the risk of loss from the chance of gain, as is done in Brockett *et al.* [1993]--where the objective is to maximize the probability of exceeding the S&P 500 and where risk is represented in the form of a chance constraint with a tail orientation in which the only choices allowed in pursuit of the objective are those which will <u>not</u> exceed a prescribed chance of falling below the T-Bill rate.

Actually the study reported here is part of a larger study in which the results from this questionnaire were used to check and clarify results reported in Brockett *et al.* [1993] and Kwon [1991] where these measures of risk and return were applied to *ex post* data from Morningstar, Inc. Here we have shown that the results from this questionnaire survey have other uses as well. For instance, our use of *ex ante* evaluations of the strategies included in the questionnaire made it possible to validate the positive *ex ante* to *ex ante* relations of risk and return on which most of the teaching and most of the work is based in modern finance, economics and decision theory. However, we also find a negative *ex post* risk to *ex post* return findings that have been the focus of attention in the strategic management literature dealing with the "Bowman paradox." See the first and last rows of Table 21. Although empirically perplexing, the latter finding is not logically inconsistent with the former.

It is important to be able to trace possible sources for these discrepancies. Help in doing this may be secured from the pairings that are displayed in Table 21. As can be seen in that Table, *ex ante* risk and *ex post* return relations are negative.

---

[29] See also Lakonishok, Shleifer and Vishny [1993] who report that although value stocks have a larger value of β and hence are more volatile then glamour stocks, it is incorrect to regard them as more "risky" because this volatility is caused largely by upward jumps rather than price declines in these stocks.

However, *ex ante* risk and *ex post* risk are positively related so it can only be the negative relation between *ex ante* return and *ex post* return which is the source of this behavior.

TABLE 21: Risk and Return Relations From Questionnaire Returns

DESCRIPTION		RELATIONS
Ex Ante Risk vs.	*Ex Ante* Return	Positive
Ex Ante Risk vs.	*Ex Post* Risk	Positive
Ex Ante Risk vs.	*Ex Post* Return	Negative
Ex Ante Return vs.	*Ex Post* Return	Negative
Ex Post Risk vs.	*Ex Post* Return	Negative*

*As reported in Brockett, *et al.* [1992].

In addition to attention to sources of trouble and possible improvements from uses of such pairings we also note the need for more attention in the research literature to the meanings assigned to risk and return by practitioners and the further need for suitable methods of measurement. To see the significance of doing this, at least as far as the research literature is concerned, we cite the following often quoted message from Aakers and Jacobson [1987:277-278]:

> "Failures to account for risk adequately will unquestionably lead to inappropriate decisions. All else being equal, if firms judge business performance only in terms of return (performance), they will place more resources than warranted in risky strategies, forego profitable opportunities, and apply misguided performance evaluations. Furthermore, if researchers do not control for risk in studies assessing the effects of strategic factors on profitability, they will obtain biased estimates of the effects on return of those strategic factors correlated with risk."

Even this call for research falls short. Omitted from consideration, for instance, is any reference to the need for attending to measures that will reflect the risk and return concepts actually used in practice. There is room for departing from usages in practice, of course, but it does not seem desirable to add further possible paradoxes without knowing that they may arise from discrepant measures and meanings between the research literature and management practice. Bringing these tasks and these possibilities to the fore is also an intended contribution of the research reported here.

13. Addendum

The need for further attention and the difficulty of doing so in the research literature is apparent from responses we have received from earlier versions of this paper. For instance, one finance specialist responded by saying that the negative correlation we report between *ex ante* and *ex post* returns is not surprising "because return is notoriously difficult to forecast." We think we are saying something different, however, when we add that the *ex ante* evaluations of risk are also positively correlated with *ex post* performance. This means that advice from experts like those we studied can be relied upon for positive association with risk and negative association with realized returns. Thus a statement that a high risk is associated with high return should be interpreted to mean that the risk is likely to be high and the return is likely to be low--with the converse also applying, *viz.*, low risk-low return advice should be interpreted to mean that low risk and high returns are likely to be realized.

These remarks are based on our study--which is confined to mutual funds, of course, as is true for Brockett *et al.* [1993]. However, they are consistent with the more general findings of Fama and French [1992] for returns to all investors in the markets for securities as well as the more recent results reported in Lakonishok, Shleifer and Vishny [1993]—*viz.*, "Value stocks are more volatile than glamour stocks as measured by Beta, but this is caused largely by larger jumps [i.e., increases] in the prices of value stocks, rather than by price declines. The higher return earned by value stock investors is therefore not a reward for accepting higher risk, but is rather a result of the systematic misjudgments of investors...."[30]

After examining our findings as reported in Table 21, yet another finance professor suggested that our discussion should be halted with the finding that *ex ante* risks and *ex ante* returns are positively correlated. Pressed for reasons, this finance professor replied that "otherwise no finance journal would publish this paper."

Now, however, the situation with respect to publication in the finance literature is beginning to be altered. At least there appears to be a transition, with accompanying turbulence, as responses are being developed to the Fama and French [1992] publication which also reported a negative relation between risk and return over a long span of years and a wide variety of industries. Findings like these have severe implications for "rational" investment theory and the theory of "efficient" markets[31] As presently formulated, they are too sweeping. Hopefully the responses that now become possible, will lead to findings of new factual material in which the practices of investment advisors (with their attendant effects on investment behavior) will be brought to the fore. Such discovery of new factual material often occurs when a widely accepted theory is brought into question with attendant relaxations of previous proscriptions as to "researchable" topics and findings. To be sure, a theory can "illuminate." It can also "eliminate"--i.e., it can eliminate facts from view and behaviors that require a broader or different theory to accommodate them.

What about our choice of mutual funds as a subject of study for management strategy? As was noted earlier in this paper, the public availability of strategy statements, the fact of very few employees and the identity of owner and customer interests for mutual funds would seem to render this an ideal choice--indeed, almost a Weberian "ideal type"--for studying strategies in both their *ex ante* formulations and *ex post* consequences. Weighing in against this choice, however, is the opinion we received from a specialist in the study of management strategy who believes that no one in strategy research would be interested in studies or findings like those, reported this paper, even though it is directed to mutual funds, which is now the largest of all U.S. industries.

We have just outlined some of the features which seem to make the study of mutual fund strategies and behavior an unusually attractive area for strategy

---

[30]This is consistent with our earlier comments which noted the shortcoming of Beta and related measurers such as variances and standard deviations as measures of risk.

[31]See "Beating the Market, Yes, It Can be Done," *The Economist* Dec. 5, 1992, pp. 21-23.

research. It is also important for its bearing on policies in business--indeed, all businesses--economics and government. At $3 trillion (and still growing), mutual funds are the biggest of all U.S. industries and, as reported in the January 18, 1993, issue of *Business Week*, p.62, "They [mutual funds] are reinventing the way we save [and invest] and are reshaping the economy!" In any case, we need research directed to determining why supposedly (ultra-rational) investors like mutual funds fail to exhibit the kinds of "learning from experience" that is supposedly associated with rational behavior. It is of course possible that "efficient markets" and like behaviors may emerge from suitably arranged institutional structures. Witness, for instance, the following findings reported in Gode and Sunder [1993],

> "We report market experiments in which human traders are replaced by "zero-intelligence" programs that submit random bids and offers. Imposing a budget constraint (i.e., not permitting traders to sell below their costs or buy above their values) is sufficient to raise the allocative efficiency of these auctions close to 100 percent. Allocative efficiency of a double auction derives largely from its structure, independent of traders' motivation, intelligence, or learning. Adam Smith's invisible hand may be more powerful than some may have thought; it can generate aggregate rationality not only from individual rationality but also from individual irrationality."

Surely still more can be done if we can (a) improve our understanding of how markets really behave as they (i) relate to the expectations of participants and (ii) their subsequent performances and (b) use this information to improve individual performances and better the pertinent institutional arrangements. This is what our research and related modelling efforts are directed to and this is especially important for individuals like those studied here who work in arrangements that allow them to become advisors to many thousand of other individuals. It is also important for understanding and evaluating the mutual funds which serve as surrogates for individual investors and have become dominating market forces. Of interest in a further development of these experiments would be a use of "experts" to help individual investors evaluate the risks associated with their possible choices. The latter (i.e., the investors) would then be free to concentrate on different return possibilities, including their timing, and how these different possibilities might mesh with their consumption and other requirements, such as maintaining liquid balances for emergencies.

So far as we know, this study, along with the study in Brockett *et al.* [1993], are the only ones explicitly designed to separate risks from returns so that each could be studied in their *ex ante*-to-*ex post* relations. This has made it possible to identify the possibility that a separate class of risk evaluators is developing to provide helpful advice in risk dimensions while letting investors choose possibilities that could meet their (generally) multiple-objective requirements for returns within the thus identified risk categories. Consistent with its phenomenal growth in recent years, it is also possible that the mutual fund industry is providing individual investors with improved abilities to make their own risk selections--with risks defined as in the preceding sections of this paper.

BIBLIOGRAPHY

Aaker, D.A., & R. Jacobson "The Role of Risk in Explaining Differences in Profitability." *Academy of Management Journal* (1987), 30: 277-296.

Albaum, G.A., G. Kozmetsky and R. Peterson "Attitudes of Newspaper Business Editors and General Public Toward Capitalism," *Journalism Quarterly* (Spring, 1984) 61, pp. 56-65.

Armstrong, J.C., & T.S. Overton, "Estimating Non-response Bias in Mail Surveys." *Journal of Marketing Research* (1977), 14: 396-402.

Baumol, W.J.; Goldfeld, S.M.; Gordon, L.A.; Koehn, M.F. *The Economics of Mutual Fund Markets: Competition vs. Regulations,* (Boston: Kluwer Academic Publishers, 1989)

Bebbington, A.C. "The Effect of Non-response in the Sample Survey with an Example." *Human Relations* (1970), 23: 169-180.

Becker, S.L. "Why an Order Effect." *Public Opinion Quarterly* (1954), 18 (Fall): 221-278.

Blume, M.E. and I. Friend "High Risk Investments: Are They Worth the Gamble?" *The Wharton Magazine* (Spring, 1977) pp. 30-33.

Bourgeois, L.J. "On the Measurement of Organizational Slack." *Academy of Management Review* (1981), 6: 29-39.

Bourgeois, L.J., & Singh, J.V. "Organizational Slack and Political Behavior Within Top Management Teams." *Academy of Management Proceedings* (1983), 43-47.

Bowman, E.H. "A Risk/Return Paradox for Strategic Management." *Sloan Management Review* (1980), 21(3): 17-31.

Boyden, D. P. (Ed.). *Gale Directory of Publications* (21st ed.). Detroit: Gale Research Inc., Book Tower (1989)

Brockett, P.L., A. Charnes, W.W. Cooper, K.H. Kwon and T.W. Ruefli "Chance Constrained Programming Models for Empirical Analyses of Mutual Fund Investment Strategies." *Decision Sciences* 23, (1993) pp. 385-408.

Bromiley, P. "Testing a Causal Model of Corporate Risk Taking and Performance," *Academy of Management Journal* (1991), 34 (1), pp. 37-59.

Byrne, R.F. "Recent Research and New Developments in Budgeting Theory and Practice: A Survey and Evaluation" (1971). Unpublished working paper, University of Pittsburgh, Graduate School of Business.

Byrne, R.F.; A. Charnes, W.W. Cooper and K.O. Kortanek "A Chance-Constrained Approach to Capital Budgeting with Portfolio Type Payback and Liquidity Constraints and Horizon Posture Controls." In R. Byrne, Charnes, A., Cooper, W. W., Gilford, D. & O. A. Davis (Eds.), *Studies in Budgeting* Amsterdam: North Holland Publishing Co. (1971).

Campbell, D.T. and P.J. Mohr P.J. "The Effect of Ordinal Position Upon Responses to Items in a Check List." *Journal of Applied Psychology* (1950), 34: 62-67.

Cantril, H. *Gauging Public Opinion*. Princeton University Press (1944).

Carpenter, E.H. and L.G. Blackwood "The Effect of Question Position on Responses to Attitudinal Questions." *Rural Sociology* (1979), 44(1): 56-72.

Charnes, A., R.L. Clarke and W.W. Cooper "Testing for Organization Slack With R. Banker's Game Theoretic Formulation of DEA." *Research in Governmental and Nonprofit Accounting* (1989), 5: 211-230.

Charnes, A.; W.W. Cooper, K.H. Kwon and T.W. Ruefli "Chance Constrained Programming and Other Approaches to Risk in Strategic Management." Paper presented at a *Conference in Honor of M. J. Gordon* in 1993, L. Gould & P. Halpern (Eds.).

Cochran, W.G. "Some Methods for Strengthening the Common χ^2 Tests." *Biometrika* (1954), 10: 417-451.

Cochran, W. G. *Sampling Techniques*, New York: John Wiley (1961).

Cooper, W. W. and Y. Ijiri (Eds.). *Kohler's Dictionary for Accountants* (6th ed.). Englewood Cliffs, N.J.: Prentice-Hall (1983).

Cowles, A., "Can Stock Market Forecasters Forecast?" *Econometrica* (1933), 1: 309-324.

Cragg, J.G. and B.G. Malkiel, "The Consensus and Accuracy of Some Predictions on the Growth of Corporate Earnings." *Journal of Finance* (1968), 23: 67-84.

Crum, R, R. Laughhunn and J.W. Payne, "Risk Preference: Empirical Evidence and Its Implications for Capital Budgeting." In F.G.J. Derkindren & R.L. Crum (Eds.), *Risk Capital Costs and Project Financing Issues in Corporate Project Selection*: Boston: Martinus Nijhoff (1981).

Cyert, R.M. and J.G. March, *A Behavioral Theory of the Firm* . Englewood Cliffs, N.J.: Prentice-Hall (1963).

Dillman, D.A., "Mail and Other Self-Administrated Questionnaires," Ch. 10 in P.H. Rossi, J.D. Wright and A.B. Anderson, eds., *Handbook of Survey Research* (New York: Academic Press, Inc., 1983).

Dubin, L.E. and L.J. Savage, *How to Gamble if You Must* . Multilith, Washington: U.S. Government Publications (1963).

Ebel, R.L., "Estimation of the Reliability of Ratings." *Psychometrika* (1951), 16(4):407-424.

Edgerton, H.A., S.H. Britt and R.D. Norman, "Objective Differences Among Various Types of Respondents to a Mailed Questionnaire." *American Sociological Review* (1947), 12: 434-444.

Elton, E.J. and M.I. Gruber, *Modern Portfolio Theory and Investment Analysis* (3rd ed.). New York: John Wiley & Sons (1987).

Fama, E.F.; and K.R. French, "The Cross-Section of Expected Stock Returns," *The Journal of Finance* (1992) pp. 427-465.

Ferber, R., "The Problem of Bias in Mail Returns: A Solution." *Public Opinion Quarterly* (1948-1949), 12: 669-676.

Ferriss, A., "A Note on Simulating Responses to Questionnaires." *American Sociological Review* (1951), 16(2): 247-249.

Fink, A., *The Survey Hankbook* Thousand Oaks, Calif., Sage Publications, Inc., 1995.

Franzen, R. andP.F. Lazarsfeld, "Mail Questionnaires as a Research Problem." *Journal of Psychology* (1945), 20: 293-320.

Friend, I., M. Blume and J. Crockett, *Mutual Funds and Other Investors* . New York: McGraw-Hill (1970).

Gode, D.K. and S. Sunder "Allocative Efficiency of Markets with Zero-Intelligence Traders: Markets as a Partial Substitute for Individual Rationality" *Journal of Political Economy*, (1993), 101, pp. 119–137.

Gooding, A.E., "Perceived Risk and Capital Asset Pricing." *Journal of Finance* (1978), 33: 1401-1421.

Gordon, M.J., *The Investment, Financing and Valuation of the Corporation*, Homewood, IL: Richard D. Irwin, Inc. (1962)

Gordon, M.J. , "Two Theories of Corporate Finance." Paper presented at a *Conference in Honor of M.J. Gordon* in 1993, L. Gould & P. Halpern (Eds.).

Grant, D.A. "The Latin Square Principle in the Design and Analysis of Psychological Experiments." *Psychological Bulletin* (1948), 45: 427 - 442.

Han, Bong-Heui, "Dispersion in Financial Analysts' Earnings Forecasts: Implications on Earnings and Risk," working paper, The University of Texas, Austin (1991).

Hulbert, M. (ed.). *The Hulbert Guide to Financial Newsletters* (3rd ed.). Chicago: Probus Publishing Co. (1989).

Jemison, D.B., "Risk and the Relationship Among Strategy, Organizational Processes and Performance." *Management Science* (1987), 33 (9), pp. 1087-1101.

Kahneman, D. and A. Tversky, "Prospect Theory: An Analysis of Decision Under Risk." *Econometrica* (1979), 47: 266-291.
Khandwalla, P.N., *The Design of Organizations*. New York: Harcourt Brace Jovanovich (1977).
Khandwalla, P.N., "Properties of Competing Organizations." In P.C. Nystrom & W.H. Starbuck (Eds.), *Handbook of Organizational Design*, vol. 1: 409-432. New York: Oxford University Press (1981).
Kwon, K.H., Chance Constrained Programming and Other Approaches to Risk in Strategic Management. Ph.D thesis. Austin, Texas: University of Texas at Austin, Graduate School of Business (1991). Also available from University Microfilms, Inc., Ann Arbor, Michigan.
Laderman, J. L., "The Best Returns for the Least Risks." *Business Week*, (February 20, 1989): 80-114.
Lakonishok, J., A. Shleifer and R. Vishny "Contrarian Investment, Extrapolation and Risk" (NBER Working Paper No. 4360) as quoted in *The NBER Digest* (Boston: National Bureau of Economic Research, Sept. 1993) p. 1.
Lawlis, G.F. and E. Lu, "Judgment of Counselling Process: Reliability, Agreement, and Error." *Psychological Bulletin* (1972), 78(1): 17-20.
Leitch, G. and J.E. Tanner, "Economic Forecast Evaluation: Profits vs. the Conventional Error Measures." *American Economic Review* (1991) 81, No. 3 pp. 580-590.
Leonard, M. and H.P. Hudson (Eds.). *Hudson's Newsletter Directory* Rhinebeck, N. Y. (1988).
MacCrimmon, K.R. and D.A. Wehrung, *Taking Risks*. New York, N.Y.: Free Press (1986).
Mao, J.C.T., "Survey of Capital Budgeting: Theory and Practice." *Journal of Finance* (1970), 25(2): 349-360.
March, J.G. and Z. Shapira, "Managerial Perspectives on Risk and Risk Taking." *Management Science* (1987), 33: 1404-1418.
Markowitz, H.M., *Portfolio Selection*. New York: John Wiley & Sons, 1959.
McWhinney, W.H., "Aspiration Levels and Utility Theory." In G. Fisk (Ed.), *The Psychology of Management Decision*: Lund, Sweden: GWK Gleerup, Publishers (1967).
Pace, C.R., "Factors Influencing Questionnaire Returns from Former University Students," *Journal of Applied Psychology* (1939), 23: 388-397.
Payne, J.W., D.J. Laughhunn and R. Crum, "Translation of Gambles and Aspiration Level Effects in Risky Choice Behavior." *Management Science* (1980), 26: 1039-1060.
Petty, J.W., II and D.F. Scott, "Capital Budgeting Practices in Large American Firms: A Restrospective Analysis and Update." In F. G.J. Derkinderen & R.L. Crum (Eds.), *Reading in Strategy for Corporate Project Selection* : Boston: Martinus Nyhoff (1981).
Pike, R.,, L. Dalgleish and K. Jackson, "Order Effects in Recognition Latency for Multi-item Probes." *Journal of Experimental Psychology* (1985), 11(2): 248-261.
Pratt, J.W., "Aversion to One Risk in the Presence of Others," *Journal of Risk and Uncertainty*, 1988, 1, pp. 395-413.
Shapira, Z., "Risk in Managerial Decision Making." Unpublished Manuscript (1986), Hebrew University.
Singh, J.V., "Performance, Slack, and Risk Taking in Strategic Decisions." *Academy of Management Journal* (1986), 29: 562-585.
Sitkin, S.B. and A. Pablo, "Reconceptualizing the Determinants of Risk Behavior," *Academy of Management Review* (1993).
Snedecor, G.W., *Statistical Methods* (3rd ed.). Ames, Iowa: Iowa State College Press (1940).

Suchman, E.A., and B. McCandless, "Who Answers Questionnaires?" *Journal of Applied Psychology* (1940), 24: 758-769.

Sunder, Shyam, "Behavior of Economic Systems under Opportunity Set Constrained Zero Intelligence Traders," *Journal of Economic Behavior and Organization*, special issue in honor of Richard M. Cyert, 1995.

The Individual Investor's Guide to Investment Publications (1st ed.). International Publishing Co., Inc.

Tinsley, H.E.A. and D.J. Weiss, "Interrater Reliability and Agreement of Subjective Judgments." *Journal of Counselling Psychology* (1975), 22(4): 358-376.

von Neumann, J. and O. Morgenstern, *Theory of Games and Economic Behavior* (2nd ed.). Princeton, N.J.: Princeton University Press (1947).

Weiss, G. , "Fixed-Income Funds: Where Safety Pays." *Business Week*, (February 26, 1990) pp. 94-109.

Zimmer, H., "Validity of Extrapolating Non-response Bias from Mail Questionnaire Follow-ups." *Journal of Applied Psychology* (1956), 40: 117-121.

APPENDIX

COVER LETTER

Dear:

The attached questionnaire forms part of a study we are undertaking on risk taking and investment behavior. Your help as an acknowledged expert in the field of open-end mutual funds could be of great service to us. We are therefore asking you to complete the attached questionnaire which we hope you will return promptly in the enclosed stamped envelope.

The enclosed questionnaire has been designed so that it should be easy to complete in a very short time. I will also be glad to answer any questions if you call me at.... As noted at the end of the questionnaire, this is my office number at.... Although the University does not accept collect calls, I will be more than happy to reimburse you for any such charges.

Many thanks for your cooperation. If you would like a copy of this study, please complete the information which is requested at the end of the questionnaire and I will be glad to send a copy as soon as it is available.

QUESTIONNAIRE PAGE 1 OF COPY ____

Your assistance as an expert in investments is solicited by asking you to complete and return this questionnaire as part of a study of risk and risk taking which is being conducted at the University of.... School of Business. A prompt response will be of great value to us. Results will be published only in aggregate form to protect the confidentiality of individual responses. To receive a copy of the completed study, check the box at the end of the questionnaire.

Questions Refer to Open-End Mutual Funds Investing in Equities.

1. Taking account of the Fund objectives shown on the left, I would rate these objectives with respect to risk and return in a typical year as follows: (Please check appropriate boxes in each row.)

STATEMENT OF OBJECTIVES	RISK					RETURN				
	Very Low	Low	Mod-erate	High	Very High	Very Low	Low	Mod-erate	High	Very High
Asses Allocation	☐	☐	☐	☐	☐	☐	☐	☐	☐	☐
Aggressive Growth	☐	☐	☐	☐	☐	☐	☐	☐	☐	☐
Balanced	☐	☐	☐	☐	☐	☐	☐	☐	☐	☐
Equity Income	☐	☐	☐	☐	☐	☐	☐	☐	☐	☐
Growth	☐	☐	☐	☐	☐	☐	☐	☐	☐	☐
Growth-Income	☐	☐	☐	☐	☐	☐	☐	☐	☐	☐
Income	☐	☐	☐	☐	☐	☐	☐	☐	☐	☐
International	☐	☐	☐	☐	☐	☐	☐	☐	☐	☐
Option Income	☐	☐	☐	☐	☐	☐	☐	☐	☐	☐
Specialty	☐	☐	☐	☐	☐	☐	☐	☐	☐	☐
Small Company	☐	☐	☐	☐	☐	☐	☐	☐	☐	☐
Financial	☐	☐	☐	☐	☐	☐	☐	☐	☐	☐
Health	☐	☐	☐	☐	☐	☐	☐	☐	☐	☐
Precious Metals	☐	☐	☐	☐	☐	☐	☐	☐	☐	☐
Natural Resources	☐	☐	☐	☐	☐	☐	☐	☐	☐	☐
Technology	☐	☐	☐	☐	☐	☐	☐	☐	☐	☐
Utilities	☐	☐	☐	☐	☐	☐	☐	☐	☐	☐

QUESTIONNAIRE PAGE 2 OF COPY ____

2. Please rank the following according to how close they come to the meanings you assign to the words risk and return. Assign a 1 to the one that comes closest, assign a 2 to the one that is next in order of closeness, and so on.

RISK Definition	Rank	RETURN Definition	Rank
Chance of Missing More Favorable Opportunities	☐	High Chance of Exceeding a Target Rate of Return (Such as the Return on the S&P 500)	☐
Chance of Making Less Than Target Rate of Return (Such as T-Bill Rate)	☐	Large Ending Value of Portfolio	☐
Chance of Loss	☐	Large Rate of Return on Assets	☐
Chance of Bankruptcy	☐	Large Capital Gains	☐
Chance of Unfavorable Event	☐	Large Dividends	☐
Large Fluctuations in Value of Portfolio	☐	Large Profit	☐

3. None of the above comes very close. My own usage of the terms risk and return can be summarized as follows. (Use other side of page if necessary).

By RISK I mean ____
____
____

By RETURN I mean ______________________________________
____
____

QUESTIONNAIRE PAGE 3 OF COPY ____

INSTRUCTIONS: Please complete this questionnaire and return it in the enclosed self-addressed, stamped envelope. Thank you very much for your help. If you have any questions, please call Professor.... Although the University ...does not accept collect calls, we will be more than happy to reimburse you for such charges. To receive a complimentary copy of the completed study, please supply the following information.

CHECK ONE:

☐ YES, I would like to receive a copy of the completed study at the following address: (Please Print)

____

Name

____

Title

____

Name of Company

____

Address

____

City State Zip Code

☐ NO, I do not wish to receive a copy of the completed study.

TABLE A-1

Comments by Editors of Financial Newsletters on their own Usages for Risk and Return

RISK	RETURN
1. Greater than U.S. Treasury, but in ranges above U.S. Treasury.	Principally, "total return." For investors interested primarily in income, the "return" is so understood.
2. The chance of encouraging investors to buy high and sell low.	Total return (dividends and capital gain reinvested) in relation to inflation - accounting for tax effects.
3. Standard deviation	Alpha (return in excess of market).
4. A calculated measure of relative chance of loss (or damage) to a valued asset.	The ultimate yield, profit or enhancement obtained through labor, rental or loan of funds, via an applied skill or management expertise, over time.
5. When 90% of capital can be lost under reasonably normal fluctuations in particular investment.	How much can be expected to make on a particular investment.
	RETURN ONLY
	"Potential" for large gain, etc.
	Total return (capital gain + income)

TABLE A-2

Comments by Brokers on their own Usages for Risk and Return

Risk and Return Questionnaire	
1. Great chance of fluctuation and/or loss of capital	Profit via growth [and] income or a contribution of both
2. Chance of loss	Total dollar return on assets
3. The client's comfort level frequently defines risk. Some people see C.D.'s and S&L's are too risky. Others are used to risk in several business ventures and are comfortable with it.	A Rate of return over at least 5 years greater than the S&P 500. The best return would have less volatility as well.
RISK-only Questionnaire	**RETURN-only Questionnaire**
Risk to me means a chance of losing parts or all of initial investment. Varying degrees of return do not denote risk to me. - Of course, as your survey seems to imply, risk is a relative term which means different things to different people. Our job as financial advisor is to determine our client's tolerance for risk in making recommendations for them in investment.	Total return to client which includes $ capital gain and $ dividends as a percentage of $ investment above purchase costs, including commission and fees. $Return = \frac{current\$NAV/unit - purch.\$NAV/unit}{purchase\$NAV/unit}$
All your choices have a *negative* overture. A relative measure of an investment against an average return of many investment categories...both above and below the average. A high rate of return can result in *both favorable and unfavorable outcomes*. While a low rate can be considered *very* risky as well (even though the perception is very low risk, i.e., passbook savings at 5%).	My own definition: [By return I mean] taking the ending value of an investment and subtracting the basis. Divide that number by the original investment and get a percentage. You may use that percentage as a rate of return of investment or then annualizes it. If this investment is part of a portfolio of investments then it is helpful to take a weighted annualized return of all investments.
Loss of capital	
When you just say risk usually I think of loss of principal. However, a more appropriate statement might be inflation risk: i.e., a T-bill is safe from loss of principal, but after taxes and inflation take its effect what is left, usually pennies to dollars. So I ask/tell my clients what happens over time if you are to just break even.	

A Multi-Dimensional Framework for Portfolio Management

Winfried Hallerbach and Jaap Spronk[1]

[1]Both Department of Finance, Erasmus University Rotterdam, P.O.Box 1738, NL-3000 DR Rotterdam, The Netherlands

1 Introduction

Given an opportunity set of financial securities, an investor faces the problem of selecting one portfolio out of the multiplicity of portfolios that can be formed. This paper presents a multi-dimensional framework that can serve as a decision aid in the portfolio selection and management process. The framework yields room for different settings of the portfolio management problem and offers an alternative to both unstructured *ad hoc* approaches and complex approaches that severely restrict the decision process. The framework offers the investor systematic guidance in the search for a portfolio that suits his tastes and preferences best, while satisfying the restrictions he faces.

After sketching relevant features of the investment decision process in section 2, we present our framework for portfolio management in section 3. Section 4 describes several ways of using the framework for portfolio selection and management in practice and section 5 concludes.

2 The investment decision

In order to support investment decisions, both the desires (preferences) of the investor and the characteristics of the investment opportunities should be adequately understood and related to each other. Unfortunately, in most models proposed for portfolio management, the real world is replaced by a simplified model-world, focusing on the 'average' investor, instead of the particular (typically non-average) investor at hand. However, the assumptions made to describe this average investor are often inadequate and may even be misleading. Consider the mean-variance framework: not only the traditional but also the most popular approach to the investment decision problem. The apparent contribution of Markowitz (1952, 1959) and Tobin (1958) is undoubtedly that they replace the classic uni-dimensional approach to investment (focusing solely on expected or mean return) by a two-dimensional approach. The formerly undefined notion of risk is formalized by identifying risk with variability of returns in a portfolio context, and operationalized by means of the (co-) variance or standard deviation. The problem, then, is to translate actual decision situation in terms of mean-variance dimensions. However, imposing this two-dimensional world may be too restrictive in practice because the mean-variance framework uses rather strong assumptions on the preferences of the investor and/or the representation of investment alternatives. Variance as a risk measure may miss its link with an investor's preference structure or with the distributions of security and portfolio returns. Information concerning mean and variance is not always sufficient to adequately discriminate between investment alternatives. Still, in many applications, the choice for mean-variance analysis seems almost natural and taken for granted, together with the restrictive nature of the underlying assumptions. In this way, the mean-variance framework becomes a *Procrustes bed*, chopping off the multi-dimensional aspects that may be perceived by the investor.

In order to elaborate the relationship between the decision context of the investor and the economic environment of the securities, we decompose the investment decision process in the following stages:

(1) *security analysis* to determine the relevant characteristics (or attributes) of the investment opportunities,
(2) *portfolio analysis* to delineate the set of non-dominated or 'efficient' portfolios,
(3) *portfolio selection* to choose the optimal portfolio from the efficient set, and
(4) *preference analysis*.

These stages are indicated by capitals in Figure 1.

In the *economic environment* box, the securities in the opportunity set are described in terms of various dimensions (attributes) in which securities are likely to differ: expected return, 'risk', maturity, income component, liquidity, manageability, taxability etc. When securities are issued by a firm (like common stocks or corporate bonds) also characteristics of the corresponding firm can be linked to the securities. This view on securities is objective in the sense that it is an 'outsider's view'. Many of these attributes are enumerated in investment text books (see also section 3).

In the *decision environment* box, the investor's profile is described, reflecting the decision context and comprising the investment objectives that the investor wishes to attain, the restrictions faced, and his tastes and preferences. Obviously,

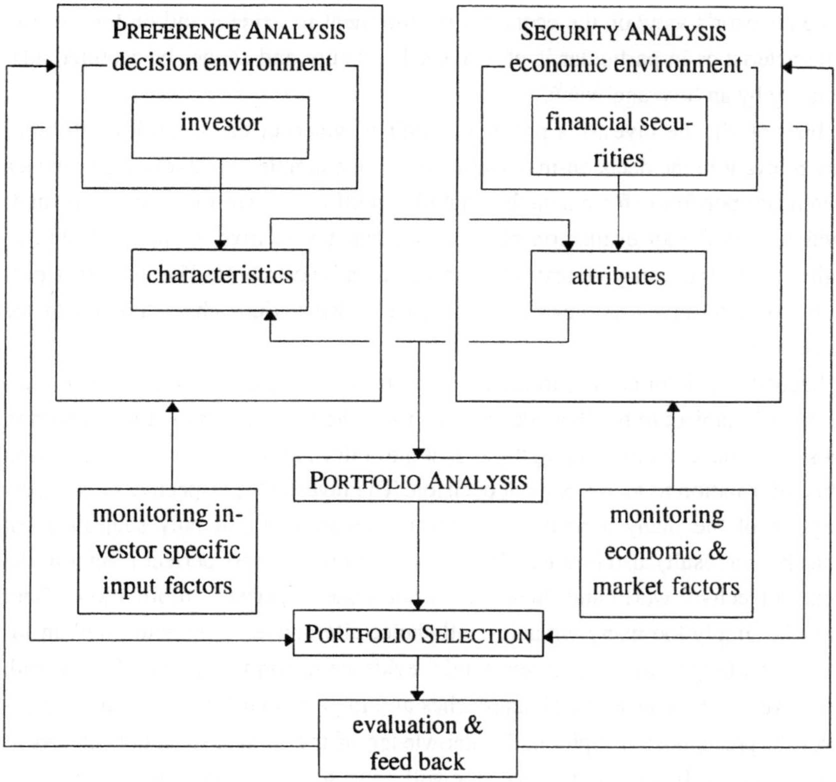

Figure 1. Global scheme of the investment process.

an investor who has to manage the portfolio of a large institution, who is severely restricted in deciding where and how to invest and who is facing tight limits on the future liquidity of his portfolio, is in quite a different position than a small private investor who is free where and how to invest but who faces high transaction costs and is very limited in his information processing capacity. In more formal terms: the investor's objective function may be multifarious and complex, and may be subject to constraints. Because of specific circumstances, the non-average investor will have to deal with externally imposed constraints. In addition, many investors impose constraints themselves, for instance because they do not like a certain class of assets or because they have only know-how with respect to certain asset categories and want to reduce the complexity of the investment problem.

The investor's view on the economic environment in general and on the security characteristics in particular is of a subjective nature and cannot be captured adequately by an 'average' view.

First of all, the investor's profile determines which of the securities' attributes are relevant in the decision-making process. For example, the investor may have a reference portfolio (for example a liability portfolio in case of a pension fund) which calls for an evaluation of security attributes relative to this portfolio. In other cases, there may be restrictions on foreign investments. The investor's profile also determines the degree of relevance of the various characteristics of the securities.

Secondly, it is important to realize that an investor's evaluation of security attributes is subject to his 'bounded rationality'. The human mind is limited both in observing data and processing these data into information, and next in translating this information to an investment decision. The investor's perspective on and perception of the many aspects of a decision situation is not only subjective but (partly as a result) also limited. The investor will not possess perfect insight in the real, 'objective' world and, hence, does not possess 'perfect' information. There may be simply too many variables and choice alternatives to monitor, and an investor is likely to use any circumstantial evidence to form a picture of the world. However, many conventional approaches and in particular the mean-variance approach, presuppose a high-quality knowledge of the joint distribution of investment returns. In reality, the investor will have some information on the future returns of possible returns but this picture will be incomplete. Not only because the possibilities to gather and process information are for every investor limited (be it that these limits differ for different investors) but also because returns depend for a large part on future prices, rates and other variables that are in principle unpredictable (e.g. because they are generated in efficient markets).

Given the relevant securities' attributes as perceived by the investor, the portfolio composition stage comprises the combination of securities into a portfolio that exhibits a constellation of attributes according to the investor's feelings or preferences. The preference structure of the investor is normally more complicated than the relatively simple utility functions assumed within the mean-variance framework. The investor may well have other objectives than financial value maximization alone. For instance, he may want to achieve a stable growth rate of the portfolio's value or a minimum pay-out ratio. Here again, however, bounded rationality will leave its traces. It would be utopian to suggest that the investor can list all available alternatives, compare them and choose the best or optimal alternative. The decision process will instead be characterized by a step-by-step search for an alternative that satisfies his requirements. Optimizing behavior is then replaced by satisficing behavior. Contributions to investment decision problems then do not entail the specification of 'optimal' decision rules, but the design of systematic search procedures that help the investor scan the feasible choice alternatives. Note in this respect that within the mean-variance framework it is not possible to 'play' with the location of self-imposed constraints (although many investors might be willing to change a locally-imposed constraint, for example when that would considerably increase expected return).

As portfolio investment is an ongoing process, the investor's profile as well as the securities profiles need to be monitored continuously. Any relevant change is incorporated in the portfolio composition process. In addition, information about the performance of the investment portfolio is fed back and the investment cycle starts again.

We conclude that it is important that the interrelationship between the decision context and the economic environment is explicitly recognized in framing financial decisions. Given the strong interdependence between investor's and securities' characteristics, the evaluation of security characteristics must take place relative to an investor's own unique circumstances. It is precisely this notion that underlies the approach we propose in section 3.

3 A general framework

The general framework for molding the investment decision process can be labeled as a 'multi-attribute approach to portfolio selection'. The notion that the portfolio investment decision for individual investors calls for a multi-attribute approach dates back to Smith (1974, p.53). Of course, there are many more examples of molding the portfolio selection problem in terms of multi-criteria decision making (see section 3.2). In comparison with the general framework described below, most (if not all) of these approaches are at best only partial. For instance, some approaches (like Arthur & Ghandforoush, 1987) fail to give room for the inherent complexity of the decision procedure given the investor's specific decision context. Other approaches (like O'Leary & O'Leary, 1987) concentrate on the beauties of a particular multiple criteria decision method, without doing full justice to the decision context and to the results and principles of financial economic theory.

The general framework proposed here consists of two stages: the formulation of a multi-attribute representation of securities and the selection of a portfolio. The multi-attribute representation of securities assumes that an investor can demarcate a set of security attributes that he considers relevant. For the investor, a financial security then represents a basket of, say, k attributes and can fully be characterized by a k-tuple of attribute values. In this view, when buying a security, an investor is actually buying an exposure to various attributes. The issue of multi-attribute portfolio selection is to balance the attributes of the individual securities on the portfolio level. That is, given the security attributes and the investor's profile (personal context), the attributes of his portfolio must be fashioned in a way that suits his particular circumstances and preferences best. The two stages will be discussed in more detail below.

3.1 Multi-attribute representation of securities

In representing securities, we distinguish between attributes that are directly return related (comprising (explicit) expected return and risk measures) and those that are indirectly return related (which are more loosely related to risk and return).

Directly return-related attributes

Most academic work assumes that securities can fully be characterized by the joint distribution of their returns. Any probability distribution can fully be described by means of its locus and its shape. An obvious attribute candidate then is the 'expected return' on a security, which refers to the location parameter of the return distribution.

The risk attached to a security's return is directly related to the shape of its distribution. The shape of a distribution can be described either by its shape parameters or by statistical moments. From the view of portfolio formation, not the (marginal) distributions of the securities must be represented, but their joint distribution. When relying on either shape parameters or moments as risk attributes, one faces three problems, each related to one of the stages in the investment decision process as described in the former section.[2] In the stage of security analysis,

[2] Additional problems are of a pure statistical nature. When relying on shape parameters one must explicitly assume that the security returns are generated by some specific distribution. Furthermore, it is required that the distribution belongs to the stable class, i.e. the set of distributions that are closed under addition. Otherwise, portfolio returns would obey distributions different from those of the securities and hence possess different shape characteristics. Alternatively, one could try to describe the distributions' shapes by means of their moments. Unfortunately, there is no one-to-one relationship between the shape of a

an *'information problem'* arises because all of the interactions between the security returns on the level of the relevant parameters or moments have to be accounted for. In the stage of portfolio analysis, a *'combination problem'* arises because the relevant security attributes must be processed and aggregated in order to obtain portfolio attributes. Finally, there is a *'criteria problem'*: in order to incorporate the probabilistic information in the decision process one must specify both the investor's tastes with respect to each of these attributes and their relative importance. It follows that the risk dimension is truly problematic in its complexity. The ambiguity of the aspect of perceived risk is an additional problem. Indeed, the very lack of adequate risk definitions may be symptomatic for the multi-dimensional nature of risk.

In order to tackle the information problem and the combination problem, we suggest a multiple factor approach for extracting risk measures. The key assumption is that the returns of the securities in the opportunity set are influenced or generated by a series of identifiable economic variables or 'factors' (cf. Chen, Roll & Ross, 1986, and Berry, Burmeister & McElroy, 1988). Each of these factors represents a dimension of the economic environment in which the security returns are generated. Depending on the specific circumstances (the profile) of the investor, a specific set of factors may be relevant. The relationship between a security's return and changes in these factors is described by a response coefficient or factor sensitivity. By means of these sensitivities, the joint distribution of security returns is linked to the joint distribution of factor changes. In this interpretation, the sensitivity coefficients can serve as risk measures. Considering factor sensitivities as relevant security attributes, we arrive at a *multi-factor representation* of security returns. As the variability of returns is linked to the variability in various identifiable economic variables, investment risk becomes an intuitive and multi-dimensional concept. In the context of this risk concept, the investor is assumed to have some idea of the securities' factor sensitivities.

The use of factor models permits replacing return variance as a uni-dimensional risk measure by multi-dimensional risk measures. These measures provide more insight in the nature of risk than a uni-dimensional, 'aggregate' risk measure. In addition to this decision-theoretic argument there is a statistical argument: the use of factor models for simplifying the representation of joint return distributions is

---

distribution and its moments. For example, zero odd-order moments are a necessary and not a sufficient condition for distribution symmetry. Cf. Barnes, Zinn & Eldred (1978) for references on this point.

indispensable for enhancing the computational tractability and practical applicability of risk measures.

Indirect return related attributes

Experiences from practice show that not all relevant information seems to be captured by explicit return and risk attributes. To fill this gap, we leave room for additional attributes that may be incorporated at the investor's discretion. These other attributes may be considered of general relevance in practice, but may also be relevant because of idiosyncrasies in the investor's personal decision context. In the latter case, the incorporation of additional attributes can be motivated from either the specific tastes and desires (goals) of the investor, from specific investment constraints he faces, or from distinctive characteristics of the investment alternatives. In short, the investor can simply indicate that there exist various other attributes with which he can discriminate between the attractiveness of various securities. For example, because of the investor's tax situation, the taxability of the portfolio components may be a relevant attribute (in which respect the portfolio's dividend yield may be important). In terms of 'liquidity' or the flexibility to revise the portfolio's composition, the marketability of the component securities may be relevant. Because of some method of performance measurement, the position with respect to some benchmark portfolio may be relevant, and so on. In addition, the investor may adhere to the notion that not all future events can be reduced to probability distributions, not even when the latter are of a subjective nature. This also implies that attributes may be considered in addition to explicit elements of return and explicit components of risk. We must seriously consider the possibility that some of these 'other' attributes act in fact as proxies for (components of) expected return and risk.

One way to concretize the potential relevance of various stock attributes is to look at the variables that appear in schemes for fundamental company analyses and industry or sector analyses.[3] A more direct way to detect dimensions in which the appraisal of securities (stocks) may differ is to look at the security analyses as conducted by investors in practice. The early study by Baker & Haslem (1974), for example, concludes that the investor's investment analysis of common stock appears to be a multi-dimensional process. In particular, they find that investors greatly differ in their perceptions of the importance of dividends, future (sales and earnings) growth expectations and financial stability.

---

[3] For these schemes, we refer to Ross, Westerfield & Jaffe (1993, Ch.2) or Reilly (1994, Chs.17&18).

With some imagination, the attributes stemming from sources as mentioned above may be labeled 'demand pull'. The data are generally available or can be obtained without many efforts, and investors may use some or all of these data in some way or another. Although the specification of relevant attributes is on the discretion of the investor himself, we can draw a clearer picture of the importance of indirect return related attributes by referring to attributes whose relevance is acknowledged through empirical study. Many of these 'validated' attributes are used and advertised by professional investors and may be marked 'supply push'.[4]

Many attributes are considered important, not only from a practical point of view, but also from an academic point of view because they represent 'anomalies'.[5] For common stocks, 'firm size' is a long-time notorious variable. Other examples are price ratios as indicators for fundamental firm value, like earnings/price, book/price (book value of common equity per share divided by market price per share), cash flow/price, sales/price and dividend/price. In the context of 'value investing' there is great renewed interest in these long time familiar attributes.[6]

In the view of (positive) financial theory, an attribute's ability to contribute to the explanation of cross-sectional return differences appears to be a convincing criterion for the selection of relevant attributes. However, an attribute will only carry a significant premium when it is 'priced' in the market. However, a non-average investor can face a set of investment opportunities that is different from the market (i.e. the average investor). Hence this investor is only interested in the relevance of this attribute in his opportunity set. Furthermore, partly connected to the

---

[4] Some examples are Goldman, Sachs & Co. (Jones, 1990, and Jones, Kohn & Melnikoff, 1990), Salomon Brothers (Sorensen, Mezrich & Thum, 1989, and Bower & Bower, 1991) and BARRA International (Arnott, Kelso, Kiscadden & Macedo, 1989, and Fogler, 1990). Note that we do not suggest that for this reason any investor should consider these attributes. Rather, the implication applies in reverse: because investors in general show a preference (or an aversion) towards some attribute, this can have a negative (positive) effect on the return on a security that 'has much' of this attribute, and vice versa.

[5] An attribute is an anomaly with respect to an asset pricing theory when that attribute possesses power to explain cross-sectional variation in expected returns in addition to the risk measures as specified by the pricing model at hand. An attribute is an anomaly with respect to the efficient market hypothesis when it can be used to forecast future returns. Detailed overviews are provided by Fama (1991) and Hawawini & Keim (1995).

[6] See for example Fama & French (1992, 1993) and Lakonishok, Schleifer & Vishny (1994).

former argument, the reward that an investor attaches to the exposure to an attribute (a 'subjective' premium) may well be different from the premium that the market as a whole attaches to that attribute (the 'objective' premium). This leads us back to the starting point that the selection of attributes depends on the personal circumstances of the investor, as summarized in his profile. In brief, there exist security attributes that are relevant in practice, despite the official view of financial theory.

As a whole, the multi-attribute representation of securities comprises a detailed and investor-specific security analysis. Preference information is used to demarcate the set of k attributes that an investor considers important. As a result, each of the securities in the opportunity set can be adequately characterized by the values that the respective k attributes take on. The selection of relevant attributes is no 'once and for all' activity. The investor's decision context and the securities' economic environment may change over time and may become 'better understood' because of 'learning effects'. As a result, the set of relevant attributes may change over time.

3.2 Multi-attribute portfolio selection

As the attributes are evaluated in a portfolio context, the first issue in portfolio analysis, then, is aggregating the security attribute scores according to their investment fractions. As most of the potential attributes considered so far are linear, this will not pose any difficulty.[7] The second issue is to confront the multi-attribute representation of securities with the investor's preference structure in order to evaluate feasible portfolios and to select a portfolio. We first discuss investor tastes and preferences about securities, expressed in scores on attributes. Next, we summarize standard approaches that are currently employed in investment practice. Finally, we present an alternative approach that does justice to the flexibility allowed in the security analysis and preference analysis.

---

[7] Note that the direct return related attributes (expected return and factor sensitivities) are linear. However, the indirect return related attributes can cause some problems. For example, individual securities' price/earnings ratios must be aggregated in *harmonic form* in order to obtain a portfolio's price/earnings ratio. It is then simpler to consider the securities' earnings/price ratios, which can be aggregated in a linear fashion to a portfolio earnings/price ratio.

Choosing between attribute exposures

The step from securities to their representation in terms of attribute scores can be justified by referring to consumer theory, where 'characteristics models' have been developed for describing consumer behavior. In this respect we especially note Lancaster (1966, p.133), whose contribution is "breaking away from the traditional approach that goods are the direct objects of utility and, instead, supposing that it is the properties or characteristics of the goods from which utility is derived."[8] These implied characteristics models opened the way for a theory of multi-attribute choice.

Transposed to the investment decision, we can assume that investors buy securities for the attributes they offer and that different securities are essentially different packages of attributes. This implies that investors choose between security attribute exposures instead of between the uni-dimensional securities or their returns. Hence, an investor's preference functional is directly specified in the multi-dimensional terms of relevant security attributes.[9] For an investor, a financial security then represents a basket of, say, k attributes and can fully be characterized by a k-tuple with scores or values that the attributes take on. In an investor's view, when buying a security, he is actually buying an exposure to various attributes. Hence, we can specify a mapping of the securities in the space spanned by the attributes:

$$\text{security } i \rightarrow \{ a_{i1}, a_{i2}, \ldots, a_{ij}, \ldots, a_{ik} \} \qquad , i \in N \qquad (1)$$

where a_{ij} is the value that attribute j takes for security i. Likewise, when composing a portfolio, the investor is actually composing an appropriate portfolio exposure to the various attributes:

$$\text{portfolio } p \rightarrow \{ a_{p1}, a_{p2}, \ldots, a_{pj}, \ldots, a_{pk} \} \qquad (2)$$

For a given portfolio, its exposure to a certain attribute can be calculated as a weighted average of the attribute exposures of the individual securities contained in this portfolio. The fractions invested in each of these securities can thus be treated as instrumental variables. Therefore, the attribute exposures can be seen as

---

[8] For a discussion and review of characteristics models, we refer to Deaton & Muellbauer (1980).

[9] Consumer theory considers decision making under certainty. Because we are dealing with situations of uncertainty or risk, we assume that some subset of attributes captures the risk aspects, so that attitudes towards risk can be reflected in the preference functional.

goal variables which are linear in the portfolio holdings. Often, the investor will try to either minimize or maximize each of these goal variables. Alternatively, the investor may strive to attain a target level or desired score on some attribute(s). Depending on the investor's insights and preferences, the relative importance of each of these goals may vary. Generally, no portfolio can be found for which each of the goal variables reaches its optimal value or for which all criteria are met. As a consequence, the investor has to evaluate the trade-offs between the various goal variables.

Alternative selection procedures

There are several routes leading to the selection of a portfolio, depending on the amount of information available on the investor's preference structure.

Assuming a large amount of preference information, the traditional utility framework could be extended to a multi-dimensional context by casting a utility function in terms of multiple portfolio attributes. Consequently, the mean-variance preference functional $Z(E_p, \sigma^2_p)$ is replaced by a 'Lancaster (1966)-type' of function $Z(a_{p1},...,a_{pk})$. In that case, an explicit optimization problem can be formulated and solved. Unfortunately, the complexity of specifying a multi-attribute preference functional is enormous and not likely to be overcome in practice. In multi-attribute utility theory, this complexity is reduced by assuming (strong) separability of the preferences. When this assumption is satisfied, a series of uni-dimensional (i.e. single attribute) utility functions can be assessed, whereafter these component functions are combined (in a linear, multiplicative or other fashion), using information about attribute trade-offs. In this way, the exposures are evaluated attribute by attribute and then combined to obtain an overall measure of desirability. Still, this places a heavy information burden on the investor. The problem here is to ex ante specify the uni-dimensional preferences for each of the attributes as well as the overall preference functional that incorporates the evaluation of a combination of attribute exposures and their trade-offs.

Another route is to cast the multi-dimensional preference functional in the form of a (linear) programming model. One way is to maximize the portfolio's exposure to one attribute (expected return, e.g.) subject to restrictions on the other attribute scores.[10] The problem with such a specification is that it is intrinsically uni-dimensional: only one attribute is optimized, while the other attributes only serve as constraints. Another way to extend the linear programming formulation to a multi-dimensional context is to use a weighted average of the various attrib-

---

[10] Cf. Sorensen & Thum's (1992) portfolio optimizer.

utes as the objective function.[11] A linear programming formulation like this can only be employed when the trade-offs between the attributes can be specified properly.

In formulating priorities and targets with respect to attributes and attribute exposures, goal programming offers more flexibility. The applicability of multiple goal programming to the portfolio problem was recognized in an early stage. In the (E,σ^2)-context, we have Lee (1972), Lee & Lerro (1973), Kumar, Philipatos & Ezell (1978), Lee & Chesser (1980), Spronk (1981) and O'Leary & O'Leary (1987). Aside from expected return and risk, some indirect return related attributes (notably dividend yield) are specified, but a truly multi-attribute representation is not pursued. In all studies but the last, risk is accounted for by Sharpe's (1967) linear approximation of portfolio return variance on the basis of the single index model, or simply by a target for a market index model beta. Of course, the number of attributes could easily be extended towards multi-dimensional risk measures and various additional indirect return related attributes. Multiple goal programming indeed has some attractive properties. It shows a close correspondence with decision making in practice, the goals are formulated as aspiration levels and there is always a solution for a well-defined problem (with a non-empty feasible region), even if some goals are conflicting. An important drawback of multiple goal programming, still, is its need for fairly detailed a priori information on the decision-maker's preferences.

Interactive programming methods, in contrast, neither require an explicit representation or specification of the decision-maker's preference function nor an explicit quantitative representation of the trade-offs among conflicting goals. By its nature, an interactive procedure progresses by seeking this information from the investor, removing the need for explicitizing the preference structure. For the investment problem as sketched in this study, we propose Interactive Multiple Goal Programming (henceforth IMGP), as developed by Spronk (1981). In this procedure, the investor reduces the set of alternatives interactively and systematically, thus conditioning the quality of the remaining portfolios. The investor has several options. He can continue until the remaining set of feasible portfolios becomes very small. Another possibility is to select a suitable portfolio from the set of portfolios satisfying the minimum requirements. In this respect, IMGP produces at each iteration a set of non-dominated portfolios. Finally, a set of feasible portfolios satisfying the minimum conditions on the goal values can be subjected to a second analysis by the investor. In his decision context, the investor may wish (or

---

[11] One example is Arthur & Ghandforoush (1987).

need) some elbow-room, thus requiring more than just one portfolio. The procedure then offers adequate flexibility to incorporate other, hard to quantify, criteria into the decision making process. IMGP incorporates all the advantages of 'traditional' goal programming, while circumventing the unnecessary burden of obtaining a 'complete' picture of the investor's preference pattern. In our opinion, this approach offers the desired degree of flexibility to be fruitfully applied to the multi-attribute portfolio selection problem. By tuning the attribute exposures, a specific portfolio profile can be obtained that matches the investor's profile. The stages of portfolio analysis and portfolio selection are no longer treated separately but are integrated. The interactive method then is no optimizer, but can better be described as a 'combinizer': it allows systematic scanning of the set of feasible portfolios and the selection of an optimal portfolio via an interactive process. In the interactive decision process, a learning process is embedded. By scanning the feasible portfolios, the investor first gets a feeling for the trade-offs that in the opportunity set exist between the exposures to the various attributes. Second, the investor can shape and adjust his preferences when confronted with the trade-offs between the attributes. It is in no way required that the investor performs the interactive process only once. He can explore the opportunity set in all dimensions, and is even advised to do so in order to get insight into the properties of the opportunity set at hand. Since the interactive procedure is path-independent, no desirable (feasible) alternatives can be missed, only insight can be gained.

4 Different ways of using the framework

A distinctive feature of the multi-attribute approach to portfolio selection is that it can accommodate an investor's specific decision context, as summarized by his goals, tastes and restrictions. The proposed framework offers ample flexibility and can be used in many different ways. In particular, the multi-dimensional risk concept that emerges from multi-factor models provides opportunities for a sophisticated management of investment risks.[12] Applications range from mere portfolio analysis to the actual selection of an optimal portfolio, from defensive to aggressive strategies and from passive to active strategies. A portfolio strategy now entails the choice of an appropriate pattern of factor sensitivities or, in general, of attribute scores.

---

[12] We distinguish between an 'Arbitrage Pricing Theory approach' and a 'multi-factor model approach'. The former is restricted to manipulating the exposure to market priced risk factors, whereas the latter considers both priced and non-priced risks.

A general application of the framework is 'scanning' a portfolio's profile. A scan entails an inventarization of its exposures to the attributes considered relevant. By determining the sensitivities for various factors, the systematic risk profile of a given portfolio can be summarized in terms of its economic factor exposures. As a portfolio is not likely to be perfectly diversified with respect to the economic risk factors, the factor risk profile must be complemented with information about non-factor risks. This can be done by incorporating a residual market factor, which represents the influences of both omitted economic factors and 'market mood' (psychological) factors on a portfolio's return behavior (Berry, Burmeister & McElroy, 1988). For large portfolios, most of the return variability can be attributed to the economic factors and the residual market factor. The scores on the indirect return related attributes complete the portfolio profile. Scanning can be supplemented by conducting sensitivity analyses, for example on the basis of economic scenarios. A scenario then entails a specific constellation of potential factor realizations. Given a portfolio's risk profile, the exposure to some factor can be combined with a hypothesized change in that factor. This yields a 'ceteris paribus' portfolio return that can be attributed to the factor.

Scanning implies a *passive* use of the framework. An *active* and *aggressive* portfolio strategy is 'tilting'. Starting from a 'normal' portfolio (i.e. a benchmark or target portfolio which is considered suitable over a longer time horizon; Kritzman, 1987), an investor may wish to deviate from this portfolio in a controlled way. By tuning the sensitivity coefficients, a specific risk profile of the investment portfolio can be chosen. For example, when the investor forecasts a decrease in the interest rate, he can decrease (in algebraic sense) the portfolio's interest rate sensitivity, while controlling its exposure to other factors. In specifying this 'factor tilt', 'factor play' or 'factor bet', interest rate risk is considered as an opportunity and the investor intends to earn excess returns from this source. A crucial condition for adopting this strategy is that the investor has superior forecasting abilities, in order to predict factor movements that are unexpected to other investors.[13] Tilting can be extended to the exposures to other attributes. As set out section 3.1, there are indications that some attributes possess predictive abilities for future excess returns. While controlling risk exposures and other attribute scores, an investor could then strive to tilt his portfolio to high dividend yield, low price/earnings ratio, low price/book ratio or small capitalization (see Sorensen & Thum, 1992, e.g.).

---

[13] The application of the active strategy is pointed out by Berry, Burmeister & McElroy (1988a,b), Sorensen, Mezrich & Thum (1989) and Sorensen & Thum (1992), among others.

In contrast to aggressive strategies, *defensive* strategies do not require factor forecasts. Here, risk is considered as a threat, not as an opportunity and defensive strategies intend to shield a portfolio's return from undesired factor influences. A *passive defensive* strategy is 'risk sterilization': a portfolio's composition is shifted in order to mitigate or negate factor exposures that are considered to be excessive. An *active defensive* strategy is hedging: a portfolio's risk exposure is structured in accordance with the economic profile of the investor, which encompasses the pattern of his expenditures, his other sources of income and the economic conditions he will face. In this way, the choice of an appropriate pattern of risk exposures is explicitly linked to the uses to which the income generated by the portfolio is to be put (see Roll & Ross, 1984). In general terms, a hedging strategy takes the form of 'matching'. Given the investor's liabilities as specified by his economic profile, the risk exposure of the assets (i.e. investment portfolio) is manipulated to absorb the risks that are incurred on the liability side. By pairwise equating the factor sensitivities of assets and liabilities, one could strive to achieve a certain degree of hedging. This form of risk management is especially relevant for institutional investors like pension funds.

The multi-factor context signifies as a conceptual framework for an integrated approach to risk management. Especially for hedging purposes, the analysis of the characteristics of liabilities and assets (investment portfolios) in one unified, consistent framework is indispensable.

5 Conclusions

This paper presents a multi-dimensional framework for portfolio management which is very general in the sense that it can accommodate any type of investor. The framework is decision-oriented and gives room for different settings of the portfolio management problem. It tries to make an optimal use of the information available without requesting the investor to formulate *ex ante* a complete expected return - covariance structure of future returns and it leaves room for a much broader class of preference structures than allowed within the Markowitz [1959] approach. Although the presented framework is suitable for a broad range of investors, it does frame the portfolio management process in a way which requires a certain amount of discipline on the side of the investor. From another perspective, the framework offers the investor systematic guidance in the search for a portfolio that meets his investment goals as close as possible while opening the possibility to systematically learn from past experiences.

We believe that the sketched methodology has a clear potential as a tool to support investors in evaluating and selecting portfolios that meet their investment goals as close as possible.

References

Arnott, R.D., C.M. Kelso, S. Kiscadden & R. Macedo, 1989, Forecasting Factor Returns: An Intriguing Possibility, *The Journal of Portfolio Management* Fall, pp. 28-35

Arthur, J.L. & P. Ghandforoush, 1987, Subjectivity and Portfolio Optimization, in: K.D. Lawrence, J.B. Guerard & G.R. Reeves (eds), *Advances in Mathematical Programming and Financial Planning*, Volume 1, JAI Press, Greenwich, Conn., pp. 171-186

Baker, H.K. & J.A. Haslem, 1974, Toward the Development of Client-Specified Valuation Models, The Journal of Finance 29/4, Sept, pp. 1255-1263

Barnes, J.W., C.D. Zinn & B.S. Eldred, 1978, A Methodology for Obtaining the Probability Density Function of the Present Worth of Probabilistic Cash Flow Profiles, *AIIE Transactions* 10/3, Sept, pp. 226-236

Berry, M.A., E. Burmeister & M.B. McElroy, 1988, Sorting Out Risks Using Known APT Factors, *Financial Analysts Journal* March-April, pp. 29-42

Bower, D.H. & R.S. Bower, 1991, The Salomon Brothers Electric Utility Model: Still a Challenge to Market Efficiency, *Financial Analysts Journal* March-April, pp. 45-54

Chen, N.-F., R. Roll & S.A. Ross, 1986, Economic Forces and the Stock Market, *The Journal of Business* 59/3, July, pp. 383-403

Deaton, A. & J. Muellbauer, 1980, *Economics and Consumer Behavior*, Cambridge University Press, Cambridge UK

Fama, E.F., 1991, Efficient Capital Markets: II, *The Journal of Finance* 46/5, Dec, pp. 1575-1617

Fama, E.F. & K.R. French, 1992, The Cross-Section of Expected Stock Returns, *The Journal of Finance* 47/2, June, pp. 427-465

Fama, E.F. & K.R. French, 1993, Common Risk Factors in the Returns on Stocks and Bonds, *Journal of Financial Economics* 33, pp. 3-56

Fogler, H.R., 1990, Common Stock Management in the 1990s, *The Journal of Portfolio Management* Winter, pp. 26-35

Hawawini, G. & D.B. Keim, 1995, On the Predictability of Common Stock Returns: World-Wide Evidence, in: R.A. Jarrow, V. Maksimovich & W.T. Ziemba (eds), *Finance*, in the Handbook Series, North-Holland

Jones, R.C., 1990, Designing Factor Models for Different Types of Stock: What's Good for the Goose Ain't Always Good for the Gander, *Financial Analysts Journal* March-April, pp. 25-30, 50

Jones, R.C., L.A. Kohn & M. Melnikoff, 1990, Quantitative Equity Investment Strategies, Goldman Sachs Asset Management, Goldman, Sachs & Co., New York NY, July

Kritzman, M., 1987, How to Build a Normal Portfolio in Three Easy Steps, *The Journal of Portfolio Management* Summer, pp. 21-23

Kumar, P.C., G.C. Philipatos & J.R. Ezell, 1978, Goal Programming and the Selection of Portfolios by Dual Purpose Funds, *The Journal of Finance* 33/1, March, pp. 303-310

Lakonishok, J., A. Schleifer & R.W. Vishny, 1994, Contrarian Investment, Extrapolation, and Risk, *The Journal of Finance* 49/5, Dec, pp. 1541-1578

Lancaster, K.J., 1966, A New Approach to Consumer Theory, *Journal of Political Economy* 74, April, pp. 132-157

Lee, S.M., 1972, *Goal Programming for Decision Analysis*, Auerbach, Philadelphia

Lee, S.M. & D.L. Chesser, 1980, Goal Programming for Portfolio Selection, *The Journal of Portfolio Management* Spring, pp. 22-26

Lee, S.M. & A.J. Lerro, 1973, Optimizing the Portfolio Selection for Mutual Funds, *The Journal of Finance* 28/5, Dec, pp. 1087-1101

Markowitz, H.M., 1952, Portfolio Selection, *The Journal of Finance* 7/1, March, pp. 77-91

Markowitz, H.M., 1959, *Portfolio Selection: Efficient Diversification of Investments*, John Wiley, New York NY

O'Leary, J.H. & D.E. O'Leary, 1987, A Multiple Goal Approach to the Choice of Pension Fund Management, in: K.D. Lawrence, J.B. Guerard & G.R. Reeves (eds), *Advances in Mathematical Programming and Financial Planning*, Volume 1, JAI Press, Greenwich, Conn., pp. 187-195

Reilly, F.K., 1994, *Investment Analysis and Portfolio Management*, Dryden Press, Fort Worth TX

Roll, R. & S.A. Ross, 1984, The Arbitrage Pricing Theory Approach to Strategic Portfolio Planning, *Financial Analysts Journal* May-June, pp. 14-26

Ross, S.A., R.W. Westerfield & J.F. Jaffe, 1993, *Corporate Finance*, Irwin, Homewood Ill.

Sharpe, W.F., 1967, A Linear Programming Algorithm for Mutual Fund Portfolio Selection, *Management Science* 13/7, pp. 499-510

Sharpe, W.F., 1987, An Algorithm for Portfolio Improvement, in: K.D. Lawrence, J.B. Guerard & G.R. Reeves (eds), *Advances in Mathematical Programming and Financial Planning*, Volume 1, JAI Press, Greenwich, Conn., pp. 155-169

Smith, K.V., 1974, The Major Asset Mix Problem of the Individual Investor, *Journal of Contemporary Business* Winter, pp. 49-62

Sorensen, E.H., J.J. Mezrich & C.Y. Thum, 1989, The Salomon Brothers U.S. Stock Risk Attribute Model, Salomon Brothers Inc., New York NY, Oct, 17 pp.

Sorensen, E.H. & C.Y. Thum, 1992, The Use and Misuse of Value Investing, *Financial Analysts Journal* March-April, pp. 51-58

Spronk, J., 1981, *Interactive Multiple Goal Programming: Applications to Financial Planning*, Martinus Nijhoff, Boston

Tobin, J., 1958, Liquidity Preference as Behavior Towards Risk, *Review of Economic Studies* 25, pp. 65-86

MCDM AND SUSTAINABLE DEVELOPMENT: THE CASE OF WATER RESOURCES PLANNING IN INDIA

Jared L. Cohon
School of Forestry and Environmental Studies, Yale University, New Haven, Connecticut 06511, USA

Sustainable development has emerged as the centerpiece of natural resources management and environmental protection. Defined as the use of natural resources today so as not to compromise the ability of future generations to use them, sustainable development is a compelling idea, and of indisputable importance. Translating it into plans and actions, however, is very difficult both because of the value conflicts inherent in sustainable development and the need for new planning and analytical methodologies.

Operationalizing sustainable development must deal with many complications, but perhaps the most difficult is the multiobjective nature of the problem. We are early in the development of suitable methodologies, but multiple criteria decision making (MCDM) is likely to play an important role. The purpose of this paper is to demonstrate the potential role of MCDM in planning for sustainable development, using the example of water resources in India.

1. Sustainable Development of Water Resources

Water is essential for all life on earth. It represents, therefore, a natural resource which must be used sustainably. Despite this fact, which is hardly a new revelation, global water issues are of serious and growing concern. (See, for example, Gleick [1993]). The problems take on many forms, depending on where you are: not enough or too much water, water of poor quality, creating widespread human disease, and the health of aquatic ecosystems. The focus of this paper is on large-scale water resources development, i.e., the planning, design, construction and management of structures for controlling natural water bodies (usually rivers). The structural components of water resource development usually include dams for storage, hydroelectric power plants and diversion and distribution systems for irrigation and other purposes. Although the era of building large dams has very likely ended in industrialized countries, this is not the case in developing nations. (Witness the Three Gorges Dam in China. See also Chaturvedi [1992] -- especially page 451 -- for a spirited defense of large-scale water developments.) Despite the environmental impacts of water structures, the desire to control water for human purposes is a powerful force that will make large-scale water plans a continuing challenge in sustainable development.

Large-scale water development fundamentally transforms the hydrology and ecology of river basins. These impacts are significant, and they have potentially far-reaching and long-term consequences. There have been many papers and books on this important topic -- e.g., Stanley and Alpus [1975], Goldsmith and Hildyard [1984], Canter [1985], and Tilzer and Serryua [1990] -- resulting in a good (but not fully adequate) knowledge base on environmental impacts of big water developments. Somewhat surprisingly, there has been relatively less progress in developing methodologies for integrating this knowledge base into water planning. (See, for example, Whipple [1996] for a discussion of the gap between water planning and environmental regulation, a closely related topic from a methodological perspective.) Yet, the development of such methodologies is essential if sustainable water development is to be accomplished.

Like all large projects, water developments pass through several stages over a relatively long period of time. Planning is the initial stage, after a problem or opportunity has been recognized. After a plan is formulated and approved, designs are created for the individual dams, power plants, diversion and distribution systems and related structures. These projects are then constructed leading to operation and management. Each stage -- planning, design, construction and management -- has its own methodological challenges, but our focus is on planning, which is the crucial stage for achieving sustainability. It is then that the development takes shape and when large-scale environmental impacts must be recognized. Although design or management alternatives may be effective in mitigating localized environmental impacts, large-scale impacts can only be dealt with -- and a sustainable course plotted -- at the planning stage.

2. Water Resources Planning Methodology

The modern, systems-oriented, model-based era of water resource planning started with the Harvard Water Program in the late 1950's. Maass et.al. [1962], which compiled the major results of the Harvard Program, presented for the first time a coherent methodology for large-scale water planning using simulation and optimization techniques. In addition, in chapters by Stephen Marglin, the idea of taking a multiobjective approach was advanced, starting an important line of research that has established water systems and water resource systems analysts as important sources of innovation in MCDM.

The literature of water resource systems analysis is very large and diverse, reflecting significant developments over the last 35 years. (See, for example, Loucks, et.al. [1981]) There have also been many papers in which MCDM methods have been developed and applied to water resource problems. There are no up-to-date reviews or compilations of MCDM and water resources. Dated reviews can be found in Cohon and Marks [1975] and Haimes et.al. [1975]. Many water resources applications are included in the bibliography prepared by White [1990].

Two recent applications are of particular interest. Ridgley and Rijsberman [1992] used the analytical hierarchy process to support decisions on the ecological restoration of the Rhine River Delta near Rotterdam in the Netherlands. Stewart and Scott [1995] developed an MCDM methodology for national water policy planning in South Africa. Both of these papers lay out methodologies that can accommodate environmental impacts in large-scale water planning problems. However, both also deal with that class of problems in which a relatively small number of alternatives is analyzed. (The scenario-based approach of Steward and Scott is an intriguing way of extending the method to a broader set of alternatives.) But, planning at the level of a river basin, especially relatively undeveloped basins of the sort one still finds in India and other developing countries, must deal with an infinity of possibilities. There are optimization models for doing this, and even many multiobjective ones, but none of them have incorporated the full range of environmental impacts in support of planning. A start on this effort is presented here.

3. Environmental Impacts of Large-Scale Water Developments in India

Environmental impacts, are, by definition, specific to a region and a problem setting. Nevertheless, the following discussion, which focuses on India, is probably transportable to a significant degree.

I served as a technical advisor to a project in India sponsored by the United Nations Development Programme (UNDP). Over a period from 1983 to 1987, a team of international experts, headed by the late Warren Hall, worked with the Systems Engineering Unit of India's Central Water Commission. Our overall goal was to support the Commission's efforts to develop expertise in water resource systems analysis. My role was to work with the Commission's staff on environmental impacts and their integration into planning models.

We interpreted "environmental impact" in broad terms to include impacts on humans and on natural resources in addition to the river basin and its ecosystem. Indeed, such a broad view is necessary for sustainable development, which embraces social systems as well as natural systems.

Our starting point was a list of potential environmental impacts prepared by the Ministry of Irrigation of India. This was supplemented through brainstorming by project participants. This was a time (the mid 1980's) before open and inclusive decision making, which, though clearly preferred and amenable to MCDM (see, in particular, Ridgley and Rijsberman [1992]) is still largely a phenomenon of the industrialized world. The kind of centralized planning pursued then in India is still the norm in developing countries. Thus, as attractive as stakeholder participation would have been, the approach that we took was the only one open to us. This would still be the case today.

The impacts are shown in Table 1. There are three categories of impacts : human impacts, impacts on flora and fauna, and impacts on other land and water uses and other resource systems. Within each category, each impact is listed, as well as the dimensions of the impact, i.e., the ways in which the impact is expressed. The dimensions can be thought of as the indicators of why we should care about an impact and as suggestive of measures. For example, the displacement of people is an impact; its primary dimension is loss of land. Mortality, morbidity, changes in nutrition and productivity are dimensions of health impacts. Table 1 also shows the potential causes of the impact, i.e., the mechanisms through which development may create the impact. For example, the submergence of land is the major mechanism through which people are displaced.

TABLE 1

Impacts	Dimensions	Mechanism
Human:		
Displacement	Loss of Land Social & Psychological Effects Special Classes (tribes)	Submergence
Loss of Cultural, Historical &/or Religious Resources	Cost to Relocate or Protect Social & Psychological Effects	Submergence
*Transportation	Change in Time and/or East of Travel	Submergence
Health	*Mortality *Morbidity Nutrition *Productivity	*Vector Habitat *Relocation of People *Water Supply *Pesticides & Fertilizers Food Production

*These impacts (or specific dimensions and their corresponding mechanisms) were not incorporated into the planning model. They were either to be the subject of separate studies during the planning process or dealt with at the design stage.

TABLE 1 (continued)

Health	*Mortality *Morbidity Nutrition *Productivity	*Vector Habitat *Relocation of People *Water Supply *Pesticides & Fertilizers Food Production
Flora and Fauna		
Destruction and Displacement of Animals and their Habitats	Endangered Species Cost Mitigation *Loss of Protein (fish)	Submergence Flow Regime *Barriers to Migration *Ecosystem Effects
Destruction of Plants	Loss of Protected Forested Land Endangered Species	Submergence Clearing

*These impacts (or specific dimensions and their corresponding mechanisms) were not incorporated into the planning model. They were either to be the subject of separate studies during the planning process or dealt with at the design stage.

TABLE 1 (continued)

Other Land and Water Uses and Other Resource Systems		
Surface Water Quality	Dissolved Oxygen Biochemical Oxygen Demand Coliform Count Dissolved Solids Other Parameters	 Flow Regime Water Use
	Reservoir Water Quality Salinity	Submergence Flow Regime
Riparian Rights	Water Supply	Flow Regime
Mineral Deposits	Loss of Access Cost of Production *Water Quality	Submergence

*These impacts (or specific dimensions and their corresponding mechanisms) were not incorporated into the planning model. They were either to be the subject of separate studies during the planning process or dealt with at the design stage.

TABLE 1 (continued)

River Basin Changes	Long-term Productivity Flooding *Landslides Erosion/Sedimentation *Micro-Climatic Changes *Induced Seismic Activity	

*These impacts (or specific dimensions and their corresponding mechanisms) were not incorporated into the planning model. They were either to be the subject of separate studies during the planning process or dealt with at the design stage.

Our strategy was to take the exhaustive list of Table 1 and to identify those impacts that are predominantly planning issues. Our rationale was that we should devote most of our effort to those impacts that can be materially and substantively altered by changes in the development plan. The impacts not incorporated into the planning model are marked with an asterisk(*) in Table 1. Some of these impacts, which are not necessarily less important, are most appropriately analyzed at the design stage and are often mitigated through design changes or through management policy. The rationale for including some impacts, while deferring others, can be further explained through example. The displacement of people was considered to be a planning issue, since reservoirs may be rejected or their sizes may be altered due to this impact. On the other hand, "transportation", another human impact, was thought to be predominantly a design issue, i.e., a reservoir site would probably not be abandoned if the lake represented a barrier to transportation. Bridges can be built or water transportation can be provided, effectively mitigating this impact at a relatively low cost. Thus, "transportation" is marked with an asterisk in Table 1.

Some impacts have multiple dimensions, only some of which were taken as planning issues. For example, landslides, micro-climatic changes and induced seismic activity are three dimensions of river basin changes that were not included. Landslides were thought to be treatable as a design issue. The latter two were deferred, both for the uncertainty that surrounded them at that time (and still does) and, in the case of seismicity, the confidence of the engineers that seismically active sites would be avoided.

Some impacts were not included in the planning model even though we considered them to be very important. A good example, is health effects, especially increased morbidity and mortality due to a higher incidence of, say, malaria brought about through changes in vector habitat or human population. Though potentially terribly important, we did not consider this something that could be addressed at the planning stage. The arguments against including them were part bureaucratic politics -- these are issues for the Ministry of Health -- and part engineering judgment -- vector habitat is created by all dams and irrigation sites; thus, the problem should be mitigated at the design stage. Neither argument is sound. There are many examples of significant human health effects from water developments, e.g., the widespread incidence of schistosomiasis around the High Aswan Dam on the Nile River. Governments should ask the question of whether the benefits of such development outweigh the costs, a definite planning issue. This was, in my view, a flaw in our work, but our MCDM approach already represented a significant stretch for the Central Water Commission. Health effects would have to be put off to the next iteration. Indeed, the iterative nature of planning and the identification of health and other impacts as deserving of future consideration were reasons for hope.

4. Quantification of Environmental Impacts

The next step was to quantify the environmental impacts with planning significance for inclusion in the planning model. There were some general principles which guided this effort. Quantification of an environmental impact entails the statement of the impact as a mathematical function of decision variables. There are three important requirements for this function. First, it must be meaningful, i.e., it must provide a measure of the impact which will be relevant to decision makers and which captures the essential nature of the impact. Second, the decision variables of which the impact is a function must be elements of the planning model if the impacts are to be integrated successfully. Third, data must be available to permit estimates of the parameters of the functions. Thus, we had to look ahead toward integration and data collection while we were at the quantification step. Furthermore, the requirements of meaningfulness, integration and data are often conflicting. Compromises must be made to derive operationally useful measures.

Our interpretation of "environment" is very broad, as our lists of impacts indicate. Many of the impacts, have, in fact, direct economic consequences and should be incorporated directly into development and benefit functions. Other impacts may be evaluated in economic terms and these should be incorporated into the cost functions as well. Such economic aspects were identified for each impact, as discussed below.

The quantification schemes which follow are quite general in the sense that they are likely to be useful in most river basin planning models. To provide context for the following discussion, we will consider one particular linear programming model of a sort that has been widely used for many years, e.g., Cohon and Marks [1973], Major and Lenton [1979], Loucks, et.al. [1981], and Vedula and Kumar [1996]. We will assume that there are m potential reservoir sites and the planning period includes n seasonal periods.

$$\text{Max } \mathbf{Z} \quad (\mathbf{C},\ \mathbf{S},\ \mathbf{X}) \tag{1}$$

s.t.

$$S_{i,t+1} - S_{it} + X_{it} - X_{i-1,t} = \text{Fit} \quad i,t \tag{2}$$

$$S_{it} - C_i \leq 0 \quad i,t \tag{3}$$

$$\mathbf{C},\ \mathbf{S},\ \mathbf{X} \geq 0 \tag{4}$$

in which

C_i	=	capacity of reservoir at i (a decision variable)
C	=	$(C_1, ..., C_m)$
S_{it}	=	storage in reservoir i at the beginning of season t (a decision variable)
S	=	$(S_{11}, ..., S_{mn})$
X_{it}	=	release from reservoir i during season t (a decision variable)
X	=	$(X_{11}, ..., X_{mn})$
F_{it}	=	natural streamflow into reservoir i during season t

The objective function is multiobjective, consisting of Z_1 (**C, S, X**), the maximization of net economic efficiency benefits, and Z_2 (**C, S, X**) .... Z_p (**C, S, X**) which are the environmental objectives, discussed below, and other objectives, e.g., distributional equity, which may be applicable to the particular situation.

The constraint in (2) insures continuity of mass, i.e., water which flows into site i (comprised of release from i-1, the site immediately upstream of i, and natural inflow) must be either stored in the reservoir at site i or released downstream. Constraint (3) allows the reservoir capacity to be determined as the largest of the seasonal storages.

This basic model can be extended in many ways to incorporate all water uses. See, for example, Major and Lenton [1979] for treatments of hydroelectric power plants and irrigation, and Loucks et.al. [1981] for flood control. Including these uses and water structures introduces additional decision variables, e.g., power plant capacities and irrigation zone sizes, and constraints, e.g., hydroelectric production from the volume and potential energy of reservoir releases and limits on arable land.

In the following discussion, of each impact, a measure is introduced, quantified in terms of the above decision variables, and discussed briefly.

4.1 Displacement of People

- Loss of land

$$\text{Min } Z_2 = \sum_i h_i(C_1) \qquad (5)$$

where h_i ( ) is a function which indicates the amount of land, currently used by people, submerged by the reservoir at site i.

Units : Hectares

Disadvantages : The measure treats all hectares equally, no matter the quality of the land or the structures on it.

- Economic cost of relocation and replacement of property.

Assess directly the cost of relocating people and compensating them fully for their loss of property. These costs, which are a function of reservoir capacity, should be added to the appropriate reservoir capacity - cost function included in Z1.

- Social and psychological effects of displacement

Even after taking into account the loss of property and its economic value, there is a further, social cost of relocation. People simply don't like to move, if it is not their idea. It is unlikely that displacement can ever be fully mitigated. Our measure is :

$$\text{Min } Z_3 = \sum_i P_i(C_1) \qquad (6)$$

where P_i ( ) is a function which indicates the number of people who would be submerged (if not moved) by reservoir i.

Units : People (who are not Tribal members)

Disadvantages : The measure does not capture individual differences (some people care more than others about being displaced). It also does not consider the segmentation of a village, i.e., moving some residents of a village may have greater impact than relocating the whole village (our measure indicates just the opposite over certain ranges). The distance of the move may also be important, especially if the move is to a culturally different area (say from one that is predominantly Hindu to one which is predominantly Muslim). This too, is not captured.

Alternative measure :

$$\text{Min } Z^1{}_3 = \sum_i V_i(C_1) \qquad (7)$$

where V_i ( ) indicates the number of villages inundated as a function of reservoir capacity. (Assume if any part of a village is submerged then the whole village is moved).

$$\text{Min } Z^{11}{}_3 \quad = \quad \Sigma_i \quad r_i P_i(C_1) \qquad (8)$$

where r_i is the population density of the area around site i. The idea here is that the more densely populated the area the more likely it is that the displaced population will be forced to move to a distant location. Thus, more social impact is counted for disruption in densely populated areas.

- Dislocation of Tribal people

$$\text{Min } Z_4 \quad = \quad \Sigma_i \quad t_i(C_1) \qquad (9)$$

where t_i ( ) indicates the number of Tribal people displaced by reservoir i, as a function of reservoir capacity.

Units : Tribal people

All of the disadvantages of the previous measure apply here, as well.

The economic costs of relocating these people should be incorporated into the reservoir cost functions.

4.2 Loss of Cultural, Historic & Religious Resources

- Economic cost or relocation or protection.

The cost of relocation or protection of these resources should be estimated as a function of reservoir capacity and incorporated into the reservoir cost function.

- Social & psychological effects.

Social and psychological effects are extremely difficult to quantify in this case. We considered minimizing the total number of such resources submerged, but decided that here, especially, the nature of the resource and its location -- neither of which is captured -- are crucial. Our conclusion was to provide a list of all such resources for each plan, confident that, should particularly important ones be involved, an appropriate political reaction would be forthcoming.

4.3 Health Impacts

The impacts on health that, for reasons noted before, seemed appropriate for treatment at the planning stage were beneficial ones. : improved nutrition and decreased incidence of some water-borne diseases. Both of these are important but difficult to quantify.

Improved nutrition will result from increased food supplies. Since this represents the basic purpose of irrigation development, we assumed food production to be an element of benefit estimation and, thus, an issue captured in objective Z_1. At a minimum (if monetary benefits cannot be assigned) an objective of maximizing food production should be considered.

Water-borne disease may be reduced through the provision of protected water supplies. However, there are problems in incorporating this effect into the planning models. The provision of water treatment and distribution facilities is not considered to be an element of the development project; they are installed, or not, by the State Governments -- another obstacle of Indian bureaucratic organization. Since the provision of water without treatment or distribution does not, by itself, secure the quality of domestic water, it is difficult to claim such a benefit for the development project.

If we assume that health benefits should be included, there are then problems in quantifying them. If data is available on incidence of water-borne disease, we can use this to estimate the reduction in the incidence due to protected water supplies. On the other hand, suppose water-borne disease is not present. The provision of secured water supply may insure that the disease will never occur. This is a benefit, but how is it to be estimated?

Our conclusion was that domestic water supply should be a reservoir purpose, but, since the security of water supplies was not considered to be a development purpose, quantification of health impacts should not be attempted. Integration into the planning models could be achieved, in part, by treating this as another aspect of providing minimum flows.

4.4 Flora and Fauna

- Reduction of endangered animal species

The most important impact of development on endangered animal species is the inundation of habitat. Our measure is,

$$\text{Min } Z_6 \quad = \quad \sum_j r_j \sum_i f_{ij}(L_i) \tag{11}$$

where f_{ij} ( ) is the area of forested land of quality j submerged by reservoir i, as a function of capacity, and r_j is a quality index.

Units : Hectares

Decision variable : L_i = storage at 3 meters below full tank level of reservoir i, as specified by the Indian Forest Act of 1980.

In developing the function f_{ij} ( ) the forested area cleared for construction should be included as a "fixed cost". However, care should be taken not to double count areas i.e., once cleared for construction, the forest cannot be lost again due to submergence.

Under the Indian Forest Act of 1980, an area equal in size to the deforested area, must be reforested. The costs to do this must be incorporated into the reservoir cost functions. Furthermore, the afforested area, now lost for irrigation, should be considered in the benefit calculations.

- Reduction of endangered plant species.

Endangered plant species are most likely to be found in forested areas. Thus, the measure proposed above also captures the reduction of endangered plant species. The quality index, rj, can be chosen to reflect the relative scarcity of species found in forested areas, if appropriate.

4.5 Impacts Related to Flow Regime

There are several potential impacts related to the maintenance of a minimum flow in the stream. These are : stream water quality, migration of salinity upstream from the mouth of a stream, salinity intrusion into groundwater due to inadequate recharge, loss of water for riparian users and provision of inputs to protected municipal water supplies.

Our measure is a simple one, which may be adequate for the planning model but necessitates additional analysis.

$$\text{Max } Z_k = X_{min,i} \qquad \forall \text{ i in E} \tag{12}$$

where E is the set of environmentally sensitive sites.

Units : m^3/sec.

where the following additional constraint is added to the model.

$$X_{min,i} - X_{i,t} \leq 0 \qquad \forall \text{ i in E}, \forall t \qquad (13)$$

Letting j be the cardinality of E, the objectives in (12) would correspond to Z_7, ..., Z_{7+j}

The provision of a minimum flow simultaneously addresses all of the impacts listed above. However, for analysis, we need only be concerned with those impacts which require the maximum minimum flow. On this basis, we felt confident in dismissing potential impacts such as loss of channel capacity and provision of drinking water for wildlife. The former problem is routinely taken care of by manual clearing of the stream channel. Wildlife needs (for drinking water) are likely to be minimal, so that providing virtually any non-zero flow should satisfy these needs.

4.6 Mineral Deposits

- Loss of access

$$\text{Min } Z_k = \sum_i M_{ik}(C_i) \text{ for } k = 7+j,.., 7+j+q \qquad (14)$$

where there are q minerals of concern and M_{ik} ( ) indicates the tons of mineral k no longer recoverable due to submergence at site i, as a function of reservoir capacity. A quality factor could be added and all minerals summed, as in the forested lands objective, reducing the q additional objectives to one.

Units : tons of mineral k

Disadvantages : This measure does not address the configuration of the mineral seam. If any part of the deposit is submerged, no matter its relation to the rest of the seam or its likelihood of recovery, it is counted as submerged.

- More expensive production.

Submergence of part or all of a mineral deposit may simply make its exploitation more costly, without making it inaccessible. For example, submergence of a portion of a coal seam may require deep-mining rather than surface-mining. To the extent that these added costs can be estimated, they should be added to reservoir cost functions.

4.7 Methodological and Data Consideration

It should be emphasized that the multiobjective linear program (MOLP) which results from the above is envisioned to be part of a larger, iterative multi-model methodology. For example, minimum flow values which are produced by the MOLP (see equation (13)) would be analyzed using water quality and ecosystem simulation models to determine the actual impacts from these surrogate values. One would expect to discover from such analysis limits on minimum flows or necessary relationships among flows at different sites. These can be reflected in new constraints in the LP.

The objectives of the previous section were selected with data requirements in mind. The necessary data is either generally available or collectible in a reasonable amount of time, even in a developing country setting.

Obtaining nondominated solutions from an MOLP is a well-studied problem for which many techniques exist. Cohon [1978] and Steuer [1986] present several possible techniques. In addition, interactive methods, such as Zionts and Wallenius [1976 and 1983], may be especially useful given the large number of objectives.

4.8 Application

The Systems Engineering Unit of the Central Water Commission applied this methodology to two sub-basins of the Mohanadi River, a large, undeveloped river whose watershed covers most of Madya Pradesh, India's largest internal state. One of the studies, published by Pendse, et.al., [1987], incorporated land submergence, dislocation of people, submergence of forests and submergence of coal deposits. The results showed both the feasibility of the approach and its utility for informing decision making at the planning stage.

5. The Future

It is surprising to me that relatively little work seems to have been done on the integration of environmental impacts into water resources planning models. Perhaps this is because the water planning bureaucracies need external pressure to do this. If this is the case, one should expect to see an upswing in this activity as the World Bank, which is the primary financier of large-scale water projects, moves toward more environmentally sensitive evaluation procedures.

In a methodological sense, MCDM is advanced enough to support integrated project planning. Of greater importance at this stage is a stronger scientific foundation for predicting environmental impacts. In particular, understanding the response of aquatic ecosystems to the disturbance of water development, is crucial for achieving sustainable water development. Also needed are bureaucratic organizations and planning procedures which permit the broader view of development that is essential for sustainability.

Acknowledgement

The research upon which this paper is based was supported by the United Nations Development Programme. I thank the many engineers of the Central Water Commission of India who contributed to the work, especially Directors Thomas and Pendse of the Systems Engineering Unit, Dr. Andrei Filotti and the members of the environmental planning group. I thank Dr. Abel Wolman for his comments on an earlier report and for his inspiration to me and generations of engineers who have worked on water problems in developing countries.

References

Canter, L. (1985). *Environmental Impact of Water Resources.* Lewis Publishers, Chelsea, Michigan.

Chaturvedi, M.C. (1992). "Irrigation and drainage -- systems policy analysis and India case study." *Journal of Water Resources Planning and Management*, July/August 118(4), 445-464.

Cohon, J.L. (1978). *Multiobjective Programming and Planning.* Academic Press, New York.

Cohon, J.L. and D.H. Marks. (1973). "Multiobjective screening models and water resource investment." *Water Resources Research*, 9(4), 826-836.

Cohon, J.L. and Marks, D.H. (1975). "A review and evaluation of multiobjective programming techniques." *Water Resources Research*, April 11(2).

Gleick, P. (1993). *Water in Crisis.* Oxford University Press, New York.

Goldsmith, H. and Hildyard, N.. (1984). *Social and Environmental Effects of Large Dams.* Sierra Club Books, San Francisco.

Haimes, Y.Y., Hall, W.A. and Freedman, H.T. (1975). *Multiobjective Optimization in Water Resources Systems.* Elsevier, New York.

Loucks, D.P., Stedinger, J.R. and Haith, D.A. (1981). *Water Resource Systems Planning and Analysis.* Prentice-Hall, Englewood Cliffs, NJ.

Maass, A., Hufschmidt, M., Dorfman, R., Thomas, H.A. Jr., Marghin, S. and Fair, G.F. (1962). *Design of Water Resource Systems*. Harvard University Press, Cambridge, Massachusetts.

Major, D.C. and Lenton, R.L. (1979). *Applied Water Resource Systems Planning*. Prentice-Hall, Englewood Cliffs, NJ.

Pendse, Y.D., Singh, R.N.P. and Sekhar, A. (1987). "Multiobjective analysis - a case study." In: *Proceedings of the Seminar on Application of Systems Analysis for Water Resources Development, 27-28 February*. Central Water Commission, New Delhi, pp. 299-319.

Ridgley, M.A. and Rijsberman, F.R. (1992). "Multicriteria evaluation in a policy analysis of a Rhine estuary." *Water Resources Bulletin*, December 28(6), 1095-1110.

Stanley, N.F. and Alpers, M.P. (eds). (1975). *Man-Made Lakes and Human Health*. Academic Press, New York.

Steuer, R. (1986). *Multiple Criteria Optimization*. Wiley, New York.

Stewart, T.J. and Scott, L. (1995). "A scenario-based framework for multicriteria decision analysis in water resources planning." *Water Resources Research* 31(11), 2835-2843.

Tilzer, M.M. and Serryua, C. (eds). (1990). *Large Lakes Ecological Structure and Function*. Springer-Verlag, New York.

Vedula, S. and Kumar, D.N. (1996). "An integrated model for optimal reservoir operation for irrigation of multiple crops." *Water Resources Research*, 32(4), 1101-1108.

Whipple, William, Jr. (1996). "Integration of water resources planning and environmental regulation." *Journal of Water Resources Planning and Management*, May/June, 122(3), 189-196.

White, D.J. (1990). "A bibliography on the applications of mathematical programming multiple-objective methods." *Journal of the Operational Research Society*, 41(8), 669-691.

Zionts, S. and Wallenius, J. (1976). "An interactive programming method for solving the multiple criteria problem." *Management Science*, 22(6), 652-663.

Zionts, S. and Wallenius, J. (1983). "An interactive multiple objective linear programming method for a class of underlying nonlinear utility functions." *Management Science*, 29(5), 519-529.

Role of BATNA and Pareto Optimality in Dyadic Multiple Issue Negotiations

Jeffrey Teich[1], Pekka Korhonen[2], Jim Phillips[3], Jyrki Wallenius[2]

[1] New Mexico State University, Department of Management, Las Cruces, New Mexico 88003, USA
[2] Helsinki School of Economics, Runeberginkatu 14-16, 00100 Helsinki, Finland
[3] Northeastern State University, Tahlequah, Oklahoma 74464, USA

Abstract. The role of the Best Alternative to a Negotiated Agreement (BATNA) has been the subject of much discussion in the literature recently. We extend this discussion concerning BATNAs and reservation prices to multiple issues. We have conducted an experiment to test several hypotheses concerning the role of BATNAs in a contextually relevant three-issue dyadic negotiation. We further improve the methodology of conducting negotiation experiments by not requesting the subjects to follow a given value point structure. We report the results of the experiment and discuss their implications.

Keywords: Negotiation, Best Alternative to a Negotiated Agreement, Multiple Issues

1. Introduction

The field of negotiation and assisted dispute resolution has received considerable attention from researchers and practitioners in recent years. Fisher and Ury in "Getting to Yes" (1981) explain that the reason people negotiate is to produce something better than can be obtained without negotiating. By "alternative to a negotiated agreement" we mean what a person would have if he or she didn't attempt a negotiation or agree to a negotiated settlement. The BATNA (**Best** Alternative To a Negotiated Agreement) becomes "the standard against which any proposed agreement should be measured" (Fisher and Ury, 1981). We share the view of Fisher and Ury (1981), Raiffa (1982), and Pinkley, Neale, and Bennett (1994) that an individual's BATNA is critical to the quality of his or her outcome of the negotiations.

It is generally believed that rational negotiators will not make concessions beyond the value of no agreement (Fisher and Ury, 1981; Raiffa, 1982; Neale and Bazerman, 1992), establishing a lower bound in negotiations. Such a lower bound may be defined by concepts such as BATNA, reservation price, minimum necessary share, status quo, and resistance point. (See Carnevale and Pruitt (1992) for a review of such limits in negotiations.) The value of no agreement obviously influences the limits, but according to Carnevale and Pruitt (1992) there are many other influences as well, such as ethical principles and principles of fairness. The lower limit terms are often used interchangeably by researchers, and they do indeed appear similar. White and Neale (1991), however, argue that they are different concepts and represent different lower bounds for negotiators. Further, Balakrishnan, Patton, and Lewis (1993) discuss the role of aspiration levels in negotiations. In a related discussion, Tietz (1983) argues that there are actually five aspiration levels in any bargaining situation. The lowest acceptable agreement is one of them.

Examples of empirical research discussing the influence of lower limits on the process and outcome of negotiations date back to Kelley, Beckman, and Fischer (1967). They controlled the lower limit through the use of a "minimum necessary share" (MNS). When the MNSs are unequal, they reported a "low man's advantage" in profit gains and greater difficulty in reaching agreements vis-a-vis equal MNSs. More recent examples of empirical research that test different aspects of lower limits on negotiations are Pinkley et al. (1994), Bazerman, Neale, Valley, Zajac, and Kim (1992), Pruitt and Syna (1985), McAlister, Bazerman, and Fader (1986). Typically the studies discuss single-issue negotiations. In the few studies that do address multiple issues, subjects are usually provided with, and told to follow, a value point structure aggregating the issues. The results seem to indicate that lower limits on negotiations do have an impact on the process and the outcome of the negotiations. Related research discusses the role of information and different views on rationality in negotiations. (See, for example, Roth and Murnighan (1982), Murnighan and Bazerman (1990), Ostmann (1992), and Valley, White, Neal, and Bazerman (1992).)

Most of the above mentioned studies discuss BATNAs and reservation prices with reference to a single issue. Most business negotiations, however, involve multiple issues. Multiple issues introduces complexity, not discussed by other authors, which can create a less than desired agreement for both sides of a negotiation. Often negotiators are not aware of their preferences among issues (i.e., tradeoffs of issues) subject to constraints placed on the negotiation by the other party. We propose to assist a single negotiator by eliciting his/her preferences and indifferences to his/her BATNA and then to present that information to the negotiator in an understandable manner. We will investigate the impact of providing this type of negotiation support in a contextually relevant 3-issue experiment. We also will study the difference between providing (possessing) uncertain BATNAs (not very well defined; for example, a possible job offer) as opposed to deterministic (certain high, medium, and low) BATNAs (for example, an actual job offer).[1] In addition, we investigate why one would agree to a settlement worse than the BATNA.

---

[1]We also define a stochastic BATNA as one in which probability distributions for alternative outcomes are assumed to be known or can be elicited. These distributions can be either discrete or continuous and can exist for packages of outcomes, or for individual issues. In the experiment discussed herein, we do not, however, consider such stochastic BATNAs.

White et al. (1991) and others have compared the different measures for a lower limit on negotiations. In this paper, we extend the discussion of BATNAs and reservation prices to the case of multiple issues. In Figure 1a we have represented the negotiators' BATNA constraints in value space. A BATNA constraint in value space is a straight line, restricting the negotiator to be above it. In other words, the value (utility) an individual gains from a negotiated outcome, should be above the value/utility of the BATNA. Referring to Figure 1b, assume a negotiator is maximizing both issues in a two-issue negotiation (scaled between 0-1) such that his/her ideal point is at the upper right corner (1,1). Many negotiators prefer to provide a reservation price separately for each issue. If that is the case, the reservation prices can be graphically represented by a vertical and horizontal "constraint" for issue one and two, respectively. For this example, assume that the person's reservation prices are 0.5 for each issue. The person's BATNA is graphically represented in Figure 1b (mapping from value space to issue space) by an indifference curve[2] which is concave to his/her ideal point. The two limiting factors taken together provide four regions for further discussion. Region A (see Figure 1b) falls above the BATNA indifference curve and both (all) reservation prices. This region should be surely acceptable to the negotiator. Region B is the possibly acceptable region that lies above the BATNA indifference surface, but below at least one reservation price. It is possibly acceptable depending on how "hard" or "dynamic" the reservation prices are. We assume here that the BATNA indifference surface is "harder" because it is less likely to change over short periods of time. Of course, the surface would shift outward if one could develop a better BATNA. Region C is the probably unacceptable region because it lies below the BATNA indifference surface, but above all of the reservation prices. Region D is surely unacceptable because it lies below the BATNA surface and at least one of the reservation prices.

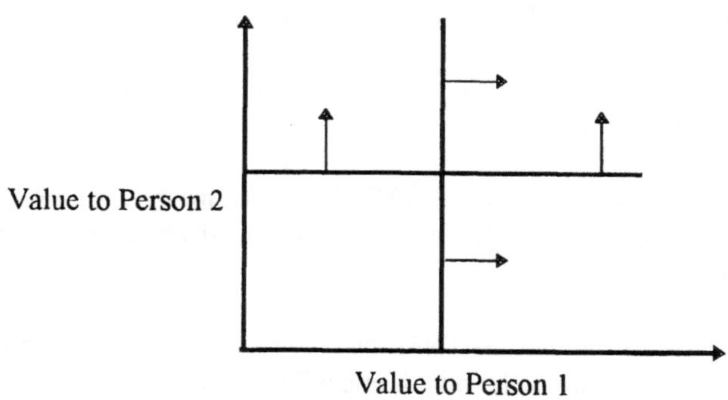

Figure 1a: BATNA Constraints in Value Space

---

[2]In three dimensions the indifference surface would look something like a satellite dish. In higher dimensions, the indifference surface would be a "hyper satellite dish". This surface acts as a sort of safety net under which a rational negotiator would not accept any negotiated settlement.

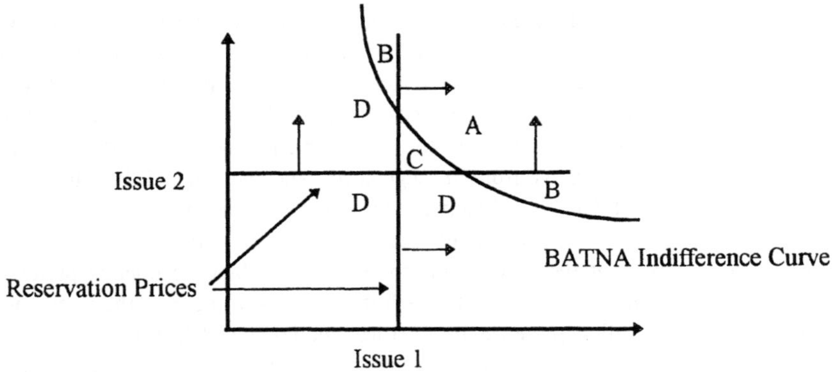

Figure 1b: Issue Space with Individual 1's BATNA Surface and Reservation Prices

This paper consists of three sections. In the second section, we describe the experiment: the hypotheses, the design, the support (the indifference surface elicitation), and the results. The third section concludes the paper. An appendix contains the instrument.

2. The Experiment

2.1 Hypotheses

Based upon the previous discussion, the following hypotheses are proposed:

H1: Supported individuals will, *on the average*, negotiate better outcomes than unsupported individuals.

H2: Supported individuals with uncertain BATNAs, *on the average*, will find the process and information more helpful than those individuals supported with deterministic BATNAs.

H3: Individuals with deterministic BATNAs, *on the average*, will achieve better outcomes than those individuals with uncertain BATNAs.

H4: Individuals provided with a deterministic BATNA are *generally*[3] able to agree to an outcome above their BATNA. Exceptions can be explained by mediated circumstances (i.e., friendship or ingratiation between the individuals).

[3] "Generally" is interpreted to mean that at least 90% of the cases in the population will behave accordingly.

H5: When both individuals are supplied with deterministic BATNAs, if the BATNAs overlap, the outcome will *generally* lie between the BATNAs; and if the BATNAs do not overlap, they will not *generally* agree to a negotiated outcome.

H6: "Negotiators will report that the BATNAs available to the other are similar to their own following the negotiation (regardless of the actual BATNA available to the other individual)" (Pinkley et al., 1994). We assume that this is *generally* true.

H7: When the BATNAs available to each member of the negotiating dyad are relatively equal (Deterministic High-High, Medium-Medium, or Low-Low), agreements will be less likely to be near Pareto-optimal than when the BATNAs are unequal (Deterministic and Low-High) (cf., Pinkley et al., 1994).

H8: "In dyads composed of negotiators with unequal BATNAs, those with a higher BATNA will receive a greater portion of the outcome when compared with those with a lower BATNA" (Pinkley et al., 1994). We assume this is true *on the average*.

H9: The highest rate of impasse will *generally* occur in dyads where both individuals have high BATNAs, a medium rate of impasse when dyads have medium BATNAs and the lowest rate of impasse in dyads where both individuals have low BATNAs (cf., Pinkley et al., 1994).

H10: If individuals have reached a non Pareto-optimal outcome, they *generally* move to the Pareto-frontier in post-settlement negotiations.

2.2 Subjects

Fifty dyads, consisting of undergraduate business school students (juniors and seniors) at New Mexico State University, were used as our experimental subjects. They all volunteered as dyads, making it possible for us to test the impact of differing types of relationships between individuals.

2.3 Task

Each dyad was requested to negotiate about a three-issue package, consisting of $10, 21 Swiss chocolates, and 2.5% of total credit (11/25 extra credit points[4]). (See the Appendix.) To make the exercise contextually relevant for the students, they were allowed to retain the negotiated settlement. Each individual was informed of his/her

[4] The relative value was the same, but the absolute value varied due to the differing point totals of the classes used in the experiment.

BATNA prior to the negotiations. Those individuals who were supported by the 7-point indifference surface corresponding to their BATNA, had to undergo the elicitation procedure described below prior to the negotiations. No time constraint was imposed on the exercise. Nor were any sanctions imposed, in case the negotiations would continue indefinitely.

2.4 Design

The design consisted of nine cells. We had two main factors: Support and BATNA, which were varied in our experiment as shown in Table 1. In this exploratory study, we have focused on the effects of the main factors. This has allowed us to keep the design reasonably simple to carry out.

Table 1: Experimental Design Indicating the Number of Dyads in Each Cell

Negotiator 2	Negotiator 1:	No Support			Support
		Low	Medium	High	Medium
No Support	Uncertain		9		7
Support	Low	4		4	
	Medium		4		
	High	4		4	
	Uncertain		7		7

Key:
No Support : The individuals are not supported
Support : The individuals are supported
Low, Medium, High } BATNA-levels when known
Uncertain : Unknown BATNA-level

The BATNA levels used in the experiment are provided in Table 2. The uncertain BATNA consisted of a choice of one of three boxes, the contents of which were unknown to the subjects. In actuality, one box contains a low level, one a medium level, and one a high level of resources as shown in Table 2.

Table 2: The BATNA-Levels Used in the Experiment

	Dollars ($)	Chocolates	Extra Credit Points
LOW	3	5	3
MEDIUM	4	9	5
HIGH	5	12	7

2.5 Performance Measures

In the experiment, we used the following performance measures:

1. The percentage of the total package that a negotiator received, and the difference in these percentages for each dyad.[5]

2. Number of times a settlement appeared Pareto-optimal or not.

3. Rate of Impasse (Tripp and Sondak, 1992).

4. Number of times the settlement dominated the BATNA in the issue space, or number of times it was dominated by the BATNA.

5. Subjective rating concerning the usefulness of the support process and information (on a 7-point scale, 7 best).

To test for Pareto-optimality, we offered the subjects 6 alternative packages, trading off (logrolling) from the negotiated settlement. These packages represented different directions from the settlement. For additional details, see the directional search in Teich, Wallenius, Wallenius, and Zionts (1993). If the subjects both preferred the corresponding trade from the settlement, then the settlement would not be Pareto-optimal.

2.6 Negotiator Support: Indifference Surface Elicitation

Keeney and Raiffa (1991) discuss a scoring scheme to evaluate individuals' BATNAs. They assume that an underlying value point structure has been elicited. The value of one's BATNA provides a lower bound for a negotiator. We feel that such a value point structure is too complicated to be assessed reliably, and at best, is a rough

[5] To illustrate, assume that individual 1 settles for $4, 7 chocolates, and 6 points, and individual 2 settles for $6, 14 chocolates, and 5 points. This means that individual 1's percentage of the total package ($10, 21, chocolates, 11 pts) is 100*(4/10+7/21+6/11)/3 = 42.6%, and individual 2's is 57.4%. Hence it follows that the difference is -14.8%.

approximation. Instead of pursuing value points, we propose an alternative approach to evaluating one's BATNA. We essentially elicit and construct an approximation of an individual's indifference surface corresponding to his/her BATNA (see, e.g., MacCrimmon and Toda, 1969, for an experimental determination of indifference curves in two dimensions). In the experiment, we perform the elicitation by using the following scheme[6]:

Step 1: Assume the existence of a BATNA

Assume a medium BATNA: [$4, 12 pts, 9 choc.].

Step 2: Construct an alternative that consists of the midpoint of each issue; call it the current alternative

A midpoint: [$5, 12.5 pts, 10.5 choc.].

Step 3: Ask the individual to compare the current alternative with the BATNA. If he or she prefers the BATNA, go to step 4. If the current alternative is preferred, go to step 5. If the individual is indifferent, go to step 6.

Assume that at the first iteration the current point is preferred. We go to step 5. At the second iteration, assume that the BATNA was preferred to a once downward adjusted alternative [$2.5, 6.25 pts, 5.25 choc.], *we go to Step 4. At the third iteration, assume that the individual prefers the BATNA over the current alternative* [$3.75, 9.375 pts, 7.875 choc.]. *Go to Step 4. At the fourth iteration assume that the individual is indifferent between the BATNA and the current alternative* [$4.375, 10.94 pts, 9.19 choc.].

Step 4: Adjust each issue upward (halfway to edge of range or to previously constructed alternative) to construct the current alternative and return to Step 3.

Now we adjust upward the current point to the midpoint of range: [$2.5, 6.25 pts, 5.25 choc.] *to* [$5, 12.5 pts, 10.5 choc.], *i.e. the next current point is* [$3.75, 9.375 pts, 7.875 choc.]; *return to Step 3. In iteration 4, we adjust up to* [$4.375, 10.94 pts, 9.19 choc.]; *Return to Step 3.*

---

[6]In this paper we assume, without loss of generality, that the individuals prefer more of each issue to less. In our experiment, this does not, however, imply a cooperative negotiation situation, since we have a fixed "pie" for each issue.

Step 5: Adjust each issue downward (halfway to edge of range or to previously constructed alternative) to construct the current alternative and return to Step 3.

The current point was preferred, adjust downward the current point: [$2.5, 6.25 pts, 5.25 choc.], and go to Step 3.

Step 6: Record this point as a point on the indifference surface. Repeat Steps 2-5 with the six different starting alternatives (in three issues), one issue at the midpoint, one at the 75th percentile, and the third at the 25th percentile.

We have found a point [$4.375, 10.94 pts, 9.19 choc.] *that lies on the same indifference curve as BATNA.*

In our experiment, we simply present these 7 points on the indifference surface as an approximation to the entire surface.

2.7 Analyses

We start our analysis by considering the following regression model, referred to as the basic model:

$$y = \mu + \alpha_1 x_{S1} + \alpha_2 x_{S2} + \beta_{L1} x_{L1} + \beta_{M1} x_{M1}$$

$$+ \beta_{H1} x_{H1} + \beta_{U1} x_{U1} + \beta_{L2} x_{L2} + \beta_{M2} x_{M2} + \beta_{H2} x_{H2} + \beta_{U2} x_{U2} + \varepsilon,$$

where

y is a difference in performance measured as explained in point 1, section 2.5

μ, α_{S1}, α_{S2}, β_{L1}, β_{M1}, β_{H1}, β_{U1}, β_{L2}, β_{M2}, β_{H2}, and β_{U2} are unknown parameters

$\varepsilon \sim N(0,\sigma^2)$; error terms are independent of each other

$$x_{Si} = \begin{cases} 0, \text{ if the } ith \text{ negotiator is not supported, } i = 1,2 \\ 1, \text{ if the } ith \text{ negotiator is supported, } i=1,2 \end{cases}$$

$$x_{Li} = \begin{cases} 0, \text{ if the } ith \text{ negotiator doesn't have a low BATNA , } i = 1,2 \\ 1, \text{ if the } ith \text{ negotiator has a low BATNA , } i=1,2 \end{cases}$$

$$x_{Mi} = \begin{cases} 0, \text{ if the } ith \text{ negotiator doesn't have a medium BATNA}, i = 1,2 \\ 1, \text{ if the } ith \text{ negotiator has a medium BATNA}, i=1,2 \end{cases}$$

$$x_{Hi} = \begin{cases} 0, \text{ if the } ith \text{ negotiator doesn't have a high BATNA}, i = 1,2 \\ 1, \text{ if the } ith \text{ negotiator has a high BATNA}, i=1,2 \end{cases}$$

$$x_{Ui} = \begin{cases} 0, \text{ if the } ith \text{ negotiator has a known BATNA}, i = 1,2 \\ 1, \text{ if the } ith \text{ negotiator has an unknown BATNA}, i=1,2 \end{cases}$$

To estimate the regression coefficients of the model, we have to fix first a regression coefficient corresponding to one of the variables x_{L1}, x_{M1}, x_{H1}, x_{U1}, and to one of the variables x_{L2}, x_{M2}, x_{H2}, x_{U2}, because these variable sets are linearly dependent with the constant variable. We set $\beta_{U1} = \beta_{U2} = 0$. In addition, we have to fix, e.g., the regression coefficient of variable x_{H2}, because our experimental design makes the rest of the variables linearly dependent. The results are given in Table 3:

Table 3: Regression Results of the Basic Model

Variable	Regr.coeff.	Std.dev.	t-values	β-coeff.
x_{S1}	-0.067	0.238	-0.28	-0.055
x_{S2}	0.045	0.225	0.20	0.041
x_{L1}	-0.376	0.379	-0.99	-0.325
x_{M1}	-0.229	0.279	-0.82	-0.259
x_{H1}	-0.173	0.379	-0.46	-0.149
x_{U1}	0.000	0.000	0.00	0.000
x_{L2}	0.230	0.223	1.03	0.199
x_{M2}	-0.186	0.279	-0.67	-0.215
x_{H2}	0.000	0.000	0.00	0.000
x_{U2}	0.000	0.000	0.00	0.000
Constant	0.248	0.316	0.78	

Variance of regression = 0.1832 df = 49
Total Sum of Squares ($SSTO_f$) = 8.9787
Residual variance = 0.1986 df_f = 42
Error Sum of Squares (SSE_f) = 8.3410
R = 0.2665
R^2 = 0.0710

First, we will test whether or not there is a relation between the dependent variable y and the independent variables x_{S1}, x_{S2}, x_{L1}, x_{M1}, x_{H1}, x_{U1}, x_{L2}, x_{M2}, x_{H2}, x_{U2}, i.e.:

H_o: $\alpha_{S1} = \alpha_{S2} = \beta_{L1} = \beta_{M1} = \beta_{H1} = \beta_{U1} = \beta_{L2} = \beta_{M2} = \beta_{H2} = \beta_{U2} = 0$

H_1: Not all coefficients are zeroes.

The F test is applied:

$$F^* = ((SSTO_f - SSE_f)/(df - df_f)) / (SSE_f/df_f)$$
$$= ((8.9787\text{-}8.3410)/(49\text{-}42))/(0.1986) = 0.4587$$

Through these tests, we will use $\alpha = 0.05$. Because $P(F > 0.4587) = 0.859 > 0.05$, we conclude H_o.

To conclude H_o means that our hypotheses H1, H2, H3, and H8[7] are not supported. Whether or not individuals are supported and level of BATNA has no statistically significant effect on the outcome. However, the type of BATNA, in general, appears at least to effect the outcome in the expected direction, even though the result is not statistically significant. Similarly, whether the BATNA is deterministic or unknown, has no effect.

Next, we consider hypotheses H4, H5, and H6, in which we have used the expression "generally". To test these hypotheses, we formulate the test problem as follows:

H_o: $p_o \geq 0.90$,

H_1: $p_o < 0.90$.

Consider hypothesis H4. First, we randomly chose the second negotiator from each dyad, and found that 5 of 35 individuals with deterministic BATNAs agreed to an outcome dominated by their BATNA.

Consider binomial variable $X_{(0.9,35)}$. Because

$$P(X_{(0.9,35)} \leq 30) = 0.2693 >> 0.05,$$

we conclude H_o, implying that generally individuals are able to agree to an outcome above their BATNA. Even though the hypothesis is not rejected, we should be surprised that 1/7 of our subjects agreed to a solution below their BATNA. We did not find any support for our conjecture that these exceptions are caused by mediated circumstances, such as friendship.

Next, we consider hypothesis H5. We found that in 15 out of 16 cases, in which BATNAs overlapped (Deterministic Low-Low, Low-High, Medium-Medium, and High-Low), the outcome did lie between the BATNAs, implying that

---

[7] Pinkley et al. (1994) found marginal support for this hypothesis.

$$P(X_{(0.9,16)} \leq 15) = 0.8147 >> 0.05.$$

Thus, we conclude H_o.

Correspondingly, in four nonoverlapping cases (Deterministic High-High), three did not agree to a negotiated outcome. For this

$$P(X_{(0.9,4)} \leq 3) = 0.3439 > 0.05.$$

Thus, we also conclude H_o. The individuals behavior is in accordance with our research hypothesis.

Data used in hypothesis H6 are based on a follow-up interview. Therefore, we can regard all individuals (70) with deterministic BATNAs as independent observations. Because 21 out of 70 considered the other individual's BATNA similar to their own, our conclusion is based on the probability

$$P(X_{(0.9,70)} \leq 21) \approx 0 << 0.05,$$

implying that we conclude H_a. This means that it is not "generally" true that the negotiator would consider the other individual's BATNA similar to their own (cf. Pinkley et al., 1994).

Consideration of hypothesis H7 is based on the frequency distribution of Pareto-optimal outcomes, when BATNAs are equal (Deterministic Low-Low (4), Medium-Medium (4), and High-High (4)), vs. BATNAs are unequal (Deterministic Low-High (4) and High-Low (4)). We conclude H_o. See Table 4. Compare this result to Pinkley et al. (1994), who found no support for this hypothesis.

Table 4: Pareto-Optimality vs. Equality of BATNAs

	Equal BATNA	Unequal BATNA	Total
Pareto-Optimal	7	18	25
Non Pareto-Optimal	5	4	9
Total	12	22	34
X^2=2.200 df=1 $P(\chi^2_{(1)} > X^2) = 0.138 > 0.05$			

Hypothesis H9 received some support, in fact stronger support than Pinkley et al. (1994), but due to the small sample size, we did not perform a rigorous statistical test. Three dyads out of 4 with high BATNAs, 0 dyads out of 4 with medium BATNAs, and 0 dyads out of 4 with low BATNAs, failed to negotiate a settlement.

Consider hypothesis H10. When testing for the Pareto-optimality of a negotiated settlement, we asked the individuals whether they would prefer six different packages to their settlement. In eleven of 50 dyads, when asked individually, both expressed willingness to make the same trade, implying non-Pareto-optimality of their settlement. Interestingly, in 8 out of the 11 such cases, the individuals refused to truly accept the corresponding Pareto improvement to their settlement (post-settlement settlement). To test whether people *generally* accept post-settlement settlements, we use the same test as for hypotheses H4, H5, and H6. Because

$$P(X_{(0.9,11)} \leq 3) = 0.00000 < 0.05,$$

we conclude H_a, implying that people obviously generally refuse to accept a Pareto improvement to the negotiated settlement.

2.8 Additional Results and Discussion

Interestingly, the support provided to the individuals was not helpful. In some cases, it seemed even harmful. We believe that this can be explained by the anchoring phenomenon (Neale and Bazerman, 1992; Buchanan and Corner, 1994). Individuals, in particular with low and medium BATNAs, seemed to anchor on the support (that is, the indifference surface corresponding to their BATNA), and therefore did not make much progress in the negotiations. In other words, prescriptively, if a negotiator has a poor BATNA, he or she should not dwell on it and rather concentrate on the negotiations or in developing a better BATNA. With a high BATNA, we still hypothesize that the support might be helpful. To conclude, knowledge of one's BATNA is important. However, too much knowledge may be harmful.

We were surprised to observe that a negotiator provided with a medium, deterministic BATNA did not perform better than individuals with an uncertain BATNA. Individuals apparently strongly disliked the uncertain BATNA, making their negotiations more successful. Kelley et al.'s (1967) discussion about the "low man's advantage" further explains this result. We suspect that a "low man's advantage" is less prevalent when an individual with an uncertain BATNA faces an individual with a high deterministic BATNA.

We speculate that the relationship between the negotiators may be an important factor related to outcome and bargaining toughness. While conducting the experiment it was observed that negotiators who did not know each other previously and, at the other extreme, those who had close friendships seemed to negotiate with a higher degree of conflict. Those with only an acquaintance type relationship were not, in general, tough negotiators and seemed interested in reducing the level of conflict (this could cause a negotiator to accept an outcome below the BATNA). Likewise, although not explicitly observed herein, if the type of relationship is "very" close, e.g., children or spouses, we predict that the level of conflict would be lowest. For a related discussion, see the dual concern model in Carnevale and Pruitt (1992, p. 539-543).

We were also surprised to observe that in 8 out of 11 cases, the individuals refused to accept the Pareto improvement to the negotiated settlement. This kind of behavior can even be said to be *generally* true.

This phenomenon contradicts traditional rationality of utility maximizing behavior. We offer two possible explanations:

- Firstly, the negotiators may simply dislike and distrust an outcome generated by computer, when they already have spent considerable time and effort in negotiating with their partner.

- Secondly, one individual's utility may be a criterion for the other individual that he or she seeks to minimize. In other words, when considering the trade, it becomes obvious that when implementing a Pareto improvement, the other individual will gain as well. The perceived gain of the other individual may be more than one's own gain, making the trade unequitable (unfair) and unacceptable (see, also Kersten, 1994).

- Thirdly, transaction costs involved in implementing the Pareto-improvement may be a factor. Perceived gains may be small relative to the costs.

This phenomenon, however, may be less likely to occur if an acknowledged expert presents the settlement alternatives to the negotiators. The expert may provide more legitimacy to the Pareto improvement. This question clearly has implications for mediation and for the interface between a Negotiation Support System and the users of such systems.

Also, work experience and/or educational level of the negotiators may have an impact on whether a Pareto improvement to an initial settlement is accepted. We speculate that those with more work experience and/or educational level would be more likely to accept such improvement. In addition, when the rewards (risks, importance etc.) are higher, individuals may be more willing to accept a Pareto improvement.

The support is likely to become more useful if we consider the principal-agent case. An example is the lawyer-client relationship. In this case, an attorney could check the acceptance of a possible settlement without consulting the client. If the indifference surface has been elicited, the client does not even have to be available for frequent consultations. An interesting future study would consist of testing the usefulness of support in the principal-agent situation. As an additional performance measure we should also consider the frequency, length, and intensity of communication between the principal and the agent.

3. Conclusion

We have investigated the role of BATNA (Best Alternative to a Negotiated Agreement) in dyadic negotiations. We have distinguished between deterministic, stochastic, and uncertain BATNAs. In a contextually relevant experiment, we have tested the impact of different BATNAs and negotiator support (in the form of an indifference surface corresponding to an individual's BATNA) to performance in negotiations. We have verified some of Pinkley et al.'s (1994) hypotheses and tested several of our own hypotheses. Our major results and their explanations can be summarized as follows:

- On the average, the type of BATNA (Deterministic Low, Deterministic Medium, Deterministic High, or Uncertain) makes no difference - at least if differences between BATNAs are not very extreme.
- On the average, individuals with an uncertain BATNA performed unexpectedly well - more precisely - not worse than individuals with a deterministic BATNA, exhibiting what Kelley et al. (1967) refer to as a "low man's advantage".
- Supported individuals do not perform better than the unsupported individuals, on the average, possibly because of anchoring on their BATNA surface.
- The individuals are generally able to agree to an outcome above their BATNA.
- If the BATNAs overlap, the outcome will generally lie between the BATNAs.
- If the BATNAs don't overlap, the individuals will generally not reach an agreement.
- The negotiators do not generally believe that their own BATNA is similar to that of the other individual.
- We did not find any difference in average performance to reach a Pareto-optimal solution, whether the BATNAs are unequal or equal.
- When a non Pareto-optimal outcome is reached through negotiations, the individuals generally retain it rather than make a Pareto-improvement (post-settlement settlement).

Some of our results were expected and some unexpected. It was surprising to us that we found no statistically significant impact from the level and type of BATNA and support on performance in negotiations. From an individual's perspective, the usefulness of negotiation support via an indifference surface elicitation has not been proven in simple dyadic negotiations. It remains a subject of future research whether this type of negotiation support is useful in: a) more complicated business negotiations, b) within individual negotiations, and c) agent-principal relationships. It was also a surprising result that the individuals were not willing to make a Pareto-improvement, even if it existed.

In our view, the contribution of our research to the field is to state and test several hypotheses related to the role of BATNAs in negotiations in an exploratory manner. In particular, we improve the methodology of conducting negotiation experiments in general by:

- using a contextually relevant multiple issue negotiation, where the subjects retain the negotiated settlement;
- not employing an imposed value (point) structure; and
- developing a simple procedure to identify Pareto-improvements to settlements, even without such value points.

Due to the richness of the negotiation field in general, and the multi-faceted role of BATNAs, in particular, many of the open ended questions should be addressed through additional experimentation.

Acknowledgement: We wish to thank Professor Sally White for providing constructive comments on an earlier draft of this paper.

References

Balakrishnan, P., Patton, C., and Lewis, P. (1993). "Toward a Theory of Agenda Setting in Negotiations", *Journal of Consumer Research*, 19, 637-654.

Buchanan, J. and Corner, J. (1994). "The Effect on Anchoring in Interactive MCDM Solution Methods", Paper presented at the XIth International Conference on Multiple Criteria Decision Making, Coimbra, Portugal, August 1-6, 1994.

Carnevale, P. & Pruitt, D. (1992). "Negotiation and Mediation", *Annual Review of Psychology*, 43, 531-582.

Fisher, R. & Ury, W. (1981). *Getting to Yes,* Boston: Houghton-Mifflin.

Keeney, R. & Raiffa, H. (1991). "Structuring and Analyzing Values for Multiple-Issue Negotiations", In H. Young (Ed.), *Negotiation Analysis.* Ann Arbor, Michigan: University of Michigan Press.

Kelley, H., Beckman, L, & Fischer, C. (1967). "Negotiating the Division of a Reward Under Incomplete Information", *Journal of Experimental Social Psychology*, 3, 361-398.

Kersten, G. (1994). "Rational Agent and Non-Efficient Compromises", Paper presented at the XIth International Conference on Multiple Criteria Decision Making, Coimbra, Portugal, August 1-6, 1994.

McAlister, L., Bazerman, M., & Fader, P. (1986). "Power and Goal Setting in Channel Negotiations", *Journal of Marketing Research*, 23, 228-236.

McCarthy, W. (1991). "The Role of Power and Principle in Getting to YES", In J. Breslin and J. Rubin (Eds.), *Negotiation Theory and Practice.* Cambridge, Mass.: Program on Negotiation Books.

McCrimmon, K. & Toda, M. (1969). "The Experimental Determination of Indifference Curves", *Review of Economic Studies*, October 1969, 433-451.

Murnighan, J. & Bazerman, M. (1990). "A Perspective on Negotiation Research in Accounting and Auditing", *Accounting Review*, 65, 642-657.

Neale, M. & Bazerman, M. (1992). "Negotiator Cognition and Rationality: A Behavioral Decision Theory Perspective", *Organizational Behavior and Human Decision Processes*, 51, 157-175.

Ostmann, A. (1992). "The Interaction of Aspiration Levels and the Social Field in Experimental Bargaining", *Journal of Economic Psychology*, 13, 233-261.

Pinkley, R., Neale, M., & Bennett, R. (1994). "The Impact of Alternatives to Settlement in Dyadic Negotiation", *Organizational Behavior and Human Decision Processes*, 57, 97-116.

Pruitt, D. & Syna, H. (1985). "Mismatching the Opponent's Offers in Negotiation", *Journal of Experimental Social Psychology*, 21, 103-113.

Raiffa, H. (1982). *The Art and Science of Negotiation,* Cambridge, Mass.: Harvard University Press.

Roth, A. & Murnighan, J. (1982). "The Role of Information in Bargaining: An Experimental Study", *Econometrica*, 50, 1123-1142.

Smith, D., Pruitt, D. & Carnevale, P. (1982). "Matching and Mismatching: The Effect of Own Limit, Other's Toughness and Time Pressure on Concession Rate in Negotiation", *Journal of Personality and Social Psychology*, 42, 876-883.

Smith, W. (1987). "Conflict and Negotiation: Trends and Emerging Issues", *Journal of Applied Social Psychology*, 17, 641-677.

Sutton, J. (1987). "Bargaining Experiments", *European Economic Review*, 31, 272-284.

Teich, J., Wallenius, H., Wallenius, J., & Zionts, S. (1993). "Identifying Pareto-Optimal Settlements for Two-Party Resource Allocation Negotiations", Working Paper No. 769, School of Management, State University of New York at Buffalo (Forthcoming in *European Journal of Operational Research*).

Tietz, R. (1983). "Aspiration Oriented Decision Making", In Lecture Notes in Economics and Mathematical Systems, No. 213, (Ed.) R. Tietz, New York, Springer-Verlag, 1-7.

Tripp, T. & Sondak, H. (1992). "An Evaluation of Dependent Variables in Experimental Negotiation Studies: Impasse Rates and Pareto Efficiency", *Organizational Behavior and Human Decision Processes*, 51, 273-295.

Valley, K., White, S., Neal, M. & Bazerman, M. (1992). "Agents as Information Brokers: The Effects of Information Disclosure on Negotiated Outcomes", *Organizational Behavior and Human Decision Processes*, 51, 220-236.

White, S. & Neale, M. (1991). "Reservation Prices, Resistance Points, and BATNAs: Determining the Parameters of Acceptable Negotiated Outcomes", *Negotiation Journal*, 7, 379-388.

Appendix: The Instrument

(Low, deterministic BATNA)

Thank you for participating in this negotiation exercise. This negotiation is real. Whatever you agree to is yours. You will be asked to negotiate over 3 issues with your counterpart. Specifically, you will negotiate over a joint resource consisting of: 10 U.S. Dollars, 21 exquisite swiss chocolates, and 11 extra credit points in MGT 325. If you fail to agree to a settlement with your counterpart in the negotiation, you will receive: 5 swiss chocolates, 3 extra credit points and 3 dollars.

Complete the following steps:

1. You may be asked to meet individually with the facilitator of the exercise prior to negotiating.

2. Negotiate with your counterpart.

3. Answer a few questions.

If you have any questions, ask the facilitator.

Multiple Criteria Scheduling on a Single Machine: A Review and a General Approach

Murat Köksalan[1,2] and Suna Köksalan Kondakci[1,2]

[1]Department of Industrial Engineering, Middle East Technical University, Ankara, Turkey
[2] Krannert School of Management, Purdue University, West Lafayette, IN 47907

Abstract

Scheduling problems have been treated as single criterion problems until recently. Many of these problems are computationally hard to solve even as single criterion problems. However, there is a need to consider multiple criteria in a real life scheduling problem in general. In recent years, many articles considering bicriteria scheduling problems, and some articles considering more than two criteria started to appear in the literature. We briefly review various categories of approaches on multiple criteria scheduling on a single machine. We then present a general approach to find the most preferred schedule in a bicriteria environment for any given nondecreasing composite function of the criteria.

1. Introduction

The problem we address is to determine the sequence of jobs waiting to be processed on a single machine. The single criterion version of this problem has been well studied for a variety of criteria. More recently, there have been many publications that consider two criteria simultaneously, and a few publications that consider more than two criteria.

In real life, these problems are indeed multiple criteria problems in nature. For example, a scheduler could be concerned with the work-in-process inventory levels as well as finishing the jobs by their due dates. The operational criteria that may represent these two concerns would usually be conflicting with each other.

An important characteristic of multiple criteria scheduling problems is their combinatorial nature. Many of the problems have been shown to be NP-hard in terms of computational complexity. Increasing the considered number of criteria makes the problem harder to solve. We believe that bicriteria scheduling problems can represent the real life problems well in many situations. Typically, one criterion can be used to represent the manufacturer's concern (such as a criterion related with inventories) and another criterion can be used to represent the customer's concern (such as a criterion related with due dates).

In Section 2 we briefly review the multiple criteria scheduling problems and present some complexity results. We present a general approach to find the best of a set of solutions for a given nondecreasing function of multiple criteria in Section 3 and make some concluding remarks in Section 4.

2. Background

Consider n jobs waiting to be processed on a single machine where each job has a deterministic processing time on the machine and a due date. Let
p_i: processing time of job i (integer)
d_i: due-date of job i (integer)
C_i: completion time of job i
T_i: tardiness of job i (T_i=Max(0,C_i-d_i))
E_i: earliness of job i(E_i=Max(0,d_i-C_i))
for i=1,...,n. To represent a scheduling problem we will use the following three field notation (Graham et al. 1979)

$$\alpha|\beta|\gamma$$

where α represents the machine environment (e.g. α=1 identifies single machine), β represents job characteristics (e.g. β=pmtn implies jobs can preempt each other, β=nmit implies no machine idle time is allowed) and γ represents performance criteria upon which schedules will be evaluated (e.g. $\gamma=\Sigma C_i$, ΣT_i implies that sum of completion times and total tardiness are used as performance criteria).

A sequence S is said to be efficient if there does not exist any other sequence S' such that $f_j(S') \leq f(S)$ for j=1,...,m and $f_k(S') < f_k(S)$ for at least one k where f_j represents the jth performance measure. Without loss of generality, we assume all criteria are to be minimized.

The ultimate goal in multiple criteria scheduling could be to generate all the efficient solutions. We distinguish between an efficient solution and an efficient sequence since there could be several distinct efficient sequences resulting with the same criteria values, that is, the same efficient solution. In the literature, the problems have been categorized in two broad headings.

1. Hierarchical scheduling where the criteria are ordered in terms of their importance and the less important criteria are only used to break ties among different sequences yielding the minimum values in the more important criteria.

This approach has been known as the Lexicographic approach in the multicriteria decision making literature.

2. Simultaneous minimization where a composite objective function of the criteria is minimized.

Since the vast majority of the publications in this area consider two criteria, we restrict our attention to bicriteria problems in most of the remainder of this paper. We briefly discuss the generalization of some of the ideas to more than two criteria later. Let Q denote the set of possible sequences and u(S) and v(S) denote the criteria values of sequence S in the first and second criteria respectively.

2.1 Hierarchical Scheduling

These problems are also called secondary criteria problems where the secondary criterion refers to the less important criterion. Let v and u respectively represent the primary and the secondary criteria. Then the problem is to

$$\begin{aligned} &\text{Min } u(S) \\ &\text{s.t. } v(S) \leq v_1 \\ &\qquad S \in Q \end{aligned}$$

where $v_1 = \text{Min}_{S \in Q}\, v(S)$. The problem is also represented as $1||F_h(u|v)$ where $F_h(u|v)$ indicates that we wish to hierarchically minimize using v as the primary criterion. We use the notation $1||F(u|v)$ to represent a problem where criterion v is bounded by an arbitrary value instead of its minimum value. Note that $1||F_h(u|v)$ is a special case of $1||F(u|v)$.

Some of the performance measures often encountered in scheduling are; sum of completion times, ΣC_i, total tardiness, ΣT_i, total earliness, ΣE_i, number of tardy jobs, ΣZ_i where Z_i is 1 if $C_i > d_i$ and 0 otherwise, maximum lateness, $L_{max} = \max_i(C_i - d_i)$, maximum earliness, $E_{max} = \max_i(E_i)$, and maximum tardiness, $T_{max} = \max_i(T_i)$.

Among the criteria, ΣC_i, ΣT_i, ΣE_i, ΣZ_i, L_{max} and E_{max}, where ΣC_i is the primary criterion and one of the remaining criteria is the secondary criterion, the problem can be solved in polynomial time. When L_{max} is the primary criterion and one of ΣC_i or E_{max} is the secondary criterion, the problem again can be solved in polynomial time. Another polynomially solvable case arises when E_{max} is the primary and L_{max} is the secondary criterion. When ΣZ_i is the primary criterion and one of ΣC_i, L_{max}, or ΣT_i is the secondary criterion, the computational complexity of the problem is not known. The remaining bicriteria combinations of the above criteria lead to NP-hard problems. A detailed complexity analysis of hierarchical problems can be found in Lee and Vairaktarakis [1993].

2.2 Simultaneous Minimization

In this case, an increasing composite function, F, of the two criteria is minimized. This problem can be represented as $1||F(u,v)$. Note that the hierarchical scheduling problem is a special case of the simultaneous minimization problem and hence if $1||F_h(u|v)$ is NP-hard then $1||F(u,v)$ is also NP-hard.

If we can generate all efficient solutions, then we can solve the above problem by evaluating each efficient solution using the function, F. Van Wassenhove and Gelders [1980] develop an algorithm to generate all the efficient solutions for ΣC_i and L_{max} criteria in polynomial time. Hoogeveen and Van de Velde [1995] find all efficient solutions for the bicriteria problem of ΣC_i and a maximum cost function, f_{max}, in polynomial time where $f_{max}=Max_i f_i(C_i)$ and f_i is any nondecreasing function of C_i. Note that L_{max} is a special case of f_{max}. Hoogeveen [1992] shows that all efficient solutions for two different maximum cost functions, f_{max} and $\hat{f}_{max}$ can be generated in polynomial time. He also shows that the same result is valid for the case where both f_i and $\hat{f}_i$ used in calculating f_{max} and $\hat{f}_{max}$ are nonincreasing (rather than nondecreasing) in C_i. E_{max}, for example, is a special case of such a function.

The computational complexities of none of the bicriteria problems where number of tardy jobs, ΣZ_i, is one of the criteria have been identified until now. All the other bicriteria combinations of the criteria mentioned so far in this paper have been shown to be NP-hard for a general nondecreasing composite function.

Although a bicriteria problem may be NP-hard with respect to a general nondecreasing composite function, it may be solved in polynomial time when the nondecreasing composite function is linear. This result is true for two different weighted sums of completion times (i.e. $\Sigma w_j C_j$ and $\Sigma \hat{w}_j C_j$) and for L_{max} and E_{max}.

For NP-hard problems, heuristic approaches can be used for generating all efficient solution or for finding the best solution.

For detailed surveys on multicriteria scheduling, the reader can refer to Gupta and Kyparisis [1985], Fry et al. [1989] and Nagar et al. [1995].

3. A General Approach

We consider a nondecreasing function of the two criteria to be minimized. We try to identify the optimal sequence while generating only a small subset of the possible solutions.

The solution to

$$\begin{aligned} (P) \quad & \text{Min } u(S) \\ & \text{s.t. } v(S) \le \tilde{v} \\ & \qquad S \in Q \end{aligned}$$

for some $\tilde{v}$ yields a weakly efficient solution. If we break ties in favor of criterion v when different sequences yield the same u value for (P), then the solution is efficient. If we solve problem (P) for each integer $\tilde{v}$ value within the range of values of criterion v (among efficient solutions), we obtain all efficient solutions. However, the complexity of problem (P) depends on u and v and it is NP-hard in many cases. When the problem is NP-hard, we may try to solve (P) approximately in order to obtain approximately efficient solutions.

Kondakci et al. [1996] developed an algorithm to find the best sequence for a nondecreasing function, F, considering the criteria, ΣC_i and T_{max}. The algorithm finds the best solution by generating only a small subset of efficient solutions. Köksalan et al. [1995] use a similar procedure for ΣC_i and E_{max}. Here, we generalize the algorithm for any two criteria. When (P) can be solved exactly, the algorithm finds the optimal solution. When (P) is solved only approximately, then the best of the approximate solutions is found.

Let u_1, and v_1, be lower bounds to the objective function values of $\text{Min}_{S \in Q}$ u(S) and $\text{Min}_{S \in Q}$ v(S), respectively. Let u_2 and v_2 be upper bounds to the objective function values of {Min u(S) s.t. $v=v_1$, $S \in Q$} and {Min v(S) s.t. $u=u_1$, $S \in Q$}, respectively. The range of values for u and v are $[u_1,u_2]$ and $[v_1,v_2]$, respectively, considering all efficient solutions.

The algorithm starts with an efficient (or approximately efficient) initial incumbent solution, and keeps the best known solution as the incumbent throughout the solution process. In each iteration it creates a dummy solution that is equivalent in terms of the F value to the incumbent. Then it generates an efficient solution in the neighborhood of the dummy solution and compares with the incumbent. The algorithm terminates when a required criterion value of the dummy solution (in order to be equivalent to the incumbent) is below the minimum possible value of that criterion.

Let u_h and v_h denote the criteria values of an initial solution and let u_I and v_I denote the criteria values of the incumbent solution.

Step 0. Let $(u_I, v_I) = (u_h, v_h)$ and $\tilde{u} = u_1$.

Step 1. Find $\tilde{v}$ such that

$F(\tilde{u}, \tilde{v}) = F(u_I, v_I)$. If $\tilde{v} \le v_1$, stop declaring the incumbent as the optimal solution.

Step 2. Solve

$$\begin{aligned} &\text{Min } u(S) \\ &\text{s.t } \quad v \leq \tilde{v} \\ &\qquad S \in Q \end{aligned}$$

and let the solution be (u^*,v^*).

Step 3. Let $\tilde{u}=u^*+1$. If $F(u^*,v^*) < F(u_l,v_l)$, let $u_l=u^*$, $v_l=v^*$. Go to step 1.

In summary, the algorithm considers the efficient (or approximately efficient) solutions starting with a lower bound on the value of u (i.e. $\tilde{u}=u_1$) and moves in the increasing direction of u (or equivalently decreasing direction of v) values. It skips some efficient (or approximately efficient) sequences in Step 1 making sure that none of the skipped sequences has a better objective function value than that of the incumbent sequence. The algorithm terminates when the $\tilde{v}$ value found in Step 1 turns out to be less than or equal to the lower bound on the v value, v_1.

Finding an Initial Solution

If a good starting solution can be found, the algorithm would skip more efficient solutions in Step 1 and converge faster. The heuristic approach of Köksalan [1993] can be used to generate the initial solution. We briefly review a simpler version of this approach below.

The approach scales the criteria values between 0 and 1 such that for a given schedule S the new criteria values are $u'=(u(S)-u_1)/(u_2-u_1)$ and $v'=(v(S)-v_1)/(v_2-v_1)$. Then, in the u',v' space it is argued that an L_p distance function from point $(u',v')=(1,1)$ joining points $(u',v')=(1,0)$ and $(u',v')=(0,1)$ well represents the locations of efficient solutions. r equally spaced dummy points on the chosen L_p distance function are evaluated in terms of the F value and the best dummy solution, say $(\overline{u'}, \overline{v'})$, is obtained. Then $\overline{v'}$ value is rescaled and $\overline{v} = v_1 + \overline{v'}(v_2-v_1)$ is found. A real efficient (or approximately efficient) solution having values u_h and v_h in the neighborhood of the best dummy solution is found by solving

$$\begin{aligned} &\text{Min } u(S) \\ &\text{s.t. } \quad v(S) \leq \overline{v} \\ &\qquad\quad S \in Q. \end{aligned}$$

Example

We demonstrate the approach for the bicriteria problem of sum of completion times, ΣC_i, and maximum tardiness, T_{max}. All efficient solutions for this problem can be generated in polynomial time. It is well known that ordering the jobs in nondecreasing order of their processing times (i.e. the shortest processing time (SPT) rule) minimizes ΣC_i and ordering the jobs in the nondecreasing order of their due dates (i.e. the earliest due date (EDD) rule) minimizes T_{max}.

Let u be ΣC_i and v be T_{max}. Then u_1=u(SPT) and v_1=v(EDD). u_2 can be obtained by breaking ties between jobs in the EDD sequence in favor of jobs having shorter processing times. Similarly, v_2 can be found by breaking ties between jobs in the SPT sequence in favor of jobs having earlier due dates.

Let $F(u,v)= 0.5\frac{(u-u_1)}{(u_2-u_1)}+0.5\frac{(v-v_1)}{(v_2-v_1)}$

and consider the 10 jobs below where job i has a processing time of p_i and due date of d_i.

Job	1	2	3	4	5	6	7	8	9	10
p_i	2	3	4	4	5	5	6	6	7	8
d_i	7	31	24	44	11	25	19	34	42	18

For simplicity we list all efficient solutions of this problem in Table 1. It can be seen that u_1=226, v_1=6, u_2=261 and v_2=32.

Using L_2 distance function from point (1,1) in the (u',v') space we obtain the arc $(u'-1)^2+(v'-1)^2=1$. Selecting r=11 equally spaced dummy points on this arc we obtain (u',v') = (0.00,1.00), (0.01,0.84), (0.05,0.69), (0.11,0.55), (0.19,0.41), (0.29,0.29), (0.41,0.19), (0.55,0.11), (0.69,0.05), (0.84,0.01), (1.00,0.00). (u',v') = (0.29,0.29) has the smallest objective function value. The T_{max} value corresponding to this dummy schedule is T=6+.29(32-6)=13.5. Solving Min u(S) st. $v(S) \leq 13.5$ $S \in Q$ yields (u_h,v_h)=(235,13).

Table 1. Efficient solutions for the example problem

No.	Job Sequence	ΣC_i	T_{max}	$F(\Sigma C_i, T_{max})$
1	1-2-3-4-5-6-7-8-9-10	226	32	0.5000
2	1-2-3-4-5-6-7-8-10-9	227	25	0.3797
3	1-2-3-4-5-6-7-10-8-9	229	19	0.2928
4	1-2-3-4-5-6-10-7-8-9	231	18	0.3022
5	1-2-3-4-5-7-10-6-8-9	233	14	0.2538
6	1-2-3-4-5-10-7-6-8-9	235	13	0.2632
7	1-2-3-5-7-10-4-6-8-9	240	12	0.3154
8	1-2-3-5-7-10-6-4-8-9	241	10	0.2912
9	1-2-3-5-10-7-6-4-8-9	243	9	0.3006
10	1-2-5-7-10-3-6-8-4-9	250	8	0.3814
11	1-3-5-7-10-6-2-8-9-4	259	7	0.4908
12	1-3-5-10-7-6-2-8-9-4	261	6	0.5000

Step 0. $(u_I,v_I)=(235,13)$ and $\tilde{u}=226$

Iteration 1

Step 1. $F(226, \tilde{v})=0.2632 \Rightarrow \quad \tilde{v}=19.7>v_I$

Step 2. Solve Min $u(S)$ st. $v(S)\leq 19.7 \quad S\in Q$

$(u^*,v^*)=(229,19)$

Step 3. $\tilde{u}=230$

$F(u^*,v^*)=0.2928>F(u_I,v_I)$

Iteration 2

Step 1. $F(230, \tilde{v})=0.2632 \Rightarrow \quad \tilde{v}=16.7>v_I$

Step 2. Solve Min $u(S)$ st. $v(S)\leq 16.7 \quad S\in Q$

$(u^*,v^*)=(233,14)$

Step 3. $\tilde{u}=234$

$F(u^*,v^*)=0.2538<F(u_I,v_I) \quad \Rightarrow \quad u_I=233,\ v_I=14$

Iteration 3

Step 1. $F(234, \tilde{v})=0.2538 \Rightarrow \quad \tilde{v}=13.3>v_I$

Step 2. Solve Min $u(S)$ st. $v(S)\leq 13.3 \quad S\in Q$

$(u^*,v^*)=(235,13)$

Step 3. $\tilde{u}=236$

$F(u^*,v^*)=0.2632>F(u_I,v_I)$

Iteration 4

Step 1. $F(236, \tilde{v})=0.2538 \Rightarrow \quad \tilde{v}=11.8>v_I$
Step 2. Solve Min u(S) st. $v(S) \leq 11.8 \quad S \in Q$
$(u^*,v^*)=(241,10)$

Step 3. $\tilde{u}=242$
$F(u^*,v^*)=0.2912>F(u_I,v_I)=0.2538$

Iteration 5

Step 1. $F(242, \tilde{v})=0.2538 \Rightarrow \quad \tilde{v}=7.3>v_I$
Step 2. Solve Min u(S) st. $v(S) \leq 7.3 \quad S \in Q$
$(u^*,v^*)=(259,7)$

Step 3. $\tilde{u}=260$
$F(u^*,v^*)=0.4908>F(u_I,v_I)=0.2538$

Iteration 6

Step 1. $F(260, \tilde{v})=0.2538 \Rightarrow \quad \tilde{v}=-6.1<v_I$
Step 4. The incumbent sequence having $(u_I,v_I)=(233,14)$ is the optimal sequence.

In this example problem we had to generate only five of the 12 efficient solutions in order to find the optimal solution.

4. Conclusions

In this paper we briefly review the bicriteria scheduling problems and discuss a general approach to find the best of the efficient or approximately efficient solutions with the aim of generating only a small subset of those solutions.

Although bicriteria problems are meaningful in scheduling, in some cases it may be necessary to consider more than two criteria. Since L_p distance functions can be defined for any number of criteria, Köksalan's [1993] heuristic approach can be easily extended to more criteria up to the point of finding a best dummy solution. Finding an efficient solution in the neighborhood of the best dummy solution, however, would be hard for more than two criteria for most criteria. This is an important area for future research. There is a need to develop good heuristic approaches for finding approximately efficient solutions for multiple criteria scheduling problems that are NP-hard.

In multiple criteria scheduling literature, the composite objective function, F, is assumed to be known. In the general multiple criteria decision making literature, on the other hand, there are many interactive approaches that assume the existence of an implicit function, F. The decision maker responds to preference questions posed by the interactive algorithms as needed and these responses are assumed to be consistent with F. The aim is to find the most preferred solution without explicitly knowing F. Similar interactive algorithms, where the decision maker's composite function is assumed to be unknown, can be developed for scheduling problems. It is more realistic to assume that a scheduler can express preference between different schedules rather than assuming that he or she can explicitly define a composite function.

References

Chen, C.L. and Bulfin, R.L. (1993). "Complexity of Single Machine, Multi-Criteria Scheduling Problems", *European Journal of Operational Research* 70, 115-125.

Fry, T.D., Armstrong, R.D. and Lewis, H. (1989). "A Framework for Single Machine Multiple Objective Sequencing Research", *Omega* 17, 595-607.

Graham,R.L., Lawler, E.L., Lenstra, J.K., and Rinnooy Kan, A.H.G. (1979). "Optimization and Approximation in Deterministic Sequencing and Scheduling", *Annals of Discrete Mathematics* 5, 287-326.

Gupta, S.K. and Kyparisis, J. (1987). "Single Machine Scheduling Research", *Omega* 15, 207-227.

Hoogeveen, J.A. (1992). "Single Machine Bicriteria Scheduling" *Unpublished Ph.D. Thesis*, CWI, Amsterdam.

Hoogeveen, J.A. and S.L. Van de Velde. (1995). "Minimizing Total Completion Time and Maximum Cost Simultaneously is Solvable in Polynomial Time", *OR Letters* 17, 205-208.

Kondakci, K.S., Azizoglu, M. and Köksalan, M. (1996). "Note: Bicriteria Scheduling for Minimizing Flow time and Maximum Tardiness", *Naval Research Logistics* 43, 929-936

Köksalan, M.M. (1993). "A Heuristic Solution to Minimizing a Function of Flowtime and Maximum tardiness," Technical Report No. 93-20, Industrial Engineering Department, Middle East Technical University.

Lee, C.Y. and Vairaktarakis, L. (1993). "Complexity of Single Machine Hierarchical Scheduling: A Survey", *Complexity in Numerical Optimization*, 269-298.

Nagar, A., Haddock, J. and Heragu, S. (1995). "Multiple and Bicriteria Scheduling: A Literature Survey", *European Journal of Operational Research* 81, 88-104.

Van Wassenhove, L.N. and Gelders, L.F. (1980). "Solving a Bicriteria Scheduling Problem", *European Journal of Operational Research* 4, 42-48.

Multiobjective Covering and Routing Problems

Brian Boffey[1] and Subhash C. Narula[1]

[1] Department of Mathematical Sciences, University of Liverpool, Liverpool L69 3BX, UK
[1] School of Business, Virginia Commonwealth University, 1015 Floyd Avenue, Richmond, Va 23284 - 4000, USA

Abstract. In this paper, we review research on point location covering problems and their development from the Maximal Covering Location Problem. With this as a basis, a particular path problem (the Maximal Covering Shortest Path Problem) is chosen to serve as a prototype combined covering - location - routing problem and possible developments are suggested. Finally, we discuss possible avenues for further research.

Key words: Cover, facility location, routing.

1 Introduction

Consider the problem of locating facilities (fire stations, police stations, etc) in a given region in such a way that the entire population of the region is within a given distance, say 2 miles, of a nearest facility. To achieve such a goal would usually be inordinately expensive. Consequently, a compromise solution is sought which provides a high level of cover yet is affordable. The Maximal Covering Location Problem (or MCLP for short) was designed for that purpose. Since its proposal in 1974, MCLP has been modified and generalized in many ways and there is now a substantial literature based around it. The model has been used in many different and varied settings and has been applied to real world problems including: locating health care centers in Colombia; analyzing archaeological settlement patterns in Mexico; locating wells and schools in Benin in relation to eradicating guinea worm disease (dracunculiasis); locating road maintenance depots in Australia; and, locating oil spill containment / removal equipment in Long Island Sound. This paper charts the evolution of MCLP based models, then looks at routing problems with a covering element and suggests some possible directions for the development of models in that area.

The location of a facility is usually influenced in some way by interactions between the facility and its 'neighborhood'. Such interactions may be essentially positive (access to markets, materials, labor, etc) or essentially negative as would be the case with a facility associated with unpleasant or toxic effluents. The *strength* of an interaction between a facility at $\boldsymbol{R}$ and a point $\boldsymbol{r}$ may be represented by a function f of $\boldsymbol{R}$ and $\boldsymbol{r}$. Typically, $f(\boldsymbol{R}, \boldsymbol{r})$ will depend only on $d(\boldsymbol{R}, \boldsymbol{r})$, the distance between $\boldsymbol{R}$ and $\boldsymbol{r}$, and will generally decrease as $d(\boldsymbol{R}, \boldsymbol{r})$ increases. (Here we shall assume that d is either Euclidean or network distance as appropriate for the interaction concerned.) Of course, independence of direction is not universally true as may be seen by considering a facility producing airborne pollutants whose spread will depend on prevailing wind patterns. Moreover, in such cases the pollution load may, for a given direction, initially increase and thereafter decrease (Boffey and Karkazis, 1995). In some cases $f(\boldsymbol{R}, \boldsymbol{r})$ may be identically zero except for a selected nearest facility. In this case f will be termed a *nearest interaction function.*

An interaction function often varies smoothly, but sometimes it may be better represented by a step function. This is the case when a precise *service distance* (or *service time*) is involved. For example, in urban fire protection, service may be considered acceptable by the insurance industry if it is available within 1.5 miles. A similar situation, but including probabilistic factors, is given in Groom (1977) who considered ambulance location in the Swansea area of South Wales. The standard quoted there was that an ambulance should be on the scene within 8 minutes of a call in at least 50% of cases, and within 20 minutes in 95% of cases. Different standards have been used by other authors.

In emergency services modelling, a demand point is said to be *covered* if it is within the service distance (time) whether or not the facility is currently free or busy. Reviews of emergency services modelling have been given by Schilling et al (1980) and ReVelle (1989) and, together with various practical aspects involved, by Marianov and ReVelle (1995). While a service distance or time is common, these are not the only possibilities; for example, observation towers (to detect forest fires) only cover those parts of the forest that are visible and this will be affected by topographic features (Goodchild and Lee, 1989).

An extensive survey of covering location problems is given in Schilling et al (1993), but only a small proportion of the papers referenced there deal *explicitly* with multicriteria problems. Here, we shall also consider problems which have *implicitly* been regarded as being multicriteria and some which, arguably, could be so regarded. This obviously means that the choice of whether or not to include a paper is subjective. However, two general rules observed were that papers on the Set Covering Problem (Section 3) are generally not referenced nor are papers on competitive location which sometimes uses a notion of 'conditional cover' depending on the locations of competitor outlets. Furthermore, references have, almost exclusively, been restricted to journal articles and books, as these are more readily available.

Section 2 gives some background terminology and notation. In the next three sections we review the literature regarding relevant work on point location starting with the basic MCLP model, moving on to capacitated, multi-server, hierarchical and dynamic problems, and then to congested, relocation and queuing models. Next, in section 6, the Maximal Covering Shortest Path Problem is formulated and possible developments of combined covering-location-routing problems are put forward. Finally, we conclude the paper with a short discussion of the state-of-the-art and possible directions for future research.

2 Terminology and Notation

Before discussing the development of covering models, it is appropriate to establish the terminology and notation that will be followed. Let H be a region in the plane containing all demand points of the system under consideration and in which facility location may take place. Usually, but not always, there will be a road (or other transportation) network $N = (V, E)$ *already* embedded in H. J will denote a set of subsets of nodes of H on which a facility may be located. Members of J may be single points (the usual case) or paths in H, or the nodes on a path through a network N in H, etc. Demand points are situated at a subset I of H with the *population* (or other *weight*) of $i \in I$ being d_i. For a large city demand may be represented by a number of demand points whose separate demands sum to the total population of the city in question. The effects of *aggregation* in relation to covering is discussed by Daskin et al (1989) and Current and Schilling (1990).

For each $v \in J$ and each interaction type k, there is an associated *cover set* C_{vk} which contains precisely those points in H which may be regarded as being covered if there were a facility at v. If v corresponds to a path in H, then C_{vk} may be defined as the union of C_{uk}, where u is a node (or perhaps any point) on path v. No restrictions are placed on C_{vk}, not even that v itself should be in C_{vk}. However, in practice C_{vk} will normally be defined in terms of a service distance or time. Moreover, in what follows, ideas will usually be presented in terms of service distance defined cover sets, it being understood that, in reality, this could be service 'time'. Also the extension to more general cover sets C_{vk} should be clear. In a similar way, each demand point will have a *neighborhood* $N_i = \{ j \mid a_{ij} = 1 \}$ such that i is *covered* if and only if there is the requisite number of facilities and types of server located in N_i.

Given these definitions, we are interested in choosing a subset $F \subset J$ at which point facilities, or a path (or paths) with given *origin* O and *destination* D, are to be located so as to:

(a) cover as much demand as possible, and

(b) minimize cost (or a surrogate such as path length).

Some concepts of multicriteria analysis will now be defined but, for a fuller account the reader may refer to Cohon (1978) or Steuer (1986). For the problem

P: maximize Z_1
maximize Z_2
subject to $\boldsymbol{x} \in \Phi$,

a feasible solution $\boldsymbol{x}$ may be discarded if there exists $\boldsymbol{y} \in \Phi$ such that

$$Z_1(\boldsymbol{y}) > Z_1(\boldsymbol{x}) \text{ and } Z_2(\boldsymbol{y}) \geq Z_2(\boldsymbol{x}) \quad \textbf{or} \quad Z_1(\boldsymbol{y}) \geq Z_1(\boldsymbol{x}) \text{ and } Z_2(\boldsymbol{y}) > Z_2(\boldsymbol{x}).$$

When (at least) one of these two conditions holds then $(Z_1(\boldsymbol{y}), Z_2(\boldsymbol{y}))$ *dominates* $(Z_1(\boldsymbol{x}), Z_2(\boldsymbol{x}))$; $\boldsymbol{y} \in \Phi$ will be termed *efficient* when $(Z_1(\boldsymbol{y}), Z_2(\boldsymbol{y}))$ is dominated by *no* $(Z_1(\boldsymbol{x}), Z_2(\boldsymbol{x}))$, $\boldsymbol{x} \in \Phi$. Furthermore, $\boldsymbol{y}$ is a *non extreme efficient solution* if there exist two feasible solutions $\boldsymbol{x}_1$ and $\boldsymbol{x}_2$ such that

$$(Z_1(\boldsymbol{y}), Z_2(\boldsymbol{y})) \leq (1-\theta)(Z_1(\boldsymbol{x}_1), Z_2(\boldsymbol{x}_1)) + \theta(Z_1(\boldsymbol{x}_2), Z_2(\boldsymbol{x}_2)), \quad 0 < \theta < 1;$$

otherwise it is an *extreme efficient solution.* The set of extreme efficient solutions can be generated by solving

$$\text{P}(w): \quad \text{maximize}_{\boldsymbol{x} \in \Phi}\; Z(w, \boldsymbol{x}) = (1-w)Z_1(\boldsymbol{x}) + w\,Z_2(\boldsymbol{x})$$

for all $0 < w < 1$. Because P has only a finite number of solutions, it is sufficient to solve P(w) for a carefully selected set $\{w_1, ..., w_t\}$ where $0 < w_1 < w_2 < ... < w_t < 1$.

3 Basic Models for Point Location

The original covering model is the classical *Set Covering Problem* (eg. Roth, 1969; Toregas et al, 1971) but this problem has much wider application (for example scheduling bus crews). Of course, complete cover will usually be expensive and so the trade off between cover and cost is of interest. This prompts the following *Bicriterion Covering Problem:*

BCP: maximize $\sum_i d_i y_i$
minimize $\sum_j c_j x_j$
subject to

$$\sum_j a_{ij} x_j - y_i \geq 0 \qquad i = 1, ..., m \tag{1}$$

$$x_j,\ y_i \in \{0, 1\} \qquad i = 1, ..., m,\ j = 1, ..., n.$$

where

$x_j = 1$ if a facility is located at site j; 0 otherwise
$y_i = 1$ if demand point i is covered; 0 otherwise
$c_j =$ cost of locating a facility at site j

d_i = demand at i
a_{ij} =1 if i would be covered by a facility located at site j; 0 otherwise.

The *cover constraints* (1) express the condition that every demand point must be covered by an opened facility or is uncovered ($y_i = 0$). It might be noted that we have formulated the problem in terms of the *cover matrix* (a_{ij}). In practice this matrix is likely to be sparse and therefore, for efficiency of implementation, a list processing approach should be adopted with the cover constraints expressed as $\Sigma_{j \in N_i} x_j - y_i \geq 0$ for all i, where N_i is the neighborhood of i. Perhaps surprisingly, BCP has been considered by only one set of authors (Church and Davis, 1992).

3.1 The Maximum Covering Location Problem

Church and ReVelle (1974) introduced the *Maximum Covering Location Problem* (or MCLP)

MCLP: maximize $\Sigma_i d_i y_i$

subject to

$$\begin{aligned} &\Sigma_j a_{ij} x_j - y_i \geq 0 && i = 1, \ldots, m \\ &\Sigma_j x_j = p && && (2) \\ &x_j, y_i \in \{0, 1\} && i = 1, \ldots, m, \; j = 1, \ldots, n. \end{aligned}$$

MCLP is the same as BCP except that one objective has been replaced by constraint (2). (White and Case (1974) formulated the *Partial Cover Problem* for which d_j=D, a constant, for all j). By solving for a range of p the trade off between cover and cost may be studied; this essentially amounts to applying the *Constraint Method* (Cohon, 1978) to solve BCP. If the cost of a facility varies with site, then (2) would be replaced by a constraint of the form $\Sigma_j k_j x_j \leq K$, where k_j is the cost at j and K is the total budget, but the appealing simplicity of MCLP is lost. In some situations it may be of interest with MCLP, or other covering problems, to study the trade off between cover and service distance; that is, considering the service distance to be variable also (Neebe, 1988). This was done by Richard et al (1990) who studied the number of fire stations to cover the rural Luxembourg province of Belgium.

A bicriterion version of MCLP was considered by

Schilling et al (1979b); the two separate objectives being 'cover of population' and 'cover of property'.

MCLP has been used by many authors in a variety of settings and to develop other multiobjective point location covering problems. Bennett et al (1982) used it to locate rural health centres, Eaton et al (1986) to locate health care units in Santo Domingo, Bell and Church (1985, 1987) to model archaeology settlement patterns, Kolesar (1980) to choose pressure measurement points for the early detection of glaucoma and Underhill (1994) and Williams and ReVelle (1996) for reserve selection for species preservation. Hodgson (1990) has shown that the 'flow-capturing' problem may be formulated as an MCLP. Pastor (1994) used MCLP to aid decision makers in selecting sites at which to open bank branches then a bicriterion problem to determine branch sizes with a view to equalizing sizes and maximizing accessibility to customers. Further applications are described in Belardo et al (1984), Dwyer and Evans (1981), Hogan (1990), Osleeb and McLafferty (1992), Rose et al (1992) and Chung (1986).

Church and ReVelle (1974) recognized that it may be necessary for a certain minimum level of service (defined by service distance T) to be available to all demand points but that, within this restriction and given the possibility of only p facilities, the cover at the higher level (service distance $S < T$) should be maximised. More precisely, this may be formulated by adding to MCLP the extra constraint $\sum_j b_{ij} x_j \geq 1$, $i = 1, \ldots, m$, where $b_{ij} = 1$ if i could be covered, relative to service distance T, by a facility located at site j, and $b_{ij} = 0$ otherwise. With this modification, the problem is known as MCLP *with Mandatory Closeness* constraints. Clearly, a multicriteria problem analogous to BCP can be formulated immediately. The effect of these closeness constraints may be achieved in a different way. First the extra (low level) constraints are modified to $\sum_j b_{ij} x_j - z_i \geq 0$ for $i = 1, \ldots, m$, where $z_i = 1$ if demand point i is covered at the lower level (that is, relative to the service distance T) and 0 otherwise. Then, the objective is modified to $\sum_i d_i y_i - M \sum_j c_j x_j$ where M is a 'very large number' to force z_i to one (assuming this is possible). This corresponds to a nearest interaction function f of the form (see Figure 1a)

$$f(\boldsymbol{R}, \boldsymbol{r}) = \begin{cases} 1 & \text{if } |\boldsymbol{R} - \boldsymbol{r}| \leq S \\ 0 & \text{if } S < |\boldsymbol{R} - \boldsymbol{r}| \leq T \\ -M & \text{if } T < |\boldsymbol{R} - \boldsymbol{r}|. \end{cases}$$

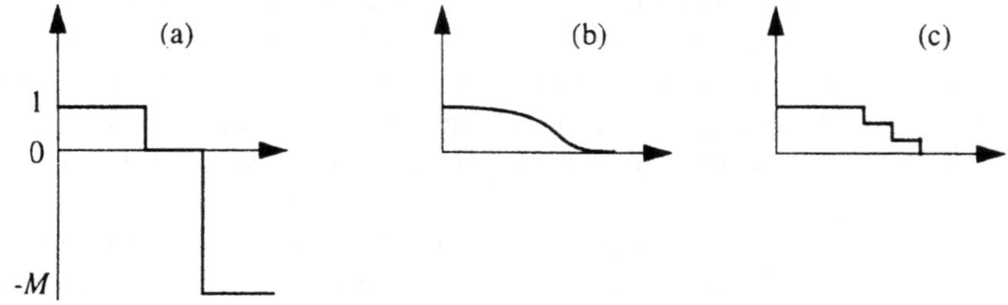

Fig. 1 Different forms of interaction function.

While a precise service distance might correspond to a standard or regulation, it may be questioned whether a sharp cut off represents reality. For, example, consider the case of fire stations and a service distance S. It is unrealistic to assume that a demand point would not receive some benefit from a fire engine attending from a distance $S+\Delta$ if Δ is small compared to S. Perhaps a nearest interaction function of the form shown in Figure 1b would be more realistic. To approximate such behavior, Church and Roberts (1983) used a piecewise linear step function (Figure 1c), and formulated a corresponding integer program. The number of steps can be varied, and 'steps up' may be included as well as 'steps down'. Ghosh and Craig (1987), in a marketing application, employed an extended MCLP model in which there were several travel time categories. Pirkul and Schilling (1991) employed a piecewise linear function in which the 'steps' (see Figure 1(c)) are replaced by a sloping line.

For firefighting, ambulance services etc, uncovered demand will in fact be serviced but the level of service will, of course, be below the desired standard. Church et al (1991) propose that a second objective be considered, namely that of minimizing the average travel distance for *uncovered* demand. They studied the trade off between demand covered and average travel distance for uncovered demand, and gave numerical experience with a 152 node problem involving the location of coffee buying centers in Busoga, Uganda.

3.2 Solution Methods for MCLP

We now briefly consider how MCLP may be solved. Church and ReVelle (1974) described three solution procedures. One heuristic procedure, the 'Greedy Adding Algorithm', starts with no facility located. At each of the p stages, a facility is added at a site that leads to the largest increase in cover. Then, at each step, a hill-climb improvement is applied with the move of a facility from its current site to an unoccupied site being considered. It is clear that this could be enhanced by substituting a modern heuristic such as tabu search (Reeves, 1993) for the hill climbs; similar remarks apply to other covering problems. A second procedure made use of integer programming with fractional solutions being resolved by branch-and-bound. For small to medium size problems there was found to be a high incidence of the linear programming relaxation being integer feasible. Galvão and ReVelle (1996) devized an effective method which employed Lagrangean relaxation together with subgradient optimization. For the *Location Set Covering Problem* (set covering with equal costs) Almiñana and Pastor (1996) used a surrogate relaxation approach based on that of Lorena and Lopes (1994) for the traditional set covering problem. It is expected that surrogate relaxation techniques, perhaps in conjunction with Lagrangean relaxation, may provide a useful tool for MCLP and possibly other network covering problems.

3.3 Extensions of MCLP

Church and Meadows (1979) assumed that facilities could be located anywhere on the network and showed that there is always an optimal solution in which every facility is sited at a point in the set R comprising demand points and points on arcs which are exactly distance S (the service distance) from some demand point. Berman (1994) observed that, while such solutions provided maximum cover, there may be other solutions providing the same cover but for which 'the maximum distance from any covered demand point to its nearest facility' is reduced. A similar result holds for other covering problems.

Several authors (Drezner 1986; Mehrez and Stulman, 1982, 1984; Megiddo et al, 1983 and Watson-Gandy, 1982) have studied covering problems in the plane, that is using Euclidean distance rather than network distance. These authors have proposed several methods for solving such problems.

Finally, MCLP may be appropriate at a particular instant but does not adequately represent a system in which there is a change in demand over time. To account for such change, Schilling (1980) introduced a multiperiod *long term dynamic* model similar to MCLP, but with the variables having an extra subscript to specify time period and the number of facilities, p_t, required in time period t increasing with t. There is also a set of *consistency conditions*

$$x_{jt} \geq x_{jt-1} \quad \text{for } j = 1, ..., n, \ \ t = 1, ..., T$$

which require that once established a facility remains in place. Their formulation was multicriteria with one objective for each time period. It may be noted that *short term* dynamic models, in which facilities (or servers) are temporarily *relocated* for operational reasons to accommodate temporary fluctuations in demand, serve a different purpose and are briefly considered in the next section.

4 Capacitated and Multi-Server Problems

The models introduced so far implicitly assume that a facility is capable of meeting any demand required of it. This will effectively be true if a facility can cover any *single* demand and if calls on a facility are rare and independent as for marine oil spills (Belardo et al, 1984), bu, 1t in many situations this is at best a reasonable approximation.

4.1 Capacitated Models

The question of major events was answered by the *Capacitated Maximal Covering Location Model* (CMCLP) of Current and Storbeck (1988) which may be formulated as:

CMCLP: maximize $\sum_i d_i y_i$

subject to

$$\begin{aligned}
&\textstyle\sum_j u_{ij} - y_i = 0 && i = 1, \ldots, m \qquad (3)\\
&\textstyle\sum_j d_i u_{ij} - Kx_j \le 0 && i = 1, \ldots, m\\
&\textstyle\sum_j x_j = p &&\\
&x_j \in \{0, 1\} && j = 1, \ldots, n.\\
&u_{ij}, y_i \ge 0 && i = 1, \ldots, m, \; j = 1, \ldots, n.
\end{aligned}$$

Note that y_i is no longer a binary variable since it is the *fraction* of demand at i that is met, and the *demand assignment variables*, u_{ij} (which are only defined when $a_{ij} = 1$) give the proportion of demand at i that is to be met from a facility at j. Also, constraints (3) must now be equalities. Of course, for operational reasons, it might be better not to have partial assignments, in which case $u_{ij} \ge 0$ is replaced by $u_{ij} \in \{0, 1\}$. Current and Storbeck (1988) do not provide a direct solution method but do mention that it could be solved, after reformulation, as a capacitated p-median problem or generalized assignment problem (for problem transformations see Church and ReVelle, 1976; Church and Weaver, 1986; Klastorin, 1979). Pirkul and Schilling (1991) observe that 'The use of a maximal response distance is meant as a tool to insure a minimum level of service to the greatest number and not as a mechanism for withholding service.' Since uncovered demand will be serviced, they argue that uncovered demand assigned to a facility should count against that facility's capacity. Using a model based on this principle, they found that the average and maximal response times may be reduced considerably for a small decrease in coverage.

Pirkul and Schilling (1989) also pose two capacitated MCLP models. The first is similar to CMCLP whereas the other requires all demand to be allocated to some facility (as would happen in practise with fire fighting and emergency medical services). They develop a Lagrangean relaxation/subgradient optimization method and report extensive computational experience with its use.

4.2 Multiserver Models

To increase capacity it may be convenient to permit more than one facility at a site by allowing x_i to be a non-negative integer rather than a binary variable. Facilities, however, frequently operate via a small number of units here to be called *servers*. Batta and Mannur (1990) permit facilities to vary in this way so as to be able to respond to emergencies of varying severity. For example, a fire station may be regarded as a facility with individual engines being the servers. With this broader view, there is no need to restrict the system to a single type of server and the associated cover sets C_{vk} may vary with k. Suppose that there are r server types, and that any call is for one server of each type. Then MCLP generalizes to

MCLP(r): maximize $\sum_i d_i y_i$

subject to

$$\begin{array}{lll} \sum_j a_{ij}^k x_j^k - y_i \geq 0 & i = 1, ..., m, \; k = 1, ..., r & (4) \\ x_j^k - z_j \leq 0 & j = 1, ..., n, \; k = 1, ..., r & \\ \sum_j x_j^k = p^k & k = 1, ..., r & \\ x_j^k, y_i \in \{0, 1\} & i = 1, ..., m, \; j = 1, ..., n, \; k = 1, ..., r. & \end{array}$$

where

$x_j^k = 1$ if a server of type k is located at site j; 0 otherwise

$y_i = 1$ if demand point i is covered; 0 otherwise

$z_j = 1$ if there is a facility located at j; 0 otherwise

$d_i =$ demand at i

$p^k =$ number of servers of type k permitted

$a_{ij}^k = 1$ if i could be covered by a server of type k located at site j; 0 otherwise.

Notice that a demand point is covered if and only if it is covered by every type of server (constraint set (4)) but this could be modified. There may also be a budget restriction $\sum_j z_j = p^s$ on the total number of facilities permitted. With this constraint MCLP(2) becomes the well known FLEET model of Schilling et al (1979a) which was proposed for fire fighting. Facilities represent fire stations; servers are of two types, namely engines (pumper companies) and trucks (ladder companies). A four objective version of FLEET requires maximization of:

cover of fires;
population cover (weighted by fire frequency);
cover of property value;
property cover (weighted by fire frequency).

Schilling et al solved, by integer programming, numerical example problems with the *single* objective of maximum population cover. As for MCLP, a high incidence of the linear programming relaxation solutions being integer feasible was found.

The Insurance Services Office (ISO) defines two types of fire coverage requirement namely: *first-due coverage* and *standard-response* coverage. The model described above relates to first-due coverage (see Marianov and ReVelle (1991) for more details). For standard-response coverage, Marianov and ReVelle (1991, 1992b) formulated a generalization of MCLP(2) which we designate FLEET(SR) and which requires each demand point to be covered by three engines and two trucks (at least).

Belardo et al (1984) studied placement of four types of containment / removal equipment to deal with possible oil spills a particular application being made to Long Island Sound. The model used was, effectively, MCLP(4) but with sites that may house equipment already in place ($z_j = 1$ for all j). Demand points were possible *events* (combinations of spill area, weather conditions and cargo type) of which 72 were identified for Long Island Sound. Also, three objectives were employed corresponding to the probability of covering *high*, *medium* and *low* economic/environmental impact spills. This provided six efficient solutions for a decision maker to consider.

4.3 Hierarchical Models

Having introduced multiple servers, it is natural to consider different types of server and / or different types of facility which may form a hierarchy of some kind. Charnes and Storbeck (1980) and Ruefli and Storbeck (1982) considered placement of *advanced life support* (ALS) vehicles carrying paramedics and *basic life support* (BLS) ambulances. Coverage of critical (life threatening) demand is best provided by ALS units and these should be placed to maximize coverage of critical demand. BLS vehicles can only serve as 'backup' for critical demand, and non critical demand must be covered by BLS units.

This multilevel problem is treated by Charnes and Storbeck (1980) and Ruefli and Storbeck (1982) in the spirit of goal programming with, for both critical and non critical demand, the goal being to maximize cover. The objective is then a combination

$$\Sigma_i c_i y_{ai}^- + \Sigma_i e_i y_{bi}^- + \Sigma_i n_i y_{ni}^-$$

where y_{ai}^-, y_{bi}^- and y_{ni}^- are, respectively, the underattainment of cover by ALS units for critical demand, by BLS units for critical demand and by BLS units for non critical demand. The coefficients are related to demand at i, and overattainment was considered to be immaterial. Eaton et al (1985) also considered ALS and BLS ambulances and, working within an MCLP framework, considered placements of ambulances in the city of Austin, Texas. They performed several computer runs and on this basis sought trade offs between response times, number of vehicles deployed, and coverage of various demand groups. Considerable interaction took place with interested parties and the final solution was both well received and resulted in very substantial cost savings.

Some systems, for example schools and health care facilities, are hierarchically related. Moore and ReVelle (1982) formulated a 2-level hierarchical extension of MCLP with regard to the health services of Honduras. The lower level facilities (clinics) are able to provide only a level one service, whereas the higher level facilities (hospitals) can provide a level two service *and* the level one service. The covering distance for level one service was permitted to be different at the two types of facility. Such a hierarchy is *successively inclusive* in that a facility provides its own level of service *and all lower levels of service.* Moore and ReVelle maximized population that was covered *at both levels*, though the trade off between the lower level and higher level services could be considered instead. Desai and Storbeck (1988) introduced a similar model which is only 'approximately successively inclusive' with linkage between sites at adjacent levels k and k+1 being linked by deviational variables d_i^+ and d_i^-, where

$$x_i^{k+1} - x_i^k - d_i^+ + d_i^- = 0,$$

and with x_i^k being 1 or 0 depending upon whether a level k facility is, or is not, located at point i.

In the context of health care, Church and Eaton (1987) considered two other types of coverage, namely 'referral-up coverage' and 'referral-down coverage'. A clinic is *referral-up covered* if a hospital is sufficiently close for a seriously ill patient to be referred there for more extensive treatment. On the other hand, a clinic is *referral-down covered* if there is a hospital sufficiently close from which to obtain medicines, equipment support staff etc, as may be needed. Church and Eaton posed separate models: one included referral-up coverage, the other referral-down coverage. This was generalized by Gerrard and Church (1994) who introduced a hierarchical model accounting for high and low level facilities together with both referrals-up and referrals-down. It was required to maximize:

population covered by a clinic;
population covered by a hospital;
number of clinics referral-down covered;
population covered by a referral-up covered clinic.

An illustrative application was given using data from the Zarzal region of Colombia.

A bicriterion model for the essential air services program was introduced by Flynn and Ratick (1988). It sought to cover small communities through air connections being available and to minimize operating costs. The model was hierarchical with the different levels being determined by aircraft type, flight frequency, number of stopovers and fares. An illustrative application of the model with four levels was given for North and South Dakota.

5 Covering and Congestion

As pointed out earlier, 'cover' is not the same as 'availability'. For emergency services it is not helpful to be 'covered' if the service is not 'available' when required. Consequently, it is important to ensure a high degree of availability which means that it is usually necessary for demand points to be covered by more than one facility. Models incorporating this feature fall into two groups: deterministic models; probabilistic and queuing models.

5.1 Deterministic Models

Hogan and ReVelle (1986) posed the BACOP2 model in which it is desired to maximize second (or *backup*) coverage while providing *primary* (first) coverage for all demand.

$$\begin{aligned}
\text{BACOP2:}\quad & \text{maximize } \textstyle\sum_i d_i y_i \\
& \text{maximize } \textstyle\sum_i d_i r_i \\
& \text{subject to} \\
& \quad \textstyle\sum_j a_{ij} x_j - y_i - r_i \geq 0, \quad i = 1, \ldots, m \\
& \quad y_i - r_i \geq 0 \qquad\qquad\qquad i = 1, \ldots, m \qquad\qquad (5) \\
& \quad \textstyle\sum_j x_j = p \\
& \quad x_j, y_i, r_i \in \{0, 1\} \qquad\quad i = 1, \ldots, m, \; j = 1, \ldots, n
\end{aligned}$$

where

$r_i = 1$ if demand point i is covered by more than one facility; 0 otherwise.

Constraint set (5) enforces the obvious requirement that i cannot be covered by a second facility if it is not already covered by a first facility. Hogan and ReVelle found that marginal reductions in primary coverage can strongly improve *backup* coverage. It should be noted that it may now be beneficial to locate more than one facility together. Various algorithms are proposed by Hall and Hochbaum (1992) for multicovering problems generally. Radar coverage provides an example where every ('demand') point must be covered by two facilities (radar stations) for the purpose of cross reference (Mehrez and Stulman, 1984); species preservation (Williams and ReVelle, 1996)) is another

Daskin and Stern (1981) solved a two level program for ambulance placement when the primary objective was to minimize the number of ambulances to provide (first) cover and the secondary objective was to maximize *excess coverage* (that is, coverage beyond first coverage, r_i now being an integer variable such that $r_i = \sum_j a_{ij} x_j - y_i$ for all i). A similar formulation was studied by Storbeck (1982). Pirkul and Schilling (1989) maximized the sum of first and second coverage, weighted by demand, subject to a limit on the number of facilities, and also imposed a constraint on each facility's workload to prevent the facility workloads being too uneven. They developed a Lagrangean relaxation / subgradient optimization approach and reported the results of various experiments. Narasimhan et al (1992) extended the above model by considering levels of coverage beyond the second.

Higher coverage (third, fourth, ...) is not valued at all in the BACOP2 model whereas Storbeck valued second, third, ... coverage equally. In practice the real value of higher coverage lies somewhere between these two extremes. Hogan and ReVelle (1986) suggested an extension of BACOP2 which treats the first, second and third levels of cover as separate objectives. Storbeck and Vohra (1988) did this using first, second, .., K-th levels of cover for some (small) integer K.

When a major event, or several smaller events, has occurred in a particular area then that area will have much reduced cover for the emergency service in question. It is common practice for the fire fighting services to *relocate* servers (engines and trucks) towards, or to, the temporarily undercovered region. Such a dynamic relocation covering problem has been studied by Kolesar and Walker (1974). The authors worked within a (location) set covering framework. When the need for relocation is detected, they solved a two level problem, first using a set covering problem to determine sites at which to relocate, then choosing between alternative optima on the basis of efficiency (that is, minimizing average response time). Chrissis et al (1982) and Gunwardane (1982) have also studied relocation, the latter within an MCLP framework.

Another view of coverage in the presence of congestion is provided by the notion of 'support coverage'. A server is *support covered* by another server if it is within a specified *support distance* R. The idea is that, if a server at i is busy when a call for its services is received then a nearby server (within distance R of i) can step in and service the call. It may be expected that R will be larger than the normal *demand* cover distance S. Moon and Chaudhry (1984) formulated the *Double Set Covering* problem which seeks to minimize the number of servers subject to all demand being covered by a server *and* every server being support covered. ReVelle et al (1996) propose a model, in the spirit of MCLP, which seeks to maximize the number of servers which are support covered subject to all demand being covered and a limitation on the number of servers to be located. They also formulated a multiobjective version to provide information on the trade off between demand covered and the number of support covered servers given a limit on the number of servers.

5.2 Probabilistic and Queuing Models

Daskin (1983), following the work of Chapman and White (1974), adopted a different view and assumed that service calls follow a binomial distribution with a *system wide* parameter q representing the probability of a facility being *busy* when called upon. Assuming (a) server busy probabilities are independent of location and (b) servers operate independently, it is readily deduced that, when a call is made from i

then one of the r covering facilities will be free with probability $1-(1-q)^r$. Hence the *marginal benefit* of increasing the number of facilities covering i from $r-1$ to r is

$$\delta_{ir} = w_i[(1-(1-q)^r) - (1-(1-q)^{r-1})] = w_i q(1-q)^{r-1}; \quad r = 1, ..., p.$$

This led Daskin (1983) to propose the *Maximum Expected Covering Location* (MEXCLP)

MEXCLP: maximize $\Sigma_i \delta_{ir} y_{ir}$

subject to

$$\Sigma_j a_{ij} x_j - \Sigma_r y_{ir} \geq 0 \quad i = 1, ..., m$$

$$\Sigma_j x_j = p$$

$$x_j, y_{ir} \in \{0, 1\} \quad i = 1, ..., m, j = 1, ..., n, r = 1, ..., p.$$

MEXCLP may also be varied by weighting the terms in the objective by the demands d_i. Daskin (1982) applied MEXCLP to ambulance location in Austin, Texas and Fujiwara et al (1987) to the problem of deploying ambulances in Bangkok. An alternative, non linear, formulation was presented by Saydam and McKnew (1985) and was solved exactly for several real life problems by means of separable programming. They also showed how lower bounds on the number of servers in demand point neighborhoods may be incorporated. Finally, Daskin et al (1988) provided an integrated discussion of excess, backup, expected cover.

An extended model, TIMEXCLP, in which variation of demand and travel times with period of day is taken account of, is given by Repede and Bernardo (1994). When the number of periods in a day is only one then the model reduces to MEXCLP. TIMEXCLP was used, together a simulation model, to study the location of emergency medical vehicles in Louisville. It was found that taking account of time variation led to a substantial increase in coverage and average response time was decreased by 36%.

Bianchi and Church (1988) observed that there is also a cost associated with ambulance stations, and introduced the MOFLEET model which might be described as a hybrid of MEXCLP and MCLP(1). Since, in general, the maximum number of stations will be less than the number of ambulances to be located, variables representing ambulance numbers at a site are taken to be integers. This led to computational difficulties which can, however, be ameliorated by restructuring in terms of binary variables. These authors studied, for two well known sets of data and a variety of ambulance numbers, the trade off between the number of stations and expected coverage. They found that it was often possible, for zero or very small reduction in coverage, to reduce the number of stations with the associated operational gains and cost reductions.

Rather than looking at expected coverage, ReVelle and Hogan (1988, 1989a) considered the location of facilities so that all demand received service with a minimum *reliability* α. That is, they located the fewest facilities subject to

$$\text{Prob}\{\text{ at least one facility is available when a call is made from node } i\ \} \geq \alpha \quad (6)$$

thus providing a probabilistic version of the Location Set Covering Problem. They translated this constraint into the requirement that at least B_i facilities must be within the service distance of i, where the constants B_i are calculated easily. However, facilities are shared between demand points and, as Pastor (1995) has shown by means of an example, it is possible that coverage may, in some circumstances, be lower than required. ReVelle and Hogan (1989b) extended their approach to the MALP model which maximizes the demand which is satisfied with reliability α thus providing a probabilistic version of MCLP.

Batta et al (1989) relaxed the Daskin independence conditions and Goldberg and Paz (1991) and Goldberg et al (1990) went even further and considered variable service times. The latter authors discussed in detail an application to ambulance deployment in Tucson, Arizona.

ReVelle and Marianov (1991) introduced a probabilistic version of FLEET which we designate PROFLEET. Its aims to locate up to p^E engines and p^T trucks at p^S stations so as to maximize demand covered by both an engine *and* a truck with reliability α (see (6)). Marianov and ReVelle (1992a) studied a very similar model, the difference from PROFLEET being a joint limit $p^E + p^T$ on the combined number of engines and trucks. In both models an integer variable y in the range $0 \leq y \leq b$ is replaced by the sum of b binary variables. This produces a constraint matrix with elements restricted to -1, 0 and 1 and the tendency for the linear programming relaxation to yield integer feasible solutions has been found to be preserved. The status of PROFLEET in relation to other models is illustrated in Figure 2.

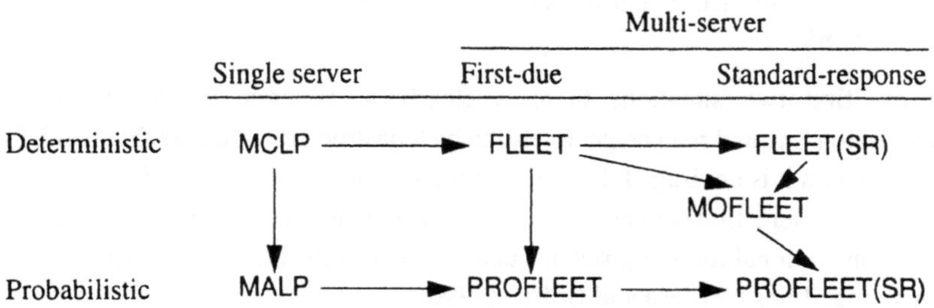

Fig. 2 Relationship between some models.

PROFLEET is a probabilistic model relevant to first-due fire protection coverage. Marianov and ReVelle (1992b) extended this in a model we designate PROFLEET(SR) which seeks to maximize demand that is standard-response covered with reliability α. Another model due to ReVelle and Snyder (1996) considered the integration of ambulance and firefighting equipment deployment. Their model was a blend of MCLP and FLEET. For a discussion of probabilistic treatment in fire fighting facility location models the reader is referred to Marianov and ReVelle (1995).

In the models described above, the probabilities of two facilities (servers) within a neighbourhood N_i being busy are assumed to be independent. This, however, is clearly not the case in practise. To avoid this problem, Marianov and ReVelle (1994) modelled the m servers within N_i by an $(M/M/m)$ queuing system with the 'number of customers' being restricted to m; that is, stacking of calls is not permitted and a call received when all servers in N_i are busy is 'lost' (though in reality it would be serviced by a server from outside N_i). In a very recent paper, Marianov and ReVelle (1996) introduced a queueing theory version of MALP also.

6 Covering by Paths

Boffey (1995) gives a review of multiobjective routing problems generally and here we shall only be interested in the small subset which include a covering aspect. Consider:

- (A) the path of an electricity transmission line sited to pass near areas of high demand;
- (B) a road constructed through an undeveloped area to pass near as many villages as possible;
- (C) bimodal delivery system; for example, a bulk newspaper delivery route which passes close to as many retail outlets as possible, or an air freight service which delivers to selected airports from whence shipment is by road;
- (D) a mobile service which makes stops at various points on an existing road network.

In the first two situations there is clearly a 'concrete extensive facility' (transmission line and road respectively) to be 'constructed'. On the other hand, for the other cases it is envisaged that an existing network is to be used. Since (C) and (D) involve 'interaction between the route taken and the surroundings', these are not simply conventional routing problems and the routes selected are regarded as being extensive facilities by Mesa and Boffey (1996).

Unlike the case of point covering problems, there has been very little research on path covering problems. Various covering tour problems which could be considered as implicitly multicriteria in nature are given by Current and Schilling (1989, 1994), Laporte and Martello (1990), Ochi et al (1995) and Gendreau et al (1996). Here we shall concentrate on MCSP/MPSP and, by analogy with the development of point covering discussed in previous sections, will describe a corresponding range of possible developments for path covering. But first, one general point. For an *O-D* path problem there will generally be a cycle (closed loop) problem and vice versa. An instance of the former can be transformed to an instance of the latter (Boffey, 1973) by 'coalescing *O* and *D*' and 'adding a dummy node' if necessary (Current and Schilling (1989). Bearing this in mind, problems will be stated as *O-D* problems.

The *Maximal Covering Shortest Path* (MCSP) problem introduced by Current et al (1985) seeks to determine a path from *O* to *D* which maximizes demand covered and minimizes cost as measured by the surrogate path length problem which may be formulated as:

MCSP: maximize $\Sigma_k d_k y_k$
minimize $\Sigma_{ij \in E} \delta_{ij} x_{ij}$
subject to=

$$\Sigma_i x_{ij} - \Sigma_k x_{jk} = \begin{cases} -1 & \text{if } j = O \\ 0 & \text{if } j \notin \{O, D\} \\ 1 & \text{if } j = D \end{cases} \tag{7}$$

$$\Sigma_k a_{kj} \Sigma_i (x_{ik} + x_{ki}) - 2y_j \geq 0 \quad \text{for } j \notin \{O, D\} \tag{8}$$

$$x_{ij}, y_k \in \{0, 1\} \qquad \text{for } ij \in E,\ k \in V \tag{9}$$

where

$y_k = 1$, if node k is covered and 0 otherwise (y_O and y_D are fixed at 1)
$x_{ij} = 1$, if arc ij is on the path chosen, and 0 otherwise
$a_{jk} = 1$, if k is covered from j, and 0 otherwise
d_k is the population (or demand) at node k
δ_{ij} = length of arc ij.

Constraint set (8), a symmetrized form of the constraint set used by Current et al (1985), determines which nodes are covered. Constraint sets (7) and (9) together ensure that the solution contains a path from node *O* to node *D*. For simplicity of exposition, we shall from now on only consider the special case, called the *Maximal Population Shortest Path* (MPSP) problem which arises when the service distance *S* is less than any arc length. This leads to constraint set (8) being simplified to

$$\Sigma_i (x_{ij} + x_{ji}) - 2y_j \geq 0 \quad \text{for } j \notin \{O, D\}. \tag{10}$$

As posed, this excludes neither *attached* nor *isolated* loops (Figure 3) which Current et al eliminated, when necessary, by using branch-and-bound. They adopted a weighting method approach (Cohon, 1978), solving the single objective problem MPSP(w) (see Section 2) for a range of values of w in the range $0 < w < 1$.

It is readily shown that a travelling salesman problem may be reduced to MPSP(w) for some w, and so MPSP(w) must be an NP-hard problem. Therefore, it is not surprizing that, for large w, the high incidence of integer feasible solutions of the linear programming relaxation experienced with MCLP does not occur. Alternatively, it is natural to use branch and bound with Lagrangean relaxation being employed to provide upper bounds. The constraints that make the problem difficult to solve are the *cover constraints* (10) and so these are incorporated into the objective. It can be shown that for

$$w < w_L = \min_{ij \in E} [2\delta_{ij}/(a_i + a_j + 2\delta_{ij})],$$

MPSP can be solved exactly by solving a shortest path problem relative to modified arc lengths. This provides a promising line of development; however, we now consider other possible path covering problems.

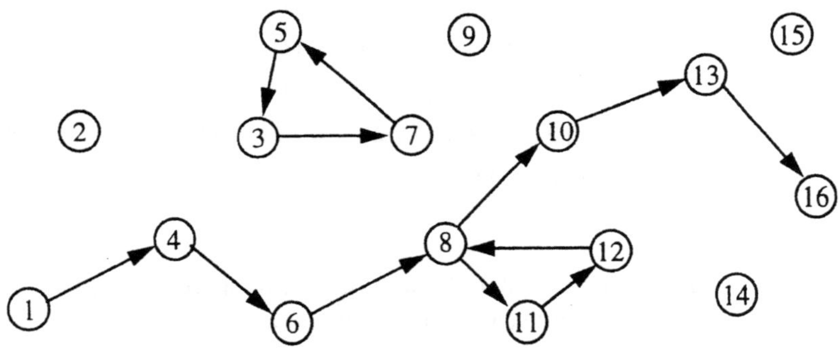

Fig. 3 Examples of an isolated loop 3-7-5-3 and an attached loop 8-11-12-8.

7 Extensions of MCSP/MPSP

In this section we shall, based on the point covering review earlier, consider possible extensions and developments of MCSP/MPSP. What constitutes a realistic development will depend on the system under consideration. Two broad classes will be considered: one typified by a mobile service (refer to examples (C) and (D) in the previous section) and the other by a physical *O-D* link (such as a transmission line (A) and road (B)).

7.1 Mobile Service

A mobile service makes *service stops* to provide the service it offers. This involves time being taken and so there is likely to be a limit on the number of service stops that can be made in a given period (eg. day). Consequently, *O-D* paths with a restriction on the number of nodes may be of interest. Additionally, or alternatively, there might be a restriction on overall path length. Such constraint(s) (which correspond to the capacity constraints of Section 4) are likely to have the consequence that a relative small proportion of demand is covered by a single *O-D* path. This raises the possibility of more facilities, that is *O-D* routes (which could, of course, correspond to different *O-D* pairs). For a single *O-D* pair this might correspond to several vehicles or to several trips by a single vehicle at different times (which is eminently reasonable if the service is only required every so often - see also Lindner-Dutton et al (1991) for a similar strategy, but relating to the carriage of hazardous materials), or something in between. Either of these possibilities can result in a node being visited by more than one facility (or server), and so the possibility of not making a service stop by one vehicle should be considered. In the same vein, the possibility of attached loops (Figure 3) may make sense with a service stop not being made on one of the visits to that node.

The preceding discussion demonstrates a situation not unlike vehicle routing problems but with service not being mandatory. On the other hand if multiple service stops are allowed but with decreasing returns then this essentially gives a model introduced in Malandraki and Daskin (1993).

Questions of long term dynamic behavior do not seem to assume importance since vehicles can be rerouted at negligible cost. Similarly, congestion and relocation also seem not to have the importance assumed in relation to emergency facilities.

7.2 Physical Link

It is likely to be undesirable for the path of the physical link be too contorted. As for the mobile service, links may be limited by length or number of nodes to ensure that they are reasonably direct (or for other reasons such as reliability in the case of electricity distribution). Again, this is likely to lead to multiple paths, but there is now no equivalent to the situation in which a mobile service traverses a set of paths in rotation. The anti looping condition of MCSP/MPSP seems to be even less justifiable now since simple 'short loops' may be replaced by *spurs* (see Figure 4). This results in a problem more like the *Hierarchical Network Design Problem* (HNDP) (Current et al, 1986; Duin and Volgenant, 1989; Pirkul et al, 1991; Sancho, 1995) but with it not being mandatory that all demand be covered.

Long term dynamic problems would certainly be relevant in some cases, for example for low/medium voltage distribution networks. On the other hand relocation does not seem to be of interest.

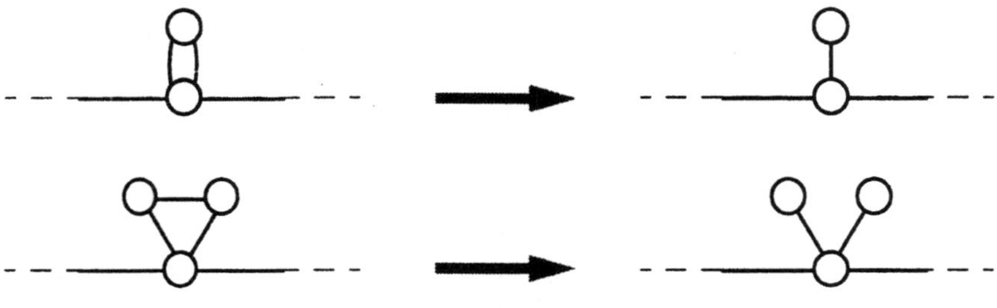

Fig. 4 Paths with loops and spurs.

8 Discussion and Conclusions

In the preceding sections, we have given a concise survey of work that has been done on covering problems with multicriteria aspects being brought out when present. Several features are apparent.

Point covering has received considerable attention and further research is likely to be directed at more detailed models including various real life features. One such feature that is suggested by Marianov and ReVelle (1995) is the question of integration between emergency services; an example of this is the siting of ambulances at fire stations (ReVelle and Snyder, 1996).

For many point covering problems, solving the integer problem via the linear programming relaxation is an effective approach in many cases though Lagrangean relaxation / subgradient optimization is, possibly, to be preferred for the more detailed models. It is felt that there is probably scope for more use of surrogate relaxation techniques, perhaps in conjunction with Lagrangean relaxation.

Path covering problems are rather harder (in a practical rather than a complexity theory sense) than point covering problems when cover is valued highly compared to cost (ie path length). However, it was argued above that, when this is the case, a single contorted path would probably be better replaced by two (or several) more direct paths. This line of development is wide open for further research. Also, since it is likely that paths will be of restricted length, it is possible that shortest path/k-shortest path methods could be useful. In this context, the method of Perko (1986) could be suitable. There is also the question of extending covering to other graphical structures and Hutson and ReVelle (1989, 1993) provide examples of this

graphical structures and Hutson and ReVelle (1989, 1993) provide examples of this for trees.

Multicriteria covering problems have generally been treated by generation techniques. For this, the weighting method has been a favorite with the constraint method being (implicitly) used when variation with the number of facilities is considered. Goal programming formulations have been proposed for some of the more basic point covering problems. Interaction with the decision maker has received relatively little prominence.

9 References

M. Almiñana and J. T. Pastor (1996) An adaptation of SH heuristic to the location set covering problem. *European Journal of Operational Research.*

R. Batta, J. M. Dolan and N. N. Krishnamurthy (1989) The maximal expected covering location problem: revisited. *Transportation Science*, **23**, 277-287.

R. Batta and N. R. Mannur (1990) Covering - location models for emergency situations that require multiple response units. *Management Science*, **36**, 16-23.

S. Belardo, J. Harrald, W. A. Wallace and J. Ward (1984) A partial covering approach to siting response resources for major maritime oil spills. *Management Science*, **30**, 1184-1196.

T. L. Bell and R. L. Church (1985) Location allocation modelling in archaeological settlement pattern research: some preliminary applications. *World Archaeology*, **16**, 354-371.

T. L. Bell and R. L. Church (1987) Location-allocation modelling in archaeology. In A. Ghosh and G. Rushton, *Spatial Analysis and Location-Allocation Models*, Van Nostrand Reinhold, New York.

V. L. Bennett, D. Eaton and R. L. Church (1982) Selecting sites for rural health workers. *Social Science and Medicine*, **16**, 63-72.

O. Berman (1994) The *p* maximal cover - *p* partial center problem. *European Journal of Operational Research*, **72**, 432-442.

G. Bianchi and R. L. Church (1988) A hybrid FLEET model for emergency medical service system design. *Social Science and Medicine*, **26**, 163-171.

T. B. Boffey (1973) A note on minimal length Hamilton path and circuit algorithms. *Operational Research Quarterly*, **24**, 437-439.

T. B. Boffey (1995) Multiobjective routing problems. *Top*, **3**, 167-203.

T. B. Boffey and J. Karkazis (1995) Location, routing and the environment. Chapter 19 (pp 453-466) in Z. Drezner, *Facility Location: a survey of applications and methods*, Springer-Verlag, New York.

S. C. Chapman and J. A. White (1974) Probabilistic formulation of emergency service facilities location problems. *ORSA/TIMS Conference*, San Juan, Puerto Rica.

A. Charnes and J. E. Storbeck (1980) A goal programming algorithm model for the siting of multilevel EMS systems. *Socio-Economic Planning Sciences*, **14**, 155-161.

J. W. Chrissis, J. W. Daves and D. M. Miller (1982) The dynamic set covering problem. *Applied Mathematics and Modelling*, **6**, 2-6.

C. H. Chung (1986) Recent applications of the Maximal Covering Location Problem (M.C.L.P.) model. *Journal of the Operational Research Society*, **37**, 735-746.

R. L. Church, J. R. Current and J. E. Storbeck (1991) A bicriterion maximal covering location formulation which considers the satisfaction of uncovered demand. *Decision Sciences*, **22**, 38-52.

R. L. Church and R. R. Davis (1992) The fixed charge maximal covering location problem. *Papers in Regional Science*, **71**, 199-215.

R. L. Church and D. J. Eaton (1987) Hierarchical location analysis using covering objectives. In *Spatial Analysis and Location - Allocation Models*, A Ghosh and G. Rushton, editors, 163-185.

R. L. Church and M. E. Meadows (1979) Location modelling utilizing maximum service distance criteria. *Geographical Analysis*, **11**, 358-373.

R. L. Church and C. S. ReVelle (1974) The maximal covering location problem. *Papers of the Regional Science Association*, **32**, 101-118.

R. L. Church and C. S. ReVelle (1976) Theoretical and computational links between the *p*-median, location set - covering, and the maximal covering location problem. *Geographical Analysis*, **8**, 406-415.

R. L. Church and K. L. Roberts (1983) Generalized coverage models and public facility location. *Papers of the Regional Science Association*, **53**, 117-135.

R. L. Church and J. R. Weaver (1986) Theoretical links between median and coverage location problems. *Annals of operations Research*, **6**, 1-19.

J. L. Cohon (1978) *Multiobjective Programming and Planning*. Academic Press, New York.

J. R. Current, C. S. ReVelle and J. L. Cohon (1985) The maximum covering / shortest path problem: a multiobjective network design and routing formulation. *European Journal of Operational Research*, **21**, 189-199.

J. R. Current, C. S. ReVelle and J. L. Cohon (1986) The hierarchical network design problem. *European Journal of Operational Research*, **27**, 57-66.

J. R. Current and D. A Schilling (1989) The covering salesman problem. *Transportation Science*, **23**, 208-213.

J. R. Current and D. A Schilling (1990) Analysis of errors due to demand aggregation in the set covering and maximal covering location problems. *Geographical Analysis*, **22**, 116-126.

J. R. Current and D. A Schilling (1994) The median tour and maximal covering tour problems: formulations and heuristics. *European Journal of Operational Research*, **73**, 114-126.

J. R. Current and J. E. Storbeck (1988) Capacitated covering models. *Environment and Planning B: Planning and Design*. **15**, 153-163.

M. S. Daskin (1982) Application of an expected covering location model to emergency service system design. *Decision Science*, **13**, 416-439.

M. S. Daskin (1983) A maximum expected covering location problem: formulation, properties and heuristic solution. *Transportation Science*, **17**, 48-70.

M. S. Daskin, A. E. Haghani, M. Khanal and C. Malandraki (1989) Aggregation effects in maximal covering models. *Annals of Operations Research*, **18**, 115-140.

M. S. Daskin, K. Hogan and C. S. ReVelle (1988) Integration of multiple, excess backup, and expected covering models. *Environment and Planning B: Planning and Design*, **40**, 15-35.

M. S. Daskin and E. H. Stern (1981) A hierarchical objective set covering model for emergency medical service vehicle deployment. *Transportation Science*, **15**, 137-152.

A. Desai and J. E. Storbeck (1988) Characterization of constraint in successively inclusive hierarchies. *Environment and Planning B: Planning and Design*, **15**, 131-141.

Z. Drezner (1986) The *p*-cover problem. *European Journal of Operational Research*, **26**, 312-313.

Z. Drezner (1995) *Facility Location: a survey of applications and methods.* Springer, New York.

C. Duin and A. Volgenant (1989) Reducing the hierarchical network design problem. *European Journal of Operational Research*, **39**, 332-344.

F. R. Dwyer and J. R. Evans (1981) A branch and bound algorithm for the list selection problem in direct mail advertising. *Management Science*, **27**, 658-667.

D. J. Eaton, M. S. Daskin D. Simmons B. Bulloch and G. Jansma (1985) Determining emergency medical service deployment in Austin, Texas. *Interfaces*, **15**, 96-108.

D. Eaton, M. Hector, V. Sanchez, R. Lantigua and J. Morgan (1986) Determining ambulance deployment in Santo Domingo, Dominican Republic. *Journal of the Operational Research Society*, **37**, 113-126.

J. Flynn and S. Ratick (1988) A multiobjective hierarchical covering model for the essential air services program. *Transportation Science*, **22**, 139-147.

O. Fujiwara, T. Makjamroen and K. K. Gupta (1987) Ambulance deployment analysis: a case study of Bangkok. *European Journal of Operational Analysis*, **31**, 9-18.

R. D. Galvão and C. S. ReVelle (1996) A Lagrangean heuristic for the maximal covering location problem. *European Journal of Operational Research*, **88**, 114-123.

M. Gendreau, G. Laporte and F. Semet (1996) The covering tour problem. *Operations Research*, To appear.

R. A. Gerrard and R. L. Church (1994) A generalized approach to modelling the hierarchical maximal covering location problem with referral. *Papers in Regional Science*, **73**, 425-453.

A. Ghosh and C. S. Craig (1987) An approach to determining optimal locations for new services. *Journal of Marketing Research*, **23**, 354-362.

A. Ghosh and G. Rushton (1987) *Spatial Analysis and Location-Allocation Models*. Van Nostrand Reinhold, New York.

J. R. Goldberg, R. Dietrich, J. M. Cheng, M. G. Mitwasi, T. Valenzuela and E. Criss (1990) Validating and applying a model for locating emergency medical vehicles in Tucson, AZ (case study). *European Journal of Operational Research*, **49**, 308-324.

J. R. Goldberg and L. Paz (1991) Locating emergency vehicle bases when service time depends on call location. *Transportation Science*, **25**, 264-280.

M. Goodchild and J. Lee (1989) Coverage problems and visibility regions on topographic surfaces. *Annals of Operations Research*, **18**, 175-186.

K. N. Groom (1977) Planning emergency ambulance services. *Journal of the Operational Research* Society, 28, 641-651.

G. Gunwardane (1982) Dynamic versions of set covering type public facility location problems. *European Journal of Operational Research*, **10**, 190-195.

N. G. Hall and D. S. Hochbaum (1992) The multicovering problem. *European Journal of Operational Research*, **62**, 323-339.

M. J. Hodgson (1990) A flow capturing location-allocation model. *Geographical Analysis*, **22**, 270-279.

K. Hogan (1980) Reducing errors in rainfall estimates through rain gauge location. *Geographical Analysis*, **22**, 270-279.

K. Hogan and C. S. ReVelle (1986) Concepts and applications of backup coverage. *Management Science*, **32**, 1434-1444.

V. A. Hutson and C. S. ReVelle (1989) Maximal direct covering tree problems. *Transportation Science*, **23**, 288-299.

V. A. Hutson and C. S. ReVelle (1993) Indirect covering tree problems on spanning tree networks. *European Journal of Operational Research*, **65**, 20-32.

T. D. Klastorin (1979) On the maximal covering location problem and the generalized assignment problem. *Management Science*, **25**, 107-111.

A. Kolen (1983) Solving covering problems and the uncapacitated plant location problem. *European Journal of Operational Research*, **12**, 266-278.

P. Kolesar (1980) Testing for vision loss in glaucoma suspects. *Management Science*, **26**, 439-450.

P. Kolesar and W. E. Walker (1974) An algorithm for the dynamic relocation of fire companies. *Operations Research*, **22**, 249-274.

G. Laporte and S. Martello (1990) The selective traveling salesman problem. *Discrete Applied Mathematics*. **26**, 193-207.

L. Lindner-Dutton, R. Batta and M. H. Karwan (1991) Equitable sequencing of a given set of hazardous materials shipments. *Transportation Sc*ience, **25**, 124-137.

L. A. N. Lorena and F. B. Lopes (1994) A surrogate heuristic for set covering problems. *European Journal of Operational Research*, **79**, 138-150.

C. Malandraki and M. S. Daskin (1993) The maximum benefit Chinese postman problem and the maximum benefit traveling salesman problem. *European Journal of Operational Research*, **65**, 218-234.

V. Marianov and C. S. ReVelle (1991) The standard response fire protection siting problem. *INFOR*, **29**, 116-129.

V. Marianov and C. S. ReVelle (1992a) A probabilistic fire protection siting model with joint vehicle reliability requirements. *Papers in Regional Science*, **71**, 217-241.

V. Marianov and C. S. ReVelle (1992b) The capacitated standard response fire protection siting problem: deterministic and probabilistic models. *Annals of Operations Research*, **40**, 203-322.

V. Marianov and C. S. ReVelle (1994) The queuing probabilistic location set covering problem and some extensions. *Socio-Economic Planning Sciences*, **28**, 167-178.

V. Marianov and C. S. ReVelle (1995) Siting emergency services. Chapter 10 in Z. Drezner, *Facility Location: a survey of applications and methods*, Springer-Verlag, New York.

V. Marianov and C. S. ReVelle (1996) The queuing maximal availability location: a model for the siting of emergency vehicles. European Journal of *Operational Research*, **93**, 110-120.

N. Megiddo, E. Zemel and S. L. Hakimi (1983) The maximum coverage location problem. *SIAM J. Alg. Disc. Method.*, **4**, 253-261.

A. Mehrez and A. Stulman (1982) The maximal covering location problem with facility placement on the entire plane. *Journal of Regional Science*, **22**, 361-365.

A. Mehrez and A. Stulman (1984) An extended continuous maximal covering location problem with facility placement. *Computers and Operations Research*, **11**, 19-23.

J. A. Mesa and T. B. Boffey. (1996) Location of extensive facilities in networks. *European Journal of Operational Research*. To appear.

I. D. Moon and S. Chaudhry (1984) An analysis of network location problems with distance constraints. *Management Science*, **30**, 290-307.

G. Moore and C. S. ReVelle (1982) The hierarchical service location problem. *Management Science*, **28**, 775-780.

S. Narasimhan, H. Pirkul and D. Schilling (1992) Capacitated emergency facility siting with multiple levels of backup. *Annals of Operations Research*, **40**, 323-338.

A. W. Neebe (1988) A procedure for locating emergency-service facilities for all possible response distances. *Journal of the Operational research Society*, **39**, 743-748.

L. S. Ochi, E. M. dos Santos, A. A. Montenegro and N. Maculan (1995) Artificial genetic algorithms for the Travelling Purchaser Problem. *Proceedings of the*

Metaheuristics International Conference, Breckenridge, Colorado. Kluwer, Norwell, Mass.

J. P. Osleeb and S. McLafferty (1992) A weighted covering model to aid in Dracunculiasis eradication. *Papers in Regional Science*, **71**, 243-257.

J. T. Pastor (1994) Bicriterion programs and managerial locations: application to the banking sector. *Journal of the Operational Research Society*, **45**, 1351-1362.

J. T. Pastor (1995) Private communication.

A. Perko (1986) Implementation of algorithms for k shortest loopless paths. *Networks*, **16**, 149-160.

H. Pirkul, J. R. Current and V. Nagarajan (1991) The hierarchical network design problem: a new formulation and solution procedures. *Transportation Science*, **25**, 175-182.

H. Pirkul and D. Schilling (1989) The capacitated maximal covering location problem with backup service. *Annals of Operations Research*, **18**, 141-154.

H. Pirkul and D. Schilling (1991) The maximal covering location problem with capacities on total workload. *Management Science*, **37**, 233-248.

C. S. Reeves (1993) *Modern Heuristics*. Blackwell-Scientific Publications, Oxford.

J. Repede and J. Bernardo (1994) Developing and validating a decision support system for locating emergency medical vehicles in Louisville, Kentucky. *European Journal ofOperational Research*, **75**, 567-581.

C. S. ReVelle (1989) Review, extension and prediction in emergency service siting models. *European Journal of Operational Research*, **40**, 58-69.

C. S. ReVelle and K. Hogan (1988) A reliability - constrained siting model with local estimates of busy fractions. *Env. and Planning B: Planning and Design*, **15**, 143-152.

C. S. ReVelle and K. Hogan (1989a) The maximum reliability location problem and a-reliable p-center problem: derivatives of the probabilistic location set covering problem. *Annals of Operations Research*, **18**, 155-174.

C. S. ReVelle and K. Hogan (1989b) The maximum availability location problem. *Transportation Science*, **23**, 192-200.

C. S. ReVelle and V. Marianov (1991) A probabilistic FLEET model with individual vehicle reliability requirements. *European Journal of Operational Research*, **53**, 93-105.

C. S. ReVelle, J. Schweitzer and S. Snyder (1996) The maximal conditional covering problem. *INFOR*. To appear.

C. S. ReVelle and S. Snyder (1996) Integrated fire and ambulance siting: a deterministic model. *Socio-Economic Planning Sciences*, **29**, 261-271.

D. Richard, H. Beguin and D. Peeters (1990) The location of fire stations in a rural environment: a case study. *Environment and Planning A*, **22**, 39-52.

G. Rose, D. W. Bennett and A. T. Evans (1992) Locating and sizing road maintenance depots. *European Journal of Operational Research*, **63**, 151-163.

R. Roth (1969) Computer solutions to minimum-cover problems. *Operations Research*, **17**, 455-465.

T. W. Ruefli and J. E. Storbeck (1982) Behaviorally linked location hierarchies. *Environment and Planning B: Planning and Design*, **9**, 257-268.

N. G. F. Sancho (1995) A suboptimal solution to a hierarchical network design problem using dynamic programming. *European Journal of Operational Research*, **83**, 237-244.

C. Saydam and M McKnew (1985) A separable programming approach to expected coverage: an application to ambulance location. *Decision Sciences*, **16**, 381-398.

D. A. Schilling (1980) Dynamic location modelling for public sector facilities: a multi-criteria approach. *Decision Sciences*, **11**, 714-725.

D. A. Schilling, D. J. Elzinga, J. Cohon, R. L. Church and C. S. ReVelle (1979a) The TEAM/FLEET models for simultaneous facility and equipment siting. *Transportation Science*, **13**, 163-175.

D. A. Schilling, C. S. ReVelle, J. Cohon and D. J. Elzinga (1979b) Some models for fire protection locational decisions. *European Journal of Operational Research*, **5**, 1-7.

D. A. Schilling, J. Vaidyanathan and R. Barkhi (1993) A review of covering problems in facility location. *Location Science*, **1**, 25-55.

R. E. Steuer (1986) *Multiple Criteria Optimization: Theory, Computation and Applications*, Wiley.

J. Storbeck (1982) Slack, natural slack and location covering. *Socio-Economic Planning Sciences*, **16**, 99-105.

J. E. Storbeck and V. Vohra (1988) A simple trade-off model for maximal and multiple coverage. *Geographical Analysis*, **20**, 220-230.

C. Toregas, R. Swain, C. S. ReVelle and L. Bergman (1971) The location of emergency service facilities. *Operations Research*, **19**, 1363-1373.

L. G. Underhill (1994) Optimal and suboptimal reserve selection algorithms. *Biological Conservation*, **20**, 85-87.

C. D. T. Watson-Gandy (1982) Heuristic procedures for the *m*-partial cover problem on a plane. *Management Science*, **11**, 149-157.

J. White and K. Case (1974) On covering problems and the central facility location problem. *Geographical Analysis*, **6**, 281-293.

J. C. Williams and C. S. ReVelle (1996) Applying mathematical programming to reserve selection. Johns Hopkins University, Internal Report.

An Evaluation of Our National Policy to Manage Nuclear Waste from Power Plants

by Ralph L. Keeney[1] and Detlof von Winterfeldt[1]

[1]Institute of Safety and Systems Management
University of Southern California
University Park
Los Angeles, CA 90089-0021

Abstract. The current national policy to manage nuclear waste from power plants is to dispose of it in a repository to be constructed at Yucca Mountain in Nevada. With many different assumptions about uncertainties and objectives, this strategy is shown to be the equivalent of $10,000 million to $50,000 million inferior to other available strategies. The implications of the analysis strongly suggest that our national policy to manage nuclear waste should be changed.

Nuclear waste management, decision analysis, nuclear waste repository

1. Managing Nuclear Waste from Power Plants

Congress passed the Nuclear Waste Policy Act in 1982 that placed the Department of Energy (DOE) in charge of identifying, constructing, and operating two underground repositories for permanent storage of nuclear waste. For each repository, DOE was to identify three final candidate sites for characterization. Characterization included digging a deep mine into the repository area to evaluate its appropriateness. In 1982, the cost estimates were $35 million for characterization and $1000 million for site construction and operation. The first repository was to accept spent fuel in 1998.

In the ensuing five years, DOE conducted studies to determine the suitability of several candidate sites. In 1986, for a first repository, DOE recommended sites in Texas, Washington, and Nevada for characterization [Department of Energy, 1986a]. By this time, the characterization costs were estimated at $1000 million per site, construction and operation had escalated to $8900 million for the cheapest site, and the start up date was delayed to 2003. In addition, many private and public groups perceived that DOE had mismanaged the nuclear waste disposal process.

The Nuclear Waste Policy Amendments Act of 1987 chose Yucca Mountain, Nevada as the site for the first repository and suspended the siting for a second repository in the eastern United States. By 1989, the estimated cost for the repository at Yucca Mountain was approximately $25,700 million of which $11,500 million were for characterization [Monitored Retrievable Storage Review Commission, 1989; Department of Energy, 1991]. The start up date for receiving spent fuel was now 2010 [Department of Energy, 1989]. These cost escalations and delays, coupled with the loss of trust and credibility in DOE's waste management process, have created a virtual impasse for the repository program. Hence, it is appropriate to revisit the fundamental question: How should the United States manage its nuclear waste generated from power plants? In particular, is it an appropriate use of over $25,000 million of financial resources to build one repository for nuclear waste? Are the benefits of deep geological storage (e.g., providing health and safety protection and safeguarding nuclear materials) worth the loss of retrievability and flexibility in managing the wastes? The analysis in this paper attempts to answer these questions by integrating numerous issues about managing nuclear waste into a common-sense framework.

2. Strategies to Manage Nuclear Waste

Strategies are the sequences of decisions made from now through 2100 to manage nuclear waste. The consequences of those strategies, and hence their relative desirability, depend both on the decisions taken and on the resolution of uncertainties over time. To clarify the strategies evaluated, it is useful to consider three features: current decisions, resolution of uncertainties over the next 100 years, and decision alternatives in 2100.

One current alternative is storing nuclear waste in an underground repository at Yucca Mountain as soon as possible. The second is an above-ground monitored retrievable storage (MRS) facility to store nuclear waste until 2100. The third is to store nuclear waste above ground at nuclear power plant sites in dry storage casks or concrete bunkers until 2100. All three alternatives would be monitored to ensure that they were safe. Indeed, there is general agreement that the monitored retrievable storage facility and on-site storage can be safely managed for a 100-year period [Monitored Retrievable Storage Review Commission, 1989; Board on Radioactive Waste Management, 1990].

As seen in Figure 1, if the initial decision is to pursue the repository at Yucca Mountain, there is uncertainty about whether the facility would be licensed. If Yucca Mountain is not licensed, then another decision must be made about whether to pursue a repository at a new site, a monitored

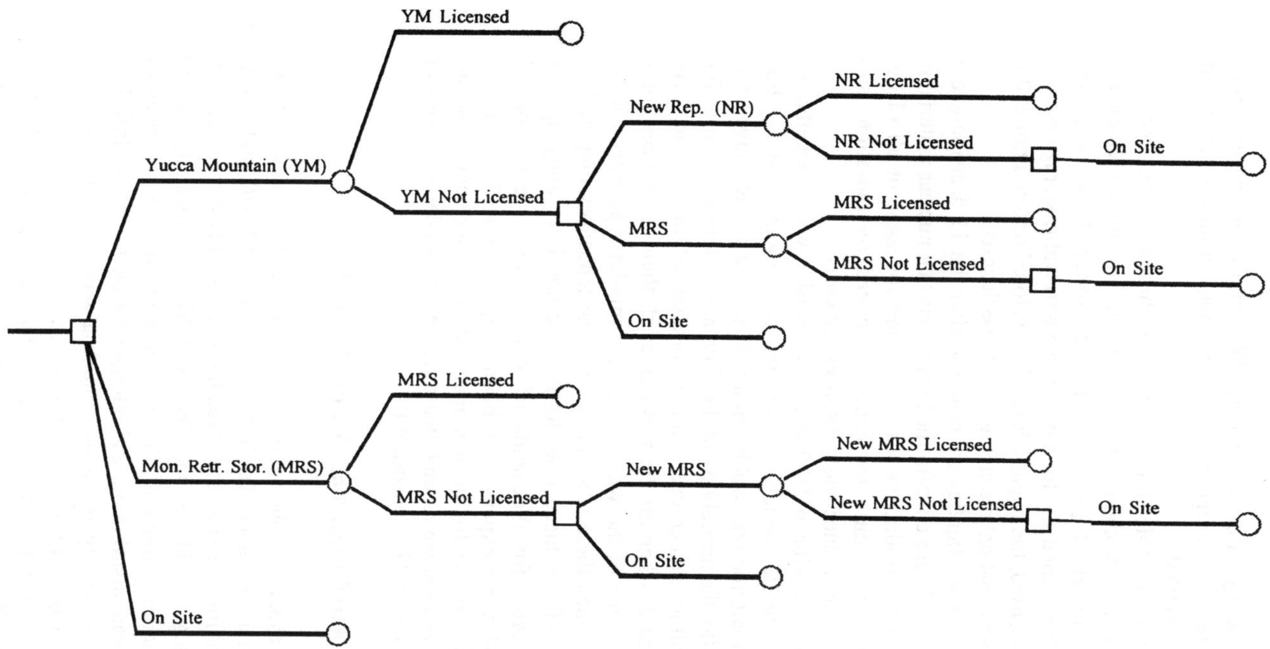

Figure 1. Current decision alternatives to manage nuclear waste

(Squares denote decisions, circles denote the resolution of uncertainty not under the control of the decision

retrievable storage facility, or on-site storage. With either another repository or a monitored retrievable storage facility, there is again uncertainty about whether it would be licensed.

If the monitored retrievable storage facility is chosen and not licensed, then a decision must be made about whether to pursue a monitored retrievable storage facility at another site or pursue on-site storage at nuclear power plants until 2100. If a second site for a monitored retrievable storage facility is chosen and not licensed, the nuclear fuel will be stored on site until 2100. In all cases, we assume that on-site storage would be licensed.

Several important uncertainties are resolved after the 11 decision-event sequences in Figure 1. These concern the likelihoods that nuclear material is stolen and/or misused, productive uses are found for some or all of the nuclear waste, there is a cure for cancers induced by radiation, and technological innovations reduce waste management costs.

If a repository exists prior to 2100, the main decision will be whether to dig up the nuclear waste, which depends on its economic value. For situations where monitored retrievable storage or on-site storage have been utilized through 2100, the desirability of the alternatives depends on uses for the waste, a possible cancer cure, and technological innovation. As illustrated in Figure 2, when all the waste is used, there is no need for a repository. When some of the waste is used, a smaller repository can be built. In addition, when there is a cancer cure, the repository may be less expensive because of a reduced need to promote absolute safety against potential cancer cases for thousands of years. Given technological innovations, it is also less expensive to build a repository in 2100. In the case when there is no use for nuclear material, no cancer cure, and no technological innovations, we assume that a repository similar to those that might be built today would be constructed.

3. Objectives, Attributes, and Value Tradeoffs

The objectives for evaluating nuclear waste management strategies are listed in Table 1. The first four objectives (pre-and postclosure health and safety, environment, social impacts) and the seventh (direct cost) have been used in previous studies [e.g., Merkhofer and Keeney, 1987]. In this analysis, four objectives were added: equity of the management strategy in terms of procedure, geographic distribution, and intergenerational burden; fulfilling the government responsibility to dispose of nuclear waste; indirect economic costs; and consequences of misuse of nuclear waste.

Each objective is operationalized by an attribute also shown in Table 1. The five attributes X_1, X_2, X_6, X_7, and X_8 have natural scales; their meaning is self explanatory. The remaining attributes utilize constructed

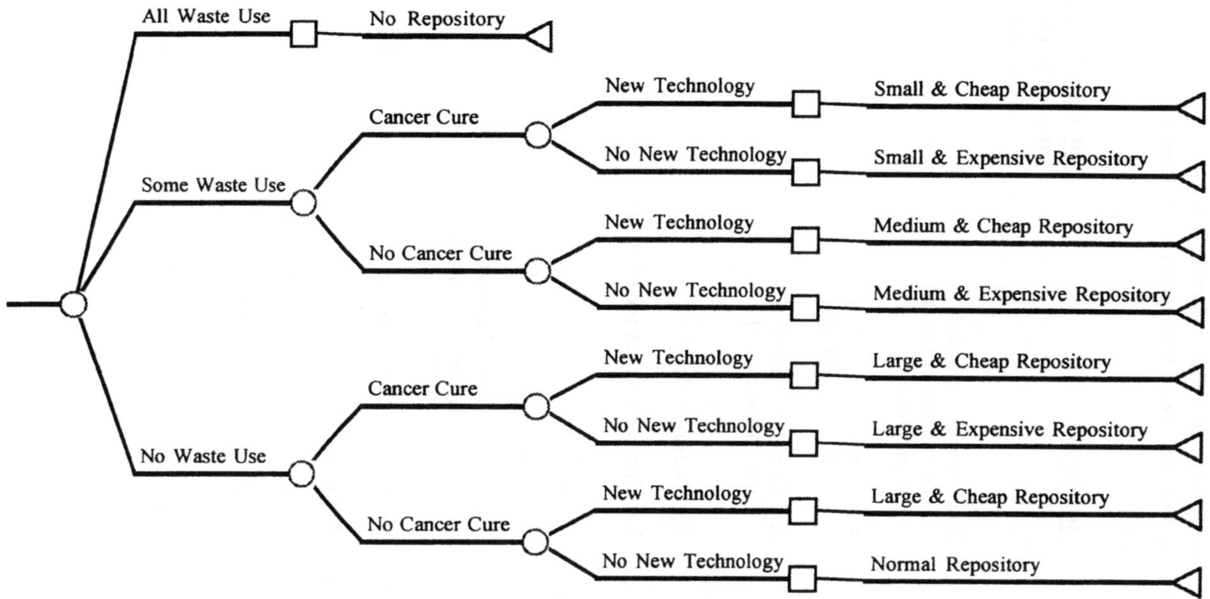

Figure 2. Decision alternatives for 2100 to manage nuclear waste, provided that an MRS or on-site storage was utilized through 2100.

(Squares denote decisions, circles denote the resolution of uncertainty not under the control of the decision makers, triangles denote endpoints of the decision tree.)

Table 1. Objectives, Attributes and Value Tradeoffs Used for Evaluating Nuclear Waste Management Strategies

Objectives	Attributes	Unit Value Tradeoff (Equiv. 1992 millions of dollars)
1. Preclosure health and safety	X_1: Number of preclosure fatalities due to cancer or accidents	4
2. Postclosure health health and safety	X_2: Number of postclosure fatalities due to cancer	1
3. Environmental impacts	X_3: Constructed scale of environmental impacts	80
4. Social impacts	X_4: Constructed scale	125
5. Equity	X_5: Component constructed scales	
5.1 Geographic	X_{51}:	1,000
5.2 Intergenerational	X_{52}:	1,000
5.3 Procedural	X_{53}:	1,000
6. Government responsibility	X_6:Years until a nuclear waste management system is operational	100
7. Direct economic costs	X_7: Millions of 1992 dollars	1
8. Indirect economic costs	X_8: Component attributes	
8.1 To states and businesses	X_{81}: millions of 1992 dollars	1
8.2 To electricity ratepayers	X_{82}: millions of 1992 dollars	1
9. Misuse of nuclear waste	X_9: Constructed scale (see Table 2)	10,000

scales to describe, mostly in words, discrete levels of performance on each corresponding objective. Table 2 illustrates the constructed attribute for the objective "misuse of nuclear waste". The attributes for the equity objectives have three levels that indicate which strategy would do best, mixed, and poorly with respect to those objectives. For example, for geographic equity, on-site storage was best, an MRS was mixed, and Yucca Mountain scored poorly. The environmental and social attribute scales are described in [Department of Energy, 1986b].

Table 2. Attribute for Implications of Theft and Misuse of Nuclear Waste

Level	Description of Attribute
0	No theft of material occurs (i.e., no missing material)
1	Theft (or loss) of material occurs but there is no evidence of subsequent misuse
2	Theft and misuse (e.g., a terrorist use) of the nuclear material occur

To evaluate and compare strategies, one must specify relative values for the various possible consequences described by each attribute and combine the values from the different attributes. This is done by constructing a value model from explicitly stated value judgments [Keeney and Raiffa, 1976]. The value model used in this analysis is technically referred to as linear additive measurable value function, and it also has the properties of a linear additive utility function [von Winterfeldt and Edwards, 1986]. It is linear because the relative values of each attribute are expressed by a linear value function, and it is additive because contributions to value from different attributes are added. The linearity and additivity are justified for four reasons: 1) the requisite value judgments are consistent with those carefully verified for the DOE analysis of repository sites [Department of Energy, 1986b]; 2) the constructed attributes that were not part of the DOE evaluation were defined so that a linear value function was reasonable; 3) the objectives chosen for the evaluation are fundamental objectives and not means objectives [Keeney, 1992]; 4) sensitivity analysis indicated that reasonable changes in these properties would not change any of the insights from the analysis.

For each attribute, the component value functions are linear, which implies that the difference in value between levels 2 and 3 of an attribute, for instance, is equivalent to the difference in value between levels 5 and 6. To integrate the implications from different objectives, one needs explicit value tradeoffs. These indicate how much a specific change in the level of one attribute is worth in terms of specific changes in the level of the second attribute. As a common unit of measure, we chose equivalent costs measured in 1992 dollars.

Because the component value functions are linear, value tradeoffs can be specified by indicating what it is worth in terms of equivalent 1992 dollars to make a change of adjacent levels of each other attribute. Specifically, we define unit value tradeoffs to indicate the value in moving one adjacent level (i.e. a unit) on an attribute in terms of equivalent 1992 dollars. These unit value tradeoffs, which are sufficient to calculate all value equivalencies between pairs of attributes, are displayed in Table 1.

The unit value tradeoffs for attributes X_1 through X_4 are adapted from the DOE repository analysis (Merkhofer and Keeney, 1987; Department of Energy, 1986b]. In that study, the unit value tradeoffs per worker and public preclosure statistical fatality were $1 million and $4 million, respectively. We chose $4 million per statistical fatality for this analysis. For postclosure fatalities, we chose $1 million per statistical fatality that might occur hundreds to thousands of years in the future. The unit value tradeoffs imply it is worth $400 million to eliminate the worst environmental impact and $500 million to eliminate the worst social impact. The DOE study used $150 and $500 million, respectively.

The unit value tradeoffs for attributes X_5 to X_9 for the base case analysis were based on discussions with individuals knowledgeable about government policy making. Regarding attribute X_7, $1 million cost equals $1 million by definition. Because these direct and indirect costs each fall on society, we chose a unit value tradeoff of $1 million for each of the indirect cost attributes, too. The unit value tradeoff for each of the equity attributes was chosen to be $1000 million and the unit value tradeoff for misuse of nuclear waste is $10,000 million. Thus, the ranges of each of the equity attributes (i.e. units) have an equivalent value of $2000 million, whereas it would be worth $20,000 million to eliminate an incidence of theft and misuse of spent nuclear fuel. The political impact of each year of delay of government fulfillment of its responsibility was assigned an equivalent value of $100 million.

4. Data and Information for Evaluating Strategies

Table 3 provides the base case uncertainties utilized in our analysis. Licensing a nuclear repository is a process based not only on technological considerations, but also on social and political considerations. Partly for this

Table 3. Base Case Uncertainties

a. Licensing

Event	Probability
Yucca Mountain repository licensed	0.6
Second repository licensed after Y.M. rejected	0.5
M.R.S. licensed after Y.M. rejected	0.7
M.R.S. initially licensed after Y.M. not pursued	0.7
Second M.R.S. licensed after first proposed MRS rejected	0.6
On-site storage licensed	1.0

b. Spent Nuclear Fuel Acceptance Dates

Storage Facility	Date	Years	Probability
Yucca Mountain repository	2010	18	0.2
	2020	28	0.8
Second repository licensed after Y.M. rejected	2025	33	0.4
	2035	43	0.6
M.R.S. licensed after Y.M. rejected	2015	23	0.7
	2025	33	0.3
M.R.S. initially licensed after Y.M. not pursued	2007	15	0.7
	2012	20	0.3
Second M.R.S. licensed after first proposed M.R.S rejected	2015	23	0.7
	2020	28	0.3
On-site storage	1997	5	1.0

c. Possible Events Over the Next 100 Years

Event	Probability
A use for all nuclear waste found by 2100	0.1
A use for some nuclear waste is found by 2100	0.2
No use for waste is found by 2100	0.7
Cure for cancer due to radiation before 2100	0.2
No cure for cancer due to radiation before 2100	0.8
Technological innovations to reduce real repository costs by 50 percent	0.7
No technological innovations for waste management	0.3

	Probability Given Storage Site		
	Repository	MRS	On-Site
Some nuclear material is stolen and misused	0.001	0.002	0.006
Some nuclear material is stolen or missing, but not used	0.005	0.01	0.03
No nuclear material is stolen or missing	0.994	0.988	0.964

reason, there is considerable uncertainty about whether the proposed Yucca Mountain repository would be licensed. To our knowledge, there are no published estimates. Privately, individuals have suggested probabilities that range from 0.1 to 0.9. For our base case analysis, we chose 0.6, which serves as an anchor for estimates that other facilities would be licensed.

We selected two possible completion dates, on-time and with significant delay, for a nuclear repository or monitored retrievable storage facility to begin accepting spent nuclear fuel. For Yucca Mountain, the probability of an on-time operation is 0.2 given the history of this alternative. For a second repository, if Yucca Mountain is rejected, we assume it is twice as likely that the facility will begin operating on time. For any monitored retrievable storage facility, we assume the probability of on-time operation is 0.7. For on-site storage, we assume that facilities could be constructed and operating five years after the alternative is chosen.

The desirability of a repository depends significantly on whether there will be a use for the waste in the next 100 years. For a base case analysis, we assign a probability of 0.1 that there is a use for all of the waste by 2100, a probability of 0.2 that there is a use for some of the waste.

It may be less expensive to build a repository in 2100 than now, due to technological innovation that may occur. We assume that the probability of technological innovations to reduce real repository costs by 50 percent is 0.7. Otherwise, there will be no technological innovation, so that the cost of a repository in 2100 would be the same in real dollars as the cost of the repository today.

Available information suggests that the probabilities of the theft and misuse of spent nuclear material are very small. Although the probabilities may be higher for above-ground storage, the Nuclear Regulatory Commission staff [Monitored Retrievable Storage Review Commission, 1989] stated that "spent fuel stored either at individual reactor sites or an MRS is a poor target for sabotage or diversion and that there is no identifiable credible threat of sabotage or diversion of fissile material. In any case, the radiological consequences of sabotage would be low." Thus, we initially assigned 0.001 as the probability that spent nuclear material would be stolen and misused during the preclosure period of time given a repository is selected. We doubled this to 0.002 when material is in a monitored retrievable storage facility, and increased the probability to 0.006 for on-site storage because there are more sites where material might be stolen. We also assume that the likelihood the material would be stolen or missing but not used is five times the likelihood that it would be stolen and misused. This gives the corresponding probabilities in Table 3c.

A complete discussion of the consequences at each endpoint of the decision tree that begins with Figure 1 and ends with Figure 2 is lengthy

[Keeney and von Winterfeldt, 1994]. However, a few remarks are useful. Pre- and postclosure fatality estimates and social and environmental impacts were adapted directly from the analysis of several potential repositories [Merkhofer and Keeney, 1987]. The estimates for equity consequences follow directly from the sequence of events and decisions. Geographic equity is best for on-site storage, worst with a repository built now. Procedural equity is best if the repository decision is delayed by 100 years, worst if repository decision is "forced" now. Intergenerational equity is best if a permanent solution is found now, worst if no process to permanently "solve the nuclear waste problem" is followed.

Table 4 provides information for estimating direct economic costs, taken partly from the report by the Monitored Retrievable Storage Review Commission [1989]. We used our judgment to estimate the costs of a repository in 2100, the cost of digging up nuclear wastes, and the benefits from using it [Keeney and von Winterfeldt, 1994]. The estimates of indirect economic consequences to states and local businesses ranged from 0 (in case of on site storage with no repository in 2100) to $1,700 million (in case of an MRS with a repository in 2100). The estimates for indirect economic consequences to the electricity ratepayers ranged from a benefit of $30,000 million (if Yucca Mountain is licensed and operates on time) to a cost of $30,000 (if two facilities are denied licenses prior to reverting to on site storage).

5. Comparing 1992 Decision Strategies

The base case analyses indicates that the best strategy for managing nuclear waste is to pursue on-site storage until 2100. At that time, no decision is necessary if all of the waste is useful. If all of the waste is not useful, a repository should be built for a cost consistent with the conditions of that time noted in Table 3.

The 1992 decision alternatives are compared in Table 5. The uncertainties and subsequent decisions that follow the current alternatives are accounted for in the consequences. Equivalent costs are calculated by multiplying each consequence by its corresponding value tradeoff from Table 1. In the base case analysis, the overall equivalent costs are $36,000 million for the repository, $18,000 million for the monitored retrievable storage facility, and $8,000 million for on-site storage. When indirect costs to electricity ratepayers or the implications of fulfilling government responsibility are deleted, or when only the set of attributes used by DOE to evaluate the repositories in 1986 are used, the equivalent cost of pursuing the repository now is at least $10,000 million higher than either pursuing the monitored retrievable storage facility or on-site storage.

Table 4. Costs Used to Construct Direct Economic Consequences of Decision Strategies

a. Costs Due to Current Generation Decisions

Item	Cost in Millions of 1992 Dollars
Repository licensed and constructed without delays	$25,700
Each year of delay in repository construction	100
Repository characterized and license rejected	10,000
Monitored-retrievable storage constructed witout delay	2,900
Each year of delay in M.R.S. construction	100
M.R.S. pursued and license rejected	2,000
On-site storage until 2100	3,000

b. Costs of a Repository Licensed and Constructed After 2100 for Different Conditions

Nuclear Waste Cancer Cure	Technological Use	Innovation	Cost in Millions of 1992 Dollars
No	None	No	$25,700
No	None	Yes	12,000
No	Some	No	20,000
No	Some	Yes	10,000
Yes	None	No	12,000
Yes	None	Yes	6,000
Yes	Some	No	10,000
Yes	Some	Yes	5,000

c. Costs Related to Use of Nuclear Waste in 2100

Item	Cost in Millions of 1992 Dollars
Dig up repository	$2,000
Economic value of nuclear waste if some of it is useful	5,000
Re-bury excess waste that is not useful	1,000
Economic value of nuclear waste if all of it is useful	10,000

Table 5. **Comparing base-case consequences and equivalent costs for the three 1992 decision alternatives: Yucca Mountain repository, a monitored retrievable storage facility, and on-site storage**
(dollar amounts are in millions)

Attribute	Pursue Yucca Mountain		Pursue M.R.S.		Pursue On-Site	
	Conse-quence	Equiv. Cost	Conse-quence	Equiv. Cost	Conse-quence	Equiv. Cost
X_1: Preclosure statistical fatalities	42.5	170	43	172	36	144
X_2: Postclosure statistical fatalities	504	504	504	504	504	504
X_3: Environmental impact level	2.8	224	2	160	1	80
X_4: Social impact level	2.68	335	2.4	300	1	125
X_{51}: Geographic equity level	1.48	1480	0.7	700	0	0
X_{52}: Intergenerational equity level	0.52	520	1.3	1300	2	2000
X_{53}: Procedural equity level	1.48	1480	0.7	700	0	0
X_6: Years to fulfill government responsibity	25.64	2564	16.05	1605	5	500
X_7: Direct economic cost	$24,857	$24,857	$14,089	$14,089	$13,454	$13,454
X_{81}: Indirect economic cost to states and business	$1340	$1340	$1550	$1550	$1200	$1200
X_{82}: Indirect economic cost to electricity ratepayers	$2,800	$2800	-$3,800	-$3800	-$10,000	-$10,000
X_9: Safeguards level	0.0132	132	0.0224	224	0.042	420
Equivalent Costs						
Total for all attributes		$36,406		$17,507		$ 8,427
All attributes except X_{82}		$33,606		$21,304		$18,427
All attributes except X_6 and X_{82}		$31,042		$19,699		$17,927
Traditional attributes: X_1, X_2, X_3, X_4, and X_7		$26,090		$15,225		$14,307

Not placing nuclear waste in a repository at this time affords us flexibility for future decisions. The value of flexibility comes about because of the ability to take advantage of any future use for nuclear waste, potential cancer cure, and technological innovations for managing the waste. The value of flexibility, calculated as the expected economic savings due to these factors, is about \$11,000 million. Even if we assume that there will be no waste use or cancer cure, the value of flexibility is still \$6,600 million.

One major reason to build a repository at Yucca Mountain is to isolate the radioactivity from future generations living thousands of years from now. However, the postclosure health effects from any possible releases of radiation material are identical for all three strategies because all three end up either with no waste because it is all used by 2100 or with a repository. Thus, potential postclosure fatalities (often stated as the reason why a repository is needed soon) cannot be used to differentiate the desirability of the strategies.

Preclosure health and safety, environmental, and social implications contribute little to overall equivalent cost. Geographical equity and intergenerational equity consequences cancel each other. If the waste is stored in a repository now, it is geographically inequitable but this generation has managed the problem, so it is equitable intergenerationally. On-site storage is equitable geographically, but we have not managed our problem so it is inequitable intergenerationally. With the monitored retrievable storage facility, the implications in terms of geographic and intergenerational equity are in between the extremes produced by the repository and on-site storage.

Major factors affecting the relative desirability of the alternatives are direct economic costs and indirect economic costs to electricity ratepayers due to the perceived quality of the management of nuclear wastes. In the base case analysis, even if these indirect costs are omitted, the alternatives of storing the waste on site or in a monitored retrievable storage facility are still much less expensive than putting it in a repository at the current time.

6. Sensitivity Analyses

Numerous sensitivity analyses support the base case conclusions [Keeney and von Winterfeldt, 1994]. Because pursuing a repository at Yucca Mountain is the current national policy to manage nuclear waste from power plants, it is instructive to make the best case for Yucca Mountain in the context of the current analysis. Table 6 shows results of sensitivity analyses that favor Yucca Mountain. First, we assume that Yucca Mountain will be licensed, which improves the equivalent cost of Yucca Mountain by \$8200 million, but does not change the equivalent cost of the MRS and on-site storage. With no technological innovation, the MRS and on-site equivalent

Table 6. Sensitivity Analysis of Conditions that Favor a Repository at Yucca Mountain

(Amounts are in millions of dollars of equivalent cost and all assumptions are base case except conditions explicitly noted)

Conditions	Yucca Mountain	M.R.S.	On-Site
Base case assumptions	$36,406	$17,507	$ 8,427
Yucca licensed for sure	28,252	17,507	8,427
Above and no technology innovation	28,252	24,722	15,645
Above and no waste use or no cancer cure	29,648	32,946	23,869
Above and theft and misuse probability multiplied by 10 for MRS and on-site	29,648	34,762	27,649
Above and cost of theft and misuse is multiplied by 10	30,278	41,748	44,449
All options licensed for sure	28,252	7,217	8,427
Above and all options operating without delay	2,652	917	8,427
Above and electricity ratepayer's benefit of Yucca Mountain being licensed and operating without delay is $40,000 million	-7,348	917	8,427

costs increase by about $7200 million each. Next we added the assumption that there is no cancer cure or waste use. This slightly increases the equivalent costs of Yucca Mountain (because of the additional postclosure health effects and the loss of expected revenue from waste use) and dramatically increases the equivalent costs of the MRS and on-site storage. However, on-site storage still comes out ahead by some $7800 million.

To tip the balance even further, we assume that the probabilities of theft and misuse remained the same for Yucca Mountain, but are higher by a factor of ten over the base case for the MRS and on-site storage. On-site storage is still preferred, but the margin narrows to $2000 million. Finally, we add the assumption that the costs of theft and misuse are increased tenfold. In this case, Yucca Mountain has the lowest equivalent costs of $30,300 million, which is lower than the MRS by $11,500 million and on-site storage by $14,200 million.

The lower part of Table 6 investigates the sensitivities of licensing and delay probabilities. First, we assumed that licensing can be assured for all options. In this case, the MRS has the lowest equivalent cost, just $1200 million lower than on-site storage, and $21,000 million lower than Yucca Mountain. In addition, if all options can be implemented without delay, both Yucca Mountain and the MRS are clear favorites over on-site storage, but the MRS still edges out Yucca Mountain by a margin of $1700 million. To make Yucca Mountain a winner, one has to assume that, in addition to the above assumptions, the indirect cost benefit to electricity ratepayers from having Yucca Mountain licensed and operating increases by at least that difference. To illustrate, we set these benefits at $40,000 million rather than the $30,000 million used in the base case. In that case, Yucca Mountain has an equivalent economic benefit of $7300 million, which is $8200 million of equivalent cost better than the MRS and $15,800 million better than on-site storage.

7. A Fund to Manage Nuclear Waste in the Future

An alternative that could be implemented now is to create a fund for society in 2100 to determine how they wish to permanently dispose of nuclear waste from our current power plants. For this analysis we assumed that spent nuclear fuel will be stored on-site until 2100. We based the size of potential funds on the historical long-term real interest rate, which depends on the inflation rate and actual return rate. Comparable data on these was available from 1926 to 1991. The average inflation rate over those 65 years was 3.19 percent as measured by the consumer price index [Department of Labor, 1992]. The compound annual growth rate on long-term government bonds, a very conservative investment, was 4.54 percent for the same period Ibbotson

Associates, 1991]. Hence, the real interest rate was 1.35 percent annually. Compounded for 108 years until 2100 at this rate, one dollar grows to 4.26 real dollars. Thus, a $6000 million fund established today would grow to $25,700 million by 2100, which is just enough to finance a repository at today's prices. A $3000 million fund would finance the expected cost of $12,500 to store waste in a repository in 2100.

While the economic implications clearly change with a nuclear waste fund, there are also some changes in the impacts of the combined on-site storage and waste fund strategy. For example, intergenerational equity improves by one level because this generation sets up the process for a permanent solution to the nuclear waste problem. Also, on-site storage with a fund would probably reflect better on the nuclear industry than on-site storage without a fund. Thus, we increased the indirect economic benefits to electricity ratepayers due to the perceived quality of nuclear waste management by an equivalent value of $10,000 million.

Using base case assumptions, the on-site storage alternative with a $6000 fund is preferable to the on-site storage alternative without a fund by $31,000 million, to the repository without a fund by $59,000 million, and to the MRS without a fund by $40,000 million. Since the entire current cost of a repository built in 2100 is covered by a $6000 million fund, the entire cost savings of $13,200 million due to technology innovation, waste use, and/or a cancer cure is attributable to the fund alternative. Most of the remaining savings are due to changes in the consequences regarding indirect cost to the nuclear industry. However, even when ignoring this attribute the savings are still $19,000 compared to on-site storage without a fund, higher when compared to the repository and the MRS without a fund. The $3000 million fund alternative is preferred to pursuing Yucca Mountain now by $38,000 million.

8. Conclusions and Recommendations

The current national strategy of attempting to store nuclear waste in a repository at Yucca Mountain appears to be a major misallocation of national resources. Under various conditions, the analysis suggests that the Yucca Mountain repository is worse than either storing the nuclear waste at the existing nuclear power plants or in a monitored retrievable storage facility by the equivalent of somewhere between $2000 million and $33,000 million. In direct economic costs alone, the repository is expected to be over $10,000 million inferior to each of the other alternatives. The repository appears to be the best strategy only when almost all uncontrollable circumstances favor it.

In most situations, the best strategy appears to be to store the nuclear waste on site for the next 100 years. This would avoid transporting nuclear waste at the current time and allow us to create an alternative for safely and soundly managing the waste over the next 100 years. Over that time period, we could involve the public and stakeholders in a legitimate manner to find a appropriate permanent solution for nuclear waste. Also, it is likely that events (e.g., technological innovations) will take place that will save considerable amounts of funds in dealing with nuclear waste. The base case expected value of this flexibility is approximately $11,000 million.

An alternative that possesses many desirable characteristics is to establish a fund at the current time to provide the financial resources 100 years from now to manage the spent nuclear fuel that would be stored between now and then at nuclear power plants. The base case analysis suggests that this on-site alternative with a $6,000 million fund is the equivalent of at least $20,000 million better than the monitored retrievable storage facility and $30,000 million better than a repository.

The insights from this analysis suggest three major recommendations:

- Stop currently pursuing Yucca Mountain as a repository to permanently store nuclear waste.

- Modify the Nuclear Waste Policy Act to allow more reasonable strategies to manage nuclear waste.

- Conduct an independent detailed analysis along the lines of the analysis presented here to help design a better national policy.

9. Acknowledgment

This report was prepared with support by the National Science Foundation under grant No. SES-8919502. A more technical and detailed description of this analysis was published in *Risk Analysis* (see Keeney and von Winterfeldt, 1994).

10. References

Board on Radioactive Waste Management (1990). "*Rethinking High-Level Radioactive Waste Disposal.*" National Academy Press, Washington, DC.

Department of Energy (1986a). "*Recommendation by the Secretary of Energy of Candidate Sites for Site Characterization for the First Radioactive-Waste Repository*" Technical Report DOE/S-0048, Office of Civilian Radioactive Waste Management, Washington, D.C.

Department of Energy (1986a). "*A Multiattribute Utility Analysis of Sites Nominated for Characterization for the First Radioactive-Waste Repository - A Decision-Aiding Methodology.*" Report No. DOE/RW-0074, Office of Civilian Radioactive Waste Management, Washington, DC.

Department of Energy (1989). "*Report to Congress on Reassessment of the Civilian Radioactive Waste Management Program*" Office of Civilian Radioactive Waste Management, Department of Energy, Washington, DC.

Department of Energy (1991). "*OCRWM Bulletin, February/March, DOE/RW-0305*" Office of Civilian Radioactive Waste Management, Department of Energy, Washington, DC.

Department of Labor (1992). "*CPI Detailed Report*" Bureau of Labor Statistics, Washington, DC.

R.L. Keeney and H. Raiffa (1976). "*Decisions with Multiple Objectives*". Wiley, New York.

D. von Winterfeldt and W. Edwards (1986). "*Decision Analysis and Behavioral Research*" Cambridge University Press.

R.L. Keeney (1992). "*Value Focused Thinking*". Harvard University Press, Cambridge, MA.

R.L. Keeney and D. von Winterfeldt, (1994). "Managing nuclear waste from power plants." *Risk Analysis,* **40**, pp. 263-279.

M.W. Merkhofer and R.L. Keeney (1986). "A multiattribute utility analysis of alternative sites for the disposal of nuclear waste." *Risk Analysis*. 7, pp. 173-194.

Monitored Retrievable Storage Review Commission (1989). "*Nuclear Waste: Is There a Need for Federal Interim Storage?*" U.S. Government Printing Office, Washington, DC.

Ibbotson Associates, Inc., (1991). "*Stocks, Bonds, Bills, and Inflation: 1991 Yearbook*" Ibbotson Associates, Chicago, Illinois.

Professor Stanley Zionts

Fifteen Years of Work with a Guru: A Lifetime or Just the Beginning

R. Ramesh
School of Management, State University of New York at Buffalo,Buffalo, NY 14260 USA

Abstract

The sixtieth birthday celebration of Stan Zionts is an important event in the lives of many - his numerous students and colleagues who have benefited enormously through his association. I have had the unique privilege of a sustained association with him for the past one and a half decades, and being the beneficiary of his mentorship, comradeship and collaboration in numerous research ventures. In this article, I summarize in a story-like fashion, our research association, accomplishments and the road ahead.

1. Introduction

The Sanskrit word guru has several connotations. Having adopted this word in the English repertoire, the American Heritage dictionary defines it as a recognized leader or guide, an acknowledged advocate, a personal spiritual teacher, the proponent of a movement or an idea. Professor Stanley Zionts, while symbolizing all the qualities attributed in these definitions and much more as far as myself and numerous others are concerned, stands tall among the so many gurus I have had in my life, worthy of a simple but poignant definition that I could attribute especially to him: Friend, Philosopher and Guide. Stan has been a symbolic referential source in virtually all of my academic pursuits and so many of my personal encounters in life. He has played a variety of roles in our interactions, guiding, philosophizing and even frolicking at times. Numerous other colleagues, I am sure, would share these sentiments and feel a deep sense of gratitude as I, for the innumerable gifts that we have received through our association with a truly remarkable man as Stan Zionts. It is indeed a pleasure to express my appreciation and gratitude to this wonderful person I have been so fortunate to have come across through an article summarizing my association with him, what I learned from him, and what he helped me to accomplish in my research career in a volume dedicated as an appreciation of the community as a whole for his numerous contributions and truly outstanding service to the profession.

I have interacted with Stan in several roles over the past fifteen years. My first encounter with him was as a student in a MCDM course that he jointly taught with Professor Mark Karwan in 1981. I was a graduate student in Industrial Engineering, a tyro of some sort. My initial impressions in the first encounter was at once frightening and encouraging. Frightening, because of the high citadel of research from which the instructors descended, encouraging because of their extremely humane approach to teaching and learning. I acquired an ability for critical thinking through those demanding engagements in class and outside, and it laid the foundations for my doctoral research. I continued my work in that course towards a dissertation under the guidance of Stan and Mark. This resulted in three years of sustained MCDM research with Stan. It was a period of immeasurable learning for me in all aspects of doctoral work and individual development. Stan played a critical role in this process, critiquing, contributing, resolving and sustaining my work and spirit. The work evolved gradually first and then took a rapid turn, and I lost control of its momentum, relying entirely on Stan's foresight, balance and vision to guide it to a successful arrival. When it finally arrived, the task of writing it all up was truly monumental, especially for a novice writer that I was. Stan stepped in again, exacting his unrelenting high standards on me. I hated him at that time for being mercilessly scrupulous in demanding the best from me, but then, I have thanked him ever since for sending me through such a grill for so long that has immensely taught me what writing is all about. Even now I fondly recall his famous phrase which would ring a bell in the memories of all his students: You should write such that even your mother would understand it. This has no derogatory connotations to a mother's ability to comprehend, but then it is also too much to expect one's mother to follow the nuances of integer programming, for instance. Certainly mine would not, but I am sure she would certainly follow what I have written in my dissertation after having gone through the drill of Stan Zionts. I wrote my 300 odd page dissertation twice in entirety, failing to satisfy him the first time around. He showed me how to write in a clear, concise and crisp fashion, and he put in endless hours poring over my manuscripts developed over a period of a year after the completion of the dissertation work. Although it delayed my defense by a year, I am extremely grateful to him for two reasons. First, it was a tremendous learning experience for me in the art of writing, and second, the effort spent on the dissertation resulted in a flawlessly smooth production of journal articles for publication. His unrelenting insistence on the highest standards of quality, be it research or writing or anything one does in life has served as a guiding beacon in all my subsequent ventures.

Upon completion of my dissertation, I joined the faculty of the School of Management at the University of Buffalo. I graduated at the fortunate time when academic jobs were aplenty even for ABDs. Stan Zionts was instrumental in my decision to stay on at Buffalo amidst offers one usually can not refuse, and enabled a smooth transition from engineering to management and integration with the business school culture. Ever since, we have worked together on numerous assignments, ventures and even practical day-to-day events in a variety of respective roles for each of us. To name a few, Stan has been a research collaborator sharing doctoral committees with me, a no-nonsense, but kind and gentle chairman, a real tough negotiator, a remarkable team-player, and many more. The mentorship of a doctoral advisor usually lasts as far as the dissertation. However, I have been very fortunate to be the beneficiary of Stan's active and sustained mentorship right through my career.

It has been a privilege and truly a great honor to write this article for this volume. Stan's contributions to the profession, colleagues and the institutions for which he had worked have been tremendous. While volumes can be written about his legacy to the field, I shall summarize in this brief article what he has helped me and several co-workers to accomplish in our research pursuits. My association of fifteen years with him seems to be a lifetime from one perspective, while in reality, it is just a beginning. I shall address our research work in three main areas: MCDM research, database systems and logistics.

2. MCDM Research

Our work in MCDM essentially spans two important domains in interactive multicriteria optimization: multicriteria linear programming and multicriteria integer programming. This work is entirely contained in my dissertation, and grew out of several prior contributions of Stan Zionts. Some of the important foundational works include the doctoral work of Bernardo Villarreal at the department of Industrial Engineering, completed under the joint guidance of Stan and Mark in 1979, and the popular method of Zionts and Wallenius [1976, 1983] for the interactive solution of multicriteria linear programming problems. My dissertation began with a focus on multicriteria integer programming using the Zionts and Wallenius method as a preference bounding step within a branch-and-bound framework. In this connection, the bicriteria integer programming problem had much to offer in terms of insights on the structural properties of the solution space, and the research swerved in the bicriteria direction. We developed the concept of convex cones as an efficient solution space reduction mechanism in the search for the solution most preferred in the feasible space by a decision maker with a nonlinear, but quasiconcave and nondecreasing utility function. A convex cone can be likened to a perfect swing of a golf club, with the positions of the handle and the golf ball identifying points in the objective function space. As the club swings from its initial position to a higher position, a convex cone wipes out from further consideration all the points in the arc space contained between the initial position and a 90 degree horizontal line from the golf handle. We were encouraged by the promised efficacy of this mechanism in reducing the search space in large chunks quickly, and investigated its structural properties. The preliminary analyses and testing confirmed this in the bicriteria problem, egging us on to test it out in more complex multicriteria problems. At this stage, we started wondering how the convex cones would impact the Zionts and Wallenius method itself, as this method was being used as a critical building block in designing an algorithm to solve the general multicriteria integer programming problems. Again, the research veered, now in the direction of muticriteria linear programming. A close investigation of the Zionts and Wallenius method quickly revealed that the method has much to gain from the use of convex cones. The structural properties of the cones and the architecture of their management within an interactive solution framework, all lent nicely and elegantly to efficient enhancements to the basic Zionts and Wallenius procedure. It took several sleepless nights of intensive programming and a few sleepy nights filled with intense dreams of points and cones in hyperspace to produce the enhanced algorithm. The algorithm worked, and again our tests confirmed the efficacy of convex cones in solving multicriteria linear programming problems as well. We were ecstatic with the results, but were equally concerned about the direction the research was moving into. The convex cones are bewitching supernatural entities capable possessing human souls. It takes tremendous courage and an indomitable self-will to extricate oneself from their shackles. Seeing our work tossing into uncharted territories in the bottomless sea of evolutionary research, Stan stepped in to direct our efforts in a purposeful way to a successful completion of the research targets. We veered back to multicriteria integer programming, only to be trapped again in the guiles of convex cones, stranded in the island of bicriteria integer programming. The efficacy of convex cones in the enhanced Zionts and Wallenius method showed that an adaptation of the method to exploit the special structural properties of the bicriteria space would considerably improve the computational performance of the algorithm in solving bicriteria problems. Surrendering our will to stay on track, we revisited the Zionts and Wallenius method, adapting it to the bicriteria problem with the support of convex cones. The intensive programming continued, and again, we were amazed at our results. The cones, along with the structural simplicity of the bicriteria problem, yielded remarkable synergies. Determined to pursue this further, we extended the enhanced Zionts and Wallenius method that has been adapted to the bicriteria problem to

solve bicriteria integer programming problems using a branch-and-bound framework. The cones provided excellent support, both within and outside the domain of the Zionts and Wallenius method in solving the integer programming problem. The results were again remarkable. Till then, we had developed comprehensive algorithms for solving multicriteria linear programming problems by the enhanced Zionts and Wallenius methos, and bicriteria integer programming problems using the enhanced and adapted Zionts and Wallenius method with cones screaming wild all over the branch-and-bound search. Nearly, two and a half years have elapsed by this time, and
my targeted research, multicriteria integer programming was nowhere in sight. I was getting seriously worried. One half of me was terribly depressed due to this, but the other half was ebullient with what we had seen and accomplished, and was egging me to go on. Fortunately, the sense of balance and direction that has been rather unique to both Stan and Mark prevailed, and they let the flow carry us through without bringing in any artificial barriers, with the hope and vision that we will arrive somewhere, safely and soundly. The big piece of the puzzle was putting it all together to confront the ultimate: the multicriteria problem. The knowledge and experience that we had gained by that time with the variations and smaller versions of the big problem supported us throughout the grand finale. We pulled together, integrated all the previous work and finally arrived at an efficient algorithm to solve the muticriteria integer programming problem.

The entire experience was truly an eye-opener for me. I learnt for the first time in my life, what true evolutionary research was all about. The sagacity, calmness, emotional support and encouragement that Stan provided throughout this process sustained my spirit and enthusiasm to continue ahead during this entire period, and taught me a great deal about research that remain fresh in my memory and outlook even till this day. Stan is a great believer in evolutionary research and has trained me to think along a pattern that is so original in him. While he would emphasize realistic targets for research, he would simply abandon them during a deliberation, but would refocus while taking stock in the intermissions. This is a simple trick in creative research, and Stan took me through each step of this process, the entire way. I am sure almost every student of his would testify to this as a legacy of Stan for all of us to keep, cherish, follow and develop.

The entire MCDM work carried out in the above program was eventually published in five journal articles and four conference proceedings. The paper in the European Journal of Operations Research [1986] presents our initial work on interactive multicriteria integer programming under a linear utility function. The paper in Annals of Operations Research [1988] develops the theory of convex cones and their structural properties important in their application in interactive multicriteria linear and integer programming. The paper in Naval Research Logistics [1989] presents the enhanced Zionts and Wallenius method with the architecture of convex cones as applied in multicriteria linear programming and empirical results. The paper in IEEE Transactions on Systems, Man and Cybernetics [1990] presents the bicriteria integer programming algorithm with convex cone enhancements and the extended and adapted Zionts and Wallenius method. Finally, the paper in Management Science [1989] presents the comprehensive approach to solving multicriteria integer programming problems interactively using preference structure representation with convex cones and the embellishments to the basic framework. The above order of journal publications roughly corresponds to the sequence of twists and turns in the evolution of our research described above. The conference papers deal with several nuggets of research results that were discovered during the entire research process.

3. Database Research

Clearly, I was rather burnt out towards the end of a nearly six year long engagement with MCDM research. I began to see everything in the world as a multicriteria problem (which is true to some degree), and needed a respite. I turned to database research, which has always been close to my heart for a very long time. I sounded Stan on this idea and he responded with great enthusiasm. It is needless to state that Stan is a true wanderer in the wonderland of research, as it is a well known fact. There may be very few areas from investment banking and market research to markov chains that Stan has not wetted his feet in. Surprisingly, database research was new to him. This made two of us, and what better combination can exist for exciting and creative research? I had been introduced to certain important optimization problems in database design and algorithms during my peregrinations in the research world. Some of these problems had hitherto been either unaddressed or simple solutions to the intriguing cases existed. In particular, optimal algorithms to perform on-line joins of relational databases have been well recognized areas of investigation in the database literature. The performance of a join algorithm is measured in terms of the CPU time required to perform a relational join, the memory requirements and long-term management issues surrounding any indices or data structures that may be used by the algorithm. Clear tradeoffs existed among these measures, and the field has always been ripe for the inundations of an operations researcher. In addition, relational join is a very practical problem, and the numerous commercial offerings of database systems often employ either brute-force methods or simple statistics guided heuristics. The database designers would normally discourage the use of joins in practical applications by designing relational structures to circumvent the possibility of large, complex joins, and the database administrators would wince when an innocent user sends in a query that would trigger millions of bytes of data joins among relations.

We were extremely fortunate to have a brilliant student to work on this problem with us. This was Ram Gopal, who is currently on the faculty of the University of Connecticut. We started out with a thorough search of the literature on join algorithms. Surprisingly, we found just three broad approaches to the join problem in over twenty years of research, but with an incredible number of variations of the basic approaches. The three approaches are: the nested-loop algorithm, join using indices and the hashing algorithms. The nested loop algorithm is a brute-force procedure requiring the least amount of run-time memory and no long term storage, but with the maximum CPU time to execute a join. Several index structures have been proposed to speed up joins, and the most important among these are the inverted indices, clustered indices and birelational join indices. These algorithms have varying degrees of run-time performance and memory requirements, and usually entail significant long-term maintenance costs for the indices. The hashing algorithms employ a pigeon-holing approach based on the divide-and-conquer principle, and partition relations based on join attributes to speed up the process. Several variations of hash algorithms exist, and most of them provide excellent run-time performance with no long term storage costs. However, their failing is their large requirements of run-time memories in order to achieve a desired level of computational performance. Motivated by these clear tradeoffs in almost every join algorithm we encountered, we set out to investigate strategies that would provide better and more efficient performance characteristics in all the measures.

We first formulated the join optimization problem as an access path optimization problem in which the objective is to minimize the number of disk pages accessed in performing a join, since the cost of data retrieval from disk accounts for most of the join processing costs. Next, we developed a model of access paths by structuring page accesses from a relation as a tree traversal process. This naturally led to a treelike data structure that can be used as a birelational join index. Consequently, the underlying optimization problem was to devise a traversal strategy on this treelike structure to minimize run-time page accesses. We considered several special cases of this data structure such as a tree for one, and developed optimal traversal strategies for them. A general traversal strategy for any kind of index structure emerged from this work, and the performance of this strategy in the general case was demonstrated using probabilistic analyses of worst-case and expected behavior. The paper in ORSA Journal on Computing

[1995] presents this research.

The general treelike access path structure can indeed be modeled as a rooted directed tree, where each node would represent a cluster of pages from a relation. In an access path, a traversal of such a tree would mean accessing all the pages contained in a node when the node is traversed. Consequently, the size of each node in terms of the number of pages it represents becomes a critical factor in run-time memory requirements. In modeling an access path as above, we developed three network structures to represent a join of two relations: page connectivity graph, cascade and block tree cascade. A page connectivity graph is a bipartite representation of the set of connected pages in two relations according to a join predicate. To reveal the structure of a join, the nodes of the bipartite graph were ordered into levels, and an isomorphic graph structure termed the cascade was obtained by a folding procedure that cascades the nodes of the page connectivity graph into levels. From a cascade, a tree structure termed the block tree cascade was derived by selectively grouping nodes at each level of the cascade into blocks. The underlying optimization problem in the general case is to devise a traversal strategy for a block tree cascade to complete the join with minimum page accesses. Again, we developed traversal strategies for block tree cascades with compact data structures for their storage and retrieval. Probabilistic analyses of worst-case and expected behavior of the algorithm showed its superior performance over the state-of-the-art algorithms. This work is described in the paper Gopal et al. [1995a].

Next, we pursued the hashing approach. Clearly, hashing had several advantages in terms of fast run-time processing, but the memory requirements turned out to be the deterrent. We began to investigate if there is a way to achieve the same or higher level of run-time performance with hashing, but with less memory requirement. The cascades provide similar run-time performance with negligible memory requirements, but then, incur the overhead of long run maintenance costs for the index structures. Therefore, the logical question was: can we combine the two approaches - hashing and cascades, to derive a new join algorithm that would yield better overall characteristics on the three performance measures? We came up with the idea of a criss-cross hashing strategy that draws from both the hashing and cascading techniques, inheriting the advantages of both. A new, but simple data structure termed page maps was developed to facilitate the criss-cross hashing strategy. The page maps aid in reducing the hashing effort and run-time memory requirements at minimal costs of long run index maintenance. The page maps also implicitly capture any of the possible inherent order in the relations, are simpler, more compact and easier to maintain than any of the traditional index structures. Extensive empirical studies and probabilistic analyses support the superior performance characteristics that were envisaged while designing this approach. This work is described in the paper Gopal et al. [1995b].

The above three papers constituted Ram Gopal's dissertation, and represented our sustained efforts over a period of three years. This sojourn in the database arena was a tremendous learning experience for all of us. In particular, this experience confirmed two of my long held, most fundamental beliefs: three heads working in concerted co-ordination are capable of synergies that one can not otherwise even imagine, and operations research has much more to offer to fields that we conventionally do not think of as recipient disciplines - database research in this case. We were delighted when the work finally reached a conclusion, and I found myself following the footsteps of my predecessor - Stan Zionts - in wandering freely across various disciplines of scientific enquiry.

4. Logistics Research

Logistics research is home turf to most operations researchers. After roaming the database world rather extensively, Stan and I decided to turn homeward in the next phase of our explorations. Everything simply fell in place when I was asked by my chairman to teach a doctoral seminar on a topic of my choice when a fresh crop of new and exciting doctoral students just arrived in our program. I sounded Stan on this idea, and he reacted with tremendous enthusiasm and support. He asked me to go ahead and organize the seminar, and indicated that he would attend every class, pitching in with his contributions. There were five doctoral students in the class: Mark, Sid, Matt, Sudip and Khalid. I organized the seminar on certain advanced topics in scheduling theory, and this laid the foundations for the doctoral dissertations of these students. Although I organized and ran the seminar, Stan was the true catalyst behind all our classroom deliberations. Having him at the forefront of the class was truly a remarkable learning experience for all of us. He would question everything said, challenge every position held, and add his tremendous insight into every discussion. Although this even slowed down our progress somewhat, we gained enormously from his participation. It was an intense exercise in clear, simple but critical thinking. Stan is a firm believer in simplicity, and would insist on breaking down any concept, however complex it may be, into its elementary components, so that one can see, feel and even relish the concept to the full extent possible. We were fortunate to repeat this exercise in another seminar course with the same participants, thanks to our chairman who saw the merit in my proposal for a repeat performance. The second seminar was centered around the logistics of planning, scheduling and control. The research programs that evolved from these interactions are: efficient lot sizing under a differential transportation cost structure for serially distributed warehouses (Mark), a unified framework for the approximation of general open and closed queuing networks with blocking (Sid), mutiprocessor cyclic scheduling (Matt), design of computer-aided process planning, scheduling and control systems (Khalid) and resource allocation and the logistics of real-time data management in distributed client/server information systems under the Distributed Computing Environment (DCE) middleware services (Sudip). While the works of Mark and Sid are essentially centered on logistics, the other three works are interdisciplinary. Matt's work deals with a shake and bake tabu search strategy for solving an intractable real-world problem, Khalid's research centers on the logistics of computer aided manufacturing, and Sudip's work addresses the logistics of information management in distributed computing environments. In the following discussion, I shall summarize the works of Mark and Sid, since they have been completed. The other research projects are currently nearing completion.

Mark's work with us is based on a real-world logistics problem at a leading paint manufacturing company at Buffalo, New York. Stan organized this contact for us during the seminar days. Our studies at this company revealed a rather intriguing but stimulating research problem in the logistics of inventory control and distribution management. The studies led to considerable model building, analysis and algorithm development. An overview of this research is as follows.

A major cost element in the logistics of distributed warehousing is transportation cost. In most practical systems, the transportation costs are volume-dependent. The unit transportation costs are usually determined differentially among intervals of shipment volumes. While the unit cost is constant over an interval, it follows a stepwise declining pattern from an interval to the next higher level of shipment volumes. This structure is analogous to that of quantity-discounted inventory systems. In this research, we consider a single product, serial warehousing system operating under the differential transportation costs and the traditional holding and ordering costs. External demand occurs at the final stage of the serial system at a constant continuous rate, and must be met on time over an infinite horizon under continuous review. The objective is to determine the ordering lot size for each warehouse such that the long-run average cost is minimized.

The above problem was encountered in the distribution logistics of the paint manufacturer. We developed a model of this decision problem and established its structural properties. We showed that the optimal solution to this problem is stationary and nested, as in the traditional models which consider only the holding and ordering costs. In particular, we addressed a two level differential transportation cost structure, which corresponds to the well known LTL/TL freight rates used by most common carriers. We developed efficient algorithms to obtain a best integer ratio and optimal powers-of-two ordering policies for the serial warehousing system. Next, we considered the general case of multilevel differential cost structures, and developed a hypercube characterization of the solution space. A hypercube presents significant structural properties that are very useful in devising search strategies. Using the hypercube model, we developed algorithms to determine best integer ratio and power-of-two policies in the general case. We carried out extensive computational investigations with the proposed algorithms with both real and simulated data, and found that the policies derived from them mostly cost within 1% of that the optimal solution. Our results demonstrated the robustness and performance efficiency of the proposed strategy. The results of this work are presented in Vroblefski et al. [1996]. We are currently extending this work to kanban-like control mechanisms that have been traditionally employed in Just-In-Time manufacturing systems to distribution networks to yield similar Just-In-Time performance in warehousing operations.

The work we did with Sid followed an evolutionary course as it had always been the case with most of our joint ventures. Sid, who always prided himself on his knowledge of french, stumbled upon a dissertation written in french on approximation techniques for open serial kanban systems during our seminar days. Stan, whose knowledge of french is legendary, jumped on this dissertation, and Stan and Sid together translated the work to me. The dissertation addressed a decomposition principle in analyzing open queuing systems. Prior to this, I had carried out research on blocking in queuing systems with other collaborators, and the ideas in the dissertation were quite interesting, although not entirely new to me. In general, blocking in queuing systems can be organized into four broad types: communication, manufacturing, minimal and general blocking. The dissertation basically addressed a fast decomposition of open queuing systems under minimal blocking, which is also known in the literature as kanban blocking. A detailed study of this work followed, and we all saw an excellent opportunity to develop a unified framework for the approximation of any type of queuing network - open or closed, and under any type of blocking.

Most manufacturing and communication systems can be modeled as queuing networks. However, many of these systems involve several complexities that make an exact analysis of the underlying queuing networks almost intractable. Consequently, approximation techniques for such analyses have become critical, and are needed to provide a robust, reliable and efficient approach to practical systems design in numerous applications. In this research, we developed a novel unified framework that works exceptionally well in approximating with a high degree of accuracy at almost insignificant computational effort, any queuing system configuration under any type of blocking. Till now, to our knowledge, the state-of-the-art in approximation research is limited to open queuing systems under minimal blocking and closed queuing systems with only two stations. The unified framework extends the state-of-the-art by significant leap, and provides excellent analytical approximations to systems that have hitherto been analyzed only through simulation. Our analyses have been based on specific decomposition strategies for each type of system under each blocking type, all fitted into an overall comprehensive decomposition framework. This work is presented in Ramesh et al. [1996].

5. Conclusion

In conclusion, it has been a rare privilege and pleasure to write this article, in celebration of the sixtieth birthday of a man who has contributed much more than anyone else in my life. I have attempted to address a very small domain of our interactions in the course of the last one and a half decade in this article. Stan has imparted his scholarly erudition, balanced wisdom and an intellectual strength to numerous colleagues and students over the years. I am indeed indebted for the many gifts of life for which he was the carrier. Looking back, the past fifteen years of our association appears rather dreamlike, highlighting what we have accomplished, and pointing to where we should go from here. I am reminded of Lewis Carrol in Alice in Wonderland, `You must run faster and faster to stay where you are'. The fifteen years is just a beginning, and there is no looking back from here anymore!

Bibliography

Gopal, D. R. Ramesh, R. and Zionts, S., "Access Path Optimization in Relational Joins, " ORSA Journal on Computing, Vol. 7, No. 3, pp. 257-268, 1995.

Ramesh, R., Karwan, M. H. and Zionts, S., "An Interactive Method for Bicriteria Integer Programming," IEEE Transactions on Systems, Man and Cybernetics, Vol. 20, No. 2, pp. 395-403, 1990.

Ramesh, R., Karwan, M. H. and Zionts, S., "Interactive Multicriteria Linear Programming: An Extension of Method of Zionts and Wallenius," Naval Research Logistics, Vol. 36, No. 3, pp. 321-335, 1989.

Ramesh, R., Karwan, M. H. and Zionts, S., "Preference Structure Representation Using Convex Cones in Multicriteria Integer Programming," Management Science, Vol. 35, No. 9, pp 1092-1105, 1989.

Ramesh, R., Karwan, M. H. and Zionts, S., "Theory of Convex Cones on Multicriteria Decision Making," Annals of Operations Research, Vol. 16, pp. 131-148, 1988.

Ramesh, R. , Karwan, M. H. and Zionts, S., "A Class of Practical Interactive Branch and Bound Algorithms for Multicriteria Integer Programming," European Journal of Operational Research, Vol. 26, No. 1, pp 161-172, 1986.

Zionts, S. and Ramesh, R., "Multicriteria Decision Making," in Encyclopedia of Operations Research, Gass, S. and Harris, C., (Eds.), Third Edition, Van Nostrand, New York, pp. 820-826, 1993.

Gopal, D. R., Ramesh, R. and Zionts, S., "A Methodology for Join Processing in Relational Databases," in the Proceedings of the Annual DSI Conference, Miami, Florida, 1991.

Ramesh, R., Karwan, M. H. and Zionts, S., "Interactive Bicriteria Integer Programming: A Performance Analysis," in Interactive Fuzzy Optimization and Mathematical Programming, (Eds.) M. Fedrizzi, J. Kacprzyk and M. Roubens, Springer-Verlag, 1991.

Ramesh, R. , Karwan, M. H. and Zionts, S., "Performance Characteristics of Three Interactive Solution Strategies for Bicriteria Integer Programming," in the Proceedings of the VIIth international Conference on MCDM, Manchester, England, 1989.

Ramesh, R., Karwin, M. H. and Zionts, s., "An Empirical Assessment and Insights on Two Multicriteria Integer Programming Algorithms," in Multiple Criteria Decision Making - Towards Interactive and Intelligent Decision Support Systems(Eds.Y. Sawaragi, K. Inoue and H. Nakayama), Springer-Verlag, 1987.

Ramesh, R., Karwan, M. H. and Zionts, S., "Degeneracy in Efficiency Testing in Bicriteria Integer Programming," in the Proceedings of the International Workshop on Methodology and Software for Interactive Decision Support, Albena, Burgaria, published by the International Institute for Applied Systems

Analysis (IIASA), Austria, 1987.

Karwan, M. H., Zionts, S., "Degeneracy in Efficiency Testing in Bicriteria Integer Programming Algorithm," in Decision Making with Multiple Objectives (Eds. Y.Y. Haimes and V. Chankong), Springer-Verlag, 1985.

Ramesh, R., Zionts, S. and Jagabandhu, S., "General Open and Closed Queuing Networks with Blocking: A Unified Framework for Approximation," working paper, School of Management, SUNY at Buffalo (1996).

Vroblefski, M., Ramesh, R. and Zionts, S., "Efficient Lot Sizing Under a Differential Transportation Cost Structure for Serially Distributed Warehouses," working paper, School of Management, SUNY at Buffalo (1996).

Gopal, D. R., Ramesh, R. and Zionts, S., "Criss-Cross High Joins: Design and Analysis," working paper, School of Management, SUNY at Buffalo (1995).

Gopal, D. R., Ramesh, R. and Zionts, S., "Cascade Graphs: Design, Analysis and Algorithms for Relational Joins," working paper, School of Management, SUNY at Buffalo, (1995).

STANLEY ZIONTS

Current as of April, 1996

School of Management
State University of New York
Buffalo, New York 14260 (USA)
Phone: 716 645 3260
Fax: 716 645 6117

Education:

B.S.	Carnegie Mellon University Electrical Engineering	1958
M.S.	Carnegie Mellon University Industrial Administration	1960
Ph.D.	Carnegie Mellon University Industrial Administration, Minor in Economics	1966

Languages:

English (Native Language). Fluent in French. Some knowledge of Mandarin Chinese.

Employment History:

1967-present School of Management, State University of New York at Buffalo Buffalo, New York 14260

Alumni Professor of Decision Support Systems, 1986-present
Professor of Management Science and Systems, 1970-present
Chairman, Department of Management Science and Systems, 1987-1991 and 1978-1981
Associate Professor of Management Science, 1967-1970
Have taught several times in the China MBA program, Dalian China, and in an MSIS program in Montpellier, France during 1991

1973-1975 Professor of Management, European Institute for Advanced Studies in Management, Brussels, Belgium

1965-1967 Program Specialist to the Joint Plant Committee, Ministry of Steel and Mines, Government of India, The Ford Foundation, Calcutta, India

1960-1965 Operations Research Technologist, U.S. Steel Applied Research Laboratory, Monroeville, Pennsylvania

Selected Professional Memberships and Activities:

The Institute for Operations Research and Management Science
Associate Editor, OPSEARCH (The Journal of the Operational Research Society of India), 1967-1973
Associate Editor, Management Science, 1980-present
Associate Editor, Naval Research Logistics, 1982-present
Associate Editor, Group Decision and Negotiation, 1991-present
Co Editor, Journal of Multicriteria Decision Analysis, 1991-present

Consulted with numerous public and private organizations and participated as an instructor or lecturer in short courses on operations research, in conjunction with several organizations.

Visiting Lecturer in Operations Research and The Management Sciences (sponsored by the Operations Research Society of America and the Institute of Management Sciences), since 1980.

President of the International Society for Multiple Criteria Decision Making (from 1979 until 1992) and Editor of FACET, the Newsletter of the predecessor organization, the Special Interest Group for Multiple Criteria Decision Making from 1981 86.

Co-editor of Special Issue of Naval Research Logistics Quarterly on Multiple Criteria Decision Making, December, 1988, Vol. 35, No. 6.

Co-editor of Special Issue of Management Science on Multiple Criteria Decision Making, November, 1984, Vol. 30, No. 10.

Co-organizer of an international conference, "Multiple Criteria Decision Making" held at CESA, Jouy en Josas, France, May, 1975.

Organizer of an international conference, "Multiple Criteria Problem-Solving Theory, Methodology, Practice," SUNYAB, August, 1977. Sponsored by the Office of Naval Research, The European Institute for Advanced Studies in Management, and SUNYAB.

Organizer of a Conference on Integer Programming, SUNYAB, June, 1978, sponsored by the National Science Foundation.

Selected Professional Memberships and Activities (continued)

Co-organizer of NATO Advanced Study Institute on Multiple Criteria Decision Making held in Istanbul, Turkey, 1987.

Professor at four of the five International Summer Schools on Multiple Criteria Decision Analysis, the most recent at Chania, Crete, Greece, July 4-16, 1994.

Helped organize a conference in honor of the 70th Birthday of Prof. Gerald L. Thompson, Carnegie Mellon University, October, 1993, and set up a fellowship in his name.

Selected Presentations

Member of the International Program Committee and Speaker at the International Workshop on Methodology and Software for Interactive Decision Support, Albena, Bulgaria, October 19 23, 1987.

"A Computer Graphics Based Approach for Multiple Criteria Linear Programming" (with P. Korhonen and J. Wallenius) IFORS 1987, Buenos Aires, Argentina, August 10-14, 1987.

Presented the Keynote Address at the Operations Research Society of South Africa Annual Conference held at the University of Zululand, South Africa, September 28-30, 1988, titled "Recent Developments in Multiple Criteria Decision Making".

Presented a plenary talk (as well as several papers) at the ninth International Conference on Multiple Criteria Decision Making sponsored by the International Society for Multiple Criteria Decision Making which was held at George Mason University, August, 1990.

Presented a tutorial on "Negotiations and Multiple Criteria Decision Making" at the ORSA/TIMS Joint National Meeting, Philadelphia, PA, October, 1990

Presented a plenary talk titled "The Aspiration Level Interactive Method of Multiple Criteria Decision Making and Some Applications" at the International Conference on Operations Research and Management Science sponsored by the Operations Research Society of the Philippines, Manila, The Philippines, December 11-14, 1990.

Member of the Organizing Committee and Speaker at the Tenth International MCDM Conference in Taipei, Taiwan sponsored by the International Society for Multiple Criteria Decision Making, July 20-24, 1992.

Member of the Organizing Committee and Speaker at the Eleventh International MCDM Conference in Coimbra, Portugal sponsored by the International Society for Multiple Criteria Decision Making, August 1-6, 1994.

Member of the Organizing Committee and Speaker at the Twelfth International MCDM Conference in Hagen, Germany sponsored by the International Society for Multiple Criteria Decision Making, June 19-23, 1995.

Also, organizer of sessions and contributor to numerous national and international meetings, and visiting lecturer in various countries including Austria, Belgium, Bulgaria, Canada, China, Finland, France, Germany, Greece, Holland, Hong Kong, Hungary, India, Israel, Italy, Japan, Norway, Philippines, Poland, Romania, Russia, South Africa, Sweden, Taiwan, Thailand, and Turkey.

Awards:

Awarded the MCDM Gold Medal by the International Society on Multiple Criteria Decision Making, Taiwan, 1992

Awarded the MCDM Presidential Service Award by the International Society on Multiple Criteria Decision Making, Taiwan, 1992

Courses Taught:

Statistics, Statistical Decision Theory, Linear Programming, Analytic Methods of Planning, Nonlinear and Integer Programming, Mathematical Analysis for Business, Management Science, Research Seminar in Operations Analysis, Dynamic Programming, Student Internship, Seminar on Multiple Criteria Decision Making, Competitive Decision Making, Introduction to Management Information Systems, Planning and Control of Operations, Management Strategy

Research Supervision:

Ph.D. students for whom I have served as major or co major professor in approximate chronological order:

Der-San Chen, Rashmi Thakkar, Kenneth Deal, Chester Lesniak, Jyrki Wallenius, Lars Nieckels, Bertil Tell, Dilip Deshpande, Zahid Khairullah, Robert M. Okello, Bernardo Villareal, Vahid Lotfi, Murat Koksalan, R. Ramesh, Moustapha Diaby, Hae Wang Chung, Steven Breslawski, U. Palekar, Yong Seok Yoon, Jeffrey Teich, Srinivas Prasad, Narasimha Reddy, Ram Gopal, and Eleazar Puente.

Current students include: Mark Vroblefski, Sridhar Jagabandhu, and Matthew Swinarski.

Grant Support Highlights:

1991-1995 National Science Foundation: "Decision Support Systems Model for U. S.-China Joint Venture Negotiations"

1986-1987 General Motors Research Laboratories: "Managerial Preference Decision Support System"

1985-1987 Office of Naval Research: "Multicriteria Integer Programming for Problems Involving Linear and Nonlinear Utility Functions"

1983-1985 National Science Foundation: "Multiple Objective Linear Programming"

1981-1982	Alcoa Foundation: "Multiple Criteria Decision Making"
1977-1978	Brookhaven Laboratories: "Multiple Criteria Analysis of Energy Planning Models"
1970-1972	Ford Foundation: "A Study of the Economic Theory of Welfare"
1964-1965	United States Steel Corporation: "Size Reduction Techniques of Linear Programming"

Current Research Interests:

Bargaining and Negotiation
Decision Support Systems
Linear and Integer Programming: Theory and Applications
Management Strategy
Multiple Criteria Decision Making
Operations Research

Publications

Books, Monographs, Proceedings, and Special Reports

Zionts, S., "Methods for Selection of an Optimum Route," Papers, Third Annual Meeting American Transportation Research Forum, December, 1962, pp. 1-12.

Reprinted in Constantin, James A., Principles of Logistics Management, Appleton Century Crofts, 1966.

Zionts, S., "Size Reduction Techniques of Linear Programming and Their Application," Ph.D. Dissertation, Carnegie Institute of Technology, Pittsburgh, PA, September, 1965.

Zionts, S., "Linear Programming," Chapter in H. B. Maynard, Handbook of Business Administration, McGraw Hill Book Company, 1967, pp. 17-72, 17-83.

Young, R. C., Zionts, S., and Bishop, A. B., "Linear Programming for Vocational Education Planning," Research and Development Series No. 93, The Center for Vocational and Technical Education, Ohio State University, December, 1973.

Chen, A.H.Y., Jen, F. C. and Zionts, S., "Borrowing and Lending Policies of a Commercial Bank," Proceedings International Meeting of the Institute of Management Science, Tel Aviv, Israel, June, 1973, pp. 779-783.

Deal, K. R., and Zionts, S., "A Differential Games Solution to the Problem of Determining the Optimal Timing of Advertising Expenditures," Proceedings. Second Annual Northeast Regional Conference of the American Institute of Decision Sciences, Kingston, Rhode Island, April, 1973.

Zionts, S., Linear and Integer Programming, a 17-chapter text on linear and integer programming, published by Prentice Hall, 1974, 514 pp.

Zionts, S., "A Deterministic Cash Management Model," in B. Jacquillat (ed.), European Finance Association. 1974 Proceedings, North Holland/American Elsevier, 1975, pp. 265-278.

Wallenius, J. and Zionts, S., "A Project on Multiple Criteria Decision Making – A Progress Report," Workshop Paper, Volume 4, IIASA Workshop on Decision Making with Multiple Conflicting Objectives, Laxenburg, Austria, October 20-24, 1975.

Thiriez, H. and Zionts, S. (eds.), Multiple Criteria Decision Making, Proceedings, May, 1975, Jouy en Josas, France, Number 130. Lecture Notes in Economics and Mathematical Systems: Operations Research, Springer Verlag, Berlin, 1976, 409 pp. + vi.

Wallenius, J. and Zionts, S., "Some Tests of an Interactive Programming Method for Multicriterion Optimization and an Attempt at Implementation," in H. Thiriez and S. Zionts (eds.), Multiple Criteria Decision Making, Jouy en Josas, France, 1975, Springer-Verlag, Berlin, 1976, pp. 319 330.

Zionts, S., "Mathematical Programming," in A. Ralston and C. L. Meek, (eds.), Encyclopedia of Computer Science, Petrocelli/Charter, New York, 1976, pp. 855-863.

Revised edition (co-authored with John Barrer) (eds. A. Ralston and E. D. Reilly), Van Nostrand Reinhold, New York, 1983, pp. 918-924.

Third edition (co-authored with R. Ramesh) (eds. A. Ralston and E. D. Reilly), Van Nostrand Reinhold, New York, 1993, pp. 820-826.

Wallenius, J. and Zionts, S., "A Research Project on Multicriterion Decision Making," Chapter 3 in Bell, D. E., Keeney, R. L., and Raiffa, H., Conflicting Objectives in Decision, Volume 1, International Series on Applied Systems Analysis, John Wiley, New York, 1977, pp. 77-96.

Zionts, S. and Deshpande, D., "A Time Sharing Computer Programming Application of a Multiple Criteria Decision Method to Energy Planning A Progress Report," in S. Zionts (ed.), Multiple Criteria Problem Solving, Proceedings, Buffalo, NY, 1977, Springer-Verlag, Berlin, 1978, pp. 549-560.

Zionts, S. (ed.), Multiple Criteria Problem Solving, Proceedings, Buffalo, NY, 1977, Number 155, Lecture Notes in Economics and Mathematical Systems, Springer-Verlag, Berlin, 1978, 567 pp. + viii.

Zionts, S., "Integer Linear Programming with Multiple Objectives," in Derek W. Bunn, Howard Thomas (eds.), "Formal Methods in Policy Formulation," Birkhauser Verlag Basel, 1978, pp. 158-167.

Wallenius, J. and Zionts, S., "Decision Making with Multiple Objectives Some Analytic Approaches," in Yuji Ijiri and Andrew B. Whinston (eds.), Quantitative Planning and Control: In Honor of William Wager Cooper, Academic Press, New York, 1979, pp. 59-68.

Zionts, S., "Multiple Criteria Decision Making for Discrete Alternatives with Ordinal Criteria," in Fiacco, A. V. and Kortanek, K. O. (eds.), Extremal Methods and Systems Analysis, Number 174, Lecture Notes in Economics and Mathematical Systems, Springer Verlag, Berlin, 1980, pp. 135-152.

Zionts, S., (Foreward) in Ellis L. Johnson, Integer Programming Facets. Subadditivity, and Duality for Group and Semi Group Problems, CBMS NSF Regional Conference Series in Applied Mathematics, Number 32, SIAM, Philadelphia, 1980.

Zionts, S. and Wallenius, J., "On Finding the Subset of Efficient Vectors of an Arbitrary Set of Vectors," in A. Prekopa (ed.), Studies on Mathematical Programming, Akademiai Kiado, Budapest, Hungary, 1980, pp. 187-200.

The following publications are in Fandel, G. and T. Gal (eds.), Multiple Criteria Decision Making: Theory and Application Proceedings, 1979, Number 177, Lecture Notes in Economics and Mathematical Systems, Springer-Verlag, Berlin, 1980.

Deshpande, D. V. and S. Zionts, "Sensitivity Analysis in Multiple Objective Linear Programming: Changes in the Objective Function Matrix," pp. 26-39.

Khairullah, Z. Y. and S. Zionts, "An Experiment with Some Algorithms for Multiple Criteria Decision Making," pp. 150-159.

Korhonen, P., J. Wallenius, and S. Zionts, "A Bargaining Model for Solving the Multiple Criteria Problem," pp. 178-188.

Villareal, B., Karwan, M. H., and Zionts, S., "An Interactive Branch and Bound Procedure for Multicriterion Integer Linear Programming," pp. 448-467.

Zionts, S., "Methods for Solving Management Problems Involving Multiple Objectives," pp. 540-558.

Karwan, M. H., Lotfi, V., Telgen, J., and Zionts, S., "A Study of Redundancy in Mathematical Programming," in J. P. Brans (ed.), Operational Research '81, North Holland, 1981, pp. 297-311.

Khairullah, Z. and Zionts, S., "An Empirical Evaluation of Some Multiple Criteria Methods for Discrete Alternatives," (abstract only) in Morse, J. N. (ed.), Organizations: Multiple Agents with Multiple Criteria, Proceedings, 1980, No. 190, Lecture Notes in Economics and Mathematical Systems, Springer-Verlag, Berlin, 1981.

Zionts, S. and Deshpande, D., "Energy Planning Using a Multiple Criteria Decision Method," in P. Nijkamp and J. Spronk, Multiple Criteria Analysis, Gower Publishing, Hampshire, England, 1981, pp. 153-162.

Koksalan, M., Karwan, M. H., and Zionts, S., "An Approach for Solving the Discrete Alternatives Multicriteria Problem, " Proceedings of the 21st IEEE Conference on Decision and Control, December, 1982.

Karwan, M. H., Lotfi, V., Telgen, J. T., and Zionts, S., Redundancy in Mathematical Programming. A State of the Art Survey, Springer-Verlag, Berlin, 1983, 278 pp.

Zionts, S., and Wallenius, J., "A Method for Identifying Redundant Constraints and Extraneous Variables in Linear Programming Problems," Chapter 3, Karwan, Lotfi, Telgen, and Zionts (1983) above.

Karwan, M. H., Koksalan, M., and Zionts, S., "An Approach for Solving the Discrete Alternatives Multicriteria Problem, " Proceedings of the 21st IEEE Conference on Decision and Control, December, 1982, Orlando (1983).

Zionts, S., "Mathematics in Business and Management," in A. Ralston and G.S. Young (eds.), The Future of College Mathematics Proceedings of a Conference/Workshop on the First Two Years of College Mathematics, Springer-Verlag, Berlin (1983), pp. 81-87.

Zionts, S., "A Report on A Project on Multiple Criteria Decision Making, 1982," in P. Hansen (ed.) Essays and Surveys on Multiple Criteria Decision Making Proceedings, Mons, 1982, Springer-Verlag, Berlin (1983), pp. 416-430.

Zionts, S. and Wallenius, J., "Recent Developments in our Approach to Multiple Criteria Decision Making," in Grauer, M., and Wierzbicki, A. P., Interactive Decision Analysis Proceedings, 1983, No. 229, Lecture Notes in Economics and Mathematical Systems, Springer-Verlag, Berlin, 1984, pp. 54-62.

Zionts, S., "Multiple Criteria Mathematical Programming: An Overview and Several Approaches," in Fandel, G., and Spronk, J., (eds.) Multiple Criteria Decision Methods and Applications, Springer-Verlag, Berlin, 1985, pp. 85-128.

Zionts, S., "Multiple Criteria Mathematical Programming: An Overview and Several Approaches," in Serafini, P. (ed.), Mathematics of Multi Objective Optimization, CISM Courses and Lectures, No. 289, International Centre for Mechanical Sciences, Springer Verlag (Vienna) 1985.

The following publications are in Haimes, Y. Y., and V. Chankong, Decision Making with Multiple Objectives Proceedings, Cleveland, Ohio, 1984 Lecture Notes in Economics and Mathematical Systems, Vol. 242, Springer-Verlag (Berlin), 1985.

Chankong, V., Y. Y. Haimes, J. Thadathil, and S. Zionts, "Multiple Criteria Optimization: A State of the Art Review", pp. 36-90.

Karwan, M. H., S. Zionts, B. Villareal, and R. Ramesh, "An Improved Interactive Multicriteria Integer Programming Algorithm," pp. 261-271.

Breslawski, S., and S. Zionts, "An Interactive Multiple Criteria Linear Programming Package", pp. 282-286.

Zionts, S., "A Report on a Project on Multiple Criteria Decision Making, 1984, " in The Institute for Systems Studies, Multicriteria Programming Problems, Proceedings of Seminar held in Moscow, April 3-4, 1984, (1985), pp. 30-39.

Zionts, S., "A Report on a Project on Multiple Criteria Decision Making, 1986, " in Athans, M., and A. H. Levis, Proceedings of the 9th MIT/ONR Workshop on C3 Systems, Monterey, CA, Dec. 1986. Laboratory for Information and Decision Systems, MIT, Cambridge, MA 02139.

Korhonen, P., Wallenius, J. and Zionts, S., "Two Interactive Procedures for Multicriterion Optimization with Multiple Decision Makers," (in Russian), S. Emelianov & O. Larichev (eds.) in Systems and Methods in Decision Support, VNIISI Publications, No. 12, (1986), pp. 4 16.

R. Ramesh, Karwan, M. H., and Zionts, S., "An Empirical Assessment and Insights on Two Multicriteria Integer Programming Algorithms" in Y. Sawaragi (Ed.) Multiple Criteria Decision Making Towards Interactive and Intelligent Decision Support Systems, Springer-Verlag, (1987).

Ramesh, R., Karwan, M. H. and Zionts, S., "Degeneracy in Efficiency Testing in Bicriteria Integer Programming", in Methodology and Software for Interactive Decision Support, Proceedings of the International Workshop held in Albena, Bulgaria, October 19-23, (1987). Lecture Notes in Economics and Mathematical Systems, Springer-Verlag, Vol. 337, 1989, pp. 99-106.

Zionts, S., "Multiple Criteria Mathematical Programming: An Updated Overview and Several Approaches, " in G. Mitra (Editor), Mathematical Models for Decision Support, Springer-Verlag, NATO ASI Series, Series F: Computer and Systems Sciences, Vol. 48, Heidelberg, 1988.

Ramesh, R., M. H. Karwan, and S. Zionts, "Performance Characteristics of Three Interactive Solution Strategies for Bicriteria Integer Programming", in Lockett, A.G., and G. Islei (eds.), Improving Decision Making in Organizations, Proceedings, Manchester, UK, August, 1988, Lecture Notes in Economics and Mathematical Systems, Vol. 335, Springer-Verlag, 1989, pp. 472-485.

Karpak, B., and S. Zionts (eds.), Multiple Criteria Decision Making and Risk Analysis Using Microcomputers, Proceedings of a NATO Advanced Study Institute, Istanbul Turkey, 1987, Springer-Verlag, Vol. F56, Heidelberg, 1989.

Zionts, S., "Multiple Criteria Mathematical Programming: An Updated Overview and Several Approaches", in Karpak, B. and S. Zionts (eds.), Multiple Criteria Decision Making and Risk Analysis Using Microcomputers, NATO ASI Series, Vol. F56, Springer Verlag, Berlin, 1989, pp. 7-60.

Jelassi, T., Kersten, G., and S. Zionts, "An Introduction to Group Decision and Negotiation Support," Readings in Multicriteria Decision Aid, edited by Carlos A. Bana e Costa, Berlin, Springer Verlag, 1990, pp. 537-568.

Zionts, S., "MCDM: Where We Have Been and Where We are Going", in M. Agustin, P. Rao, and A. Pil (eds.), Proceedings: International Conference on Operations Research / Management Science: Techniques and Applications, December 11-14, 1990, Manila, The Philippines, pp. I-9 I-25, 1990.

Zionts, S., "Negotiations and MCDM: Their Interrelationships", in P. Korhonen, A. Lewandowski, and J. Wallenius (eds.), Multiple Criteria Decision Support: Proceedings, Helsinki, Finland, 1989, Publication 356, Lecture Notes in Economics and Mathematical Systems, Springer-Verlag, Berlin, 1991, pp. 377-386.

Gopal, R. G., Ramesh, R., and Zionts, S., "A Methodology for Join Processing in Relational Databases", in Melnyk, S. A., and Narasimhan, R., 1991 Proceedings Decision Sciences Institute, November 24-26, 1991, Miami Beach, Florida, pp. 809-811.

Zionts, S., "The State of Multiple Criteria Decision Making: Past, Present, and Future", in A. Goicoechea, L. Duckstein, and S. Zionts, Proceedings, Ninth International Conference on Multiple Criteria Decision Making: Fairfax, Virginia, 1990 Springer-Verlag, Berlin, 1992, pp. 33-44.

Goicoechea, A., L. Duckstein, and S. Zionts, eds., Multiple Criteria Decision Making: Proceedings of the Ninth International Conference: Theory and Applications in Business., Industry. and Government, Springer-Verlag, Berlin, 1992.

Zionts, S., "Multiple Criteria Decision Making and Negotiating, Some Observations," in Y. Ijiri, (ed.), Creative and Innovative Approaches to the Science of Management, Quorum Books, Westport, Connecticut, 1993; pp. 423-439.

Zionts, S., "Multiple Criteria Decision Making: THE CHALLENGE THAT LIES AHEAD", in Tzeng, G. H., Wang, H. F., Wen, U. P, and Yu, P. L., Multiple Criteria Decision Making, Proceedings of the Tenth International Conference: Expand and Enrich the Domains of Thinking and Application, Springer Verlag, Berlin, 1994, pp. 17-26.

Ramesh, R., and Zionts, S., "Multiple Criteria Decision Making", Chapter in Gass, S. I., and Harris, C. M., Encyclopedia of MS/OR, to be published by Kluwer Academic Publishers, 1995

Zhao, Lianxing, and S. Zionts, "Engine Selection and Multicriterion Decision Making", in Qian, F. P., and W. O. Lively, (eds.) The Age of Project Management - China and the World, Proceedings of the 1st International Symposium on Project Management, Xi'an, China, Sept. 25-27, 1995, NPU Press, Xi'an, China, 1995, pp. 518-522.

Zionts, S., "Decision Making: Some Experiences, Myths, and Observations", Forthcoming in G. Fandel, T. Gal (Eds.): Multiple Criteria Decision Making, Proceedings of the Twelfth International Conference: Twenty Years of East-West Cooperation in MCDM, Springer-Verlag, 1996.

J. E. Aronson and Zionts, S., OPERATIONS RESEARCH: METHODS. MODELS AND APPLICATIONS, Proceedings of a Conference in Honor of Gerald L. Thompson on the Occasion of his 70th Birthday, to be published by Greenwood Publishers, 1996.

S. Zionts, "Some Interesting Aspects of Linear Programming", Forthcoming as Chapter Two in J. E. Aronson and Zionts, S., OPERATIONS RESEARCH: METHODS. MODELS AND APPLICATIONS, Proceedings of a Conference in Honor of Gerald L. Thompson on the Occasion of his 70th Birthday, to be published by Greenwood Publishers, 1996.

Ramesh, R., Karwan, M. H., and Zionts, S., "Interactive Bicriteria Integer Programming: A Performance Analysis", Forthcoming in Interactive Fuzzv Optimization and Mathematical Programming, M. Fedrizzi, J. Kacprzyk, and M. Roubens, to be published by Springer-Verlag.

Zionts, S., "Some Thoughts on MCDM: Myths and Ideas", Forthcoming in J. Climaco (ed.) *XI International Conference on MCDM Proceedings*.

Journal Articles and Notes

Mediate, A. and Zionts, S., "Note on a Paper by Hanssmann, " Operations Research, Vol. 9, No. 6, 1961, pp. 900 901.

Bruce, H. J. and Zionts, S., "Research - The Forgotten 'R' in Transportation," The Transportation Journal, Vol. 2, No. 3, 1963, pp. 28-33.

Glover, F. and Zionts, S., "A Note on the Additive Algorithm of Balas, " Operations Research, Vol. 13, No. 4, 1965, pp. 546-549.

Thompson, G. L., Tonge, F. M., and Zionts, S., "Methods for Removing Nonbinding Constraints and Extraneous Variables from Linear Programming Problems," Management Science, Vol. 12, No. 7, 1966, pp. 588-608.

Zionts, S., "A Bargaining Model for Allocating Steel Production," CORSI Bulletin (Operations Research Society of India), 1967, pp. 69-78.

Rao, M. R. and Zionts, S., "Allocation of Transportation Units to Alternative Trips - A Column Generation Scheme with Out-of-Kilter Sub-Problems," Operations Research, Vol. 16, No. 1, 1968, pp. 52-63.

Contini, B. and Zionts, S., "Restricted Bargaining for Organizations with Multiple Objectives," Econometrica, 1968, Vol. 36, No. 2, pp. 397-414.

Zionts, S., "Programming with Linear Fractional Functionals," Naval Research Logistics Quarterly, Vol. 15, No. 3, 1968, pp. 449-451.

Zionts, S., "On an Algorithm for the Solution of Mixed Integer Programming Problems," Management Science (Theory), Vol. 15, No. 1, 1968, pp. 113-116.

Zionts, S., "Implicit Enumeration Using Bounds on Variables: A Generalization of Balas' Additive Algorithm for Solving Linear Programs with Zero One

Variables," CORSI Bulletin (Operational Research Society of India), Seminar Number 1968-69, Vol. 1, pp. 7-21.

Zionts, S., "The Criss Cross Method for Solving Linear Programming Problems," Management Science (Theory), Volume 15, No. 7, 1969, pp. 426-445.

Zionts, S., "Toward a Unifying Theory for Integer Linear Programming, Operations Research, Vol. 17, No. 2, 1969, pp. 359-367.

Reprinted in Spanish as "Hacia Una Teoria Que Unifique La Programacion Lineal En Enteros, " Investigacion Operacional (published by the School of Mathematics, University of Havana, Havana, Cuba), No. 7, 1970, pp. 1-9.

Southwick, L., Jr. and Zionts, S., "Managing Incentives in a Poverty Reduction Framework," Decision Sciences, Vol 1, Nos. 3 and 4, 1970, pp. 371-396.

Chen, A.H.Y., Jen, F. C. and Zionts, S., "The Optimal Portfolio Revision Policy," The Journal of Business, Vol. 44, No. 1, 1971, pp. 51-61.

Reprinted in E. J. Elton and M. Gruber, (eds.), Security Evaluation and Portfolio Analysis, Prentice Hall, Inc., (Englewood Cliffs, NJ), 1972, pp. 434-447.

Chen, A.H.Y., Jen, F. C. and Zionts, S., "Portfolio Models with Stochastic Cash Demands," Management Science, November, 1972, Vol. 19, No. 3, pp. 319-332.

Zionts, S., "Implicit Enumeration in Integer Programming, " Naval Research Logistics Quarterly, March, 1972, Vol. 19, No. 1, pp. 165-182.

Zionts, S., (Book Review), Hammer, P. L., Ivaneseu and Rudeanu, S., "Boolean Methods in Operations Research," Econometrica, Vol. 40, 1972, pp. 777-778.

Zionts, S., "Some Empirical Tests of the Criss Cross Method, Management Science, December, 1972, Vol. 19, No. 4, pp. 406-410.

Chen, D. S. and Zionts, S., "An Exposition of the Group Theoretic Approach to Integer Programming," OPSEARCH, June, 1972, Vol. 9, No. 2, pp. 75-102.

Zionts, S. (Note), "Toward a Unifying Theory for Integer Linear Programming Some Comments on a Paper by Rubin," Management Science, March, 1973, Vol. 19, No. 7, p. 837.

Zionts, S. (Book Review), Greenberg, H., "Integer Programming," Mathematical Programming, Vol. 4, No. 2, 1973, pp. 234-237.

Southwick, L., Jr. and Zionts, S., "The Use of Tax Incentives for Employment of Underutilized Resources," Management Science, December, 1973, (Part I, Vol. 20, No. 4, pp. 449-459.)

Chen, A.H.Y., Jen, F. C. and Zionts, S., "The Joint Determination of Portfolio and Transaction Demands for Money," Journal of Finance, March, 1974, Vol. 29, No. 1, pp. 175-186.

Zionts, S. (Book Review), Garfinkel, R. S. and Nemhauser, G. L., "Integer Programming," Journal of Business, Vol. 47, No. 1, 1974, pp. 118-119.

Southwick, L., Jr. and Zionts, S., "An Optimal Control Theory Approach to the Education Investment Decision," Operations Research, November-December, 1974, Vol. 22, No. 6, pp. 1156-1179.

Southwick, L., Jr. and Zionts, S., "Resource Allocation in a Local Election Campaign," Interfaces, 1975, Vol. 6, No. 1, pp. 53-63.

Zionts, S. and Wallenius, J., "An Interactive Programming Method for Solving the Multiple Criteria Problem," Management Science, 1976, Vol. 22, No. 6, pp. 652-663.

Chen, D. S. and Zionts, S., "Comparison of Some Algorithms for Solving the Group Theoretic Integer Programming Problem," Operations Research, November-December, 1976, Vol. 24, No. 6, pp. 1120-1128.

Kendall, K. E. and Zionts, S., "Solving Integer Programming Problems by Aggregating Constraints," Operations Research, March-April, 1977, Vol. 25, No. 2, pp. 346-351.

Zionts, S., "Integer Linear Programming with Multiple Objectives," Annals of Discrete Mathematics, Vol. 1, 1977, pp. 551-562.

Reprinted in Hammer, P. L., Johnson, E. L., Korte, B. H. and Nemhauser, G. L. (eds.), Studies in Integer Programming, North Holland, Amsterdam, 1977.

Zionts, S., "A Survey of Multiple Criteria Integer Programming Methods," The Annals of Discrete Mathematics, Vol. 5, 1979, pp. 389-398.

Reprinted in Hammer, P. L., Johnson, E. L., and Korte B. H. (eds.), Discrete Optimization II, North-Holland, Amsterdam, 1979.

Zionts, S., "MCDM - If not a Roman Numeral, then what?" Interfaces, Vol. 9, 1979, pp. 94-101.

A rewritten version of the above appeared in Zionts, S., "What, MCDM is not a Roman Numeral?," in E. Turban, Cases and Readings in Management Science, BPI Publications, Plano, Texas, 1982.

Zionts, S. and Wallenius, J., "Identifying Efficient Vectors: Some Theory and Computational Results," Operations Research, Vol. 28, No. 3, Part 2, 1980, pp. 788-793.

Zionts, S., "A Multiple Criteria Method for Choosing Among Discrete Alternatives," European Journal of Operations Research, Vol. 7, No. 2, June, 1981, pp. 143-147.

Bahl, H. C., and Zionts, S., "A Noniterative Multiproduct Multiperiod Planning Method," Operations Research Letters, Vol. 1, No. 6, pp. 219-221, (1982).

Zionts, S., and Wallenius, J., "An Interactive Multiple Objective Linear Programming Method for a Class of Underlying Nonlinear Utility Functions, " Management Science, Vol. 29, No. 5, May 1983, pp. 519-529.

Reprinted in Grauer, M., Lewandowski, A., and Wierzbicki, A., (eds.) Multiobjective and Stochastic Optimization Proceedings of an IIASA Task Force Meeting, Laxenburg, Austria, pp. 311-334.

Karwan, M. H., Lotfi, V., Zionts, S., and Telgen, J., "Redundancy in Mathematical Programming," Mathematical Programming Society Committee on Algorithms Newsletter, No. 8, 1983, pp. 23-29.

Zionts, S., (Book Review), Goicoechea, A., Hansen, D. R., and Duckstein, L. "Multiobjective Decision Analysis with Engineering and Business Applications", IIE Transactions, Vol. 15, 1983, pp. 183-184.

Zionts, S. (Book Review), White, D. J., "Optimality and Efficiency." European Journal of Operational Research, Vol. 14, No. 3, 1983, page 344.

Koksalan, M., Karwan, M. H., and Zionts, S., "An Improved Method for Solving Multiple Criteria Problems Involving Discrete Alternatives," IEEE Transactions on Systems, Man, and Cybernetics, Vol. 14, No. 1, January 1984, pp. 24-34.

Korhonen, P., Wallenius, J., and Zionts, S., "Solving the Discrete Multiple Criteria Problem Using Convex Cones," Management Science, Vol. 30, No. 11, 1984, pp. 1336-1345.

Spronk, J., and Zionts, S., (Introduction), "Multiple Criteria Decision Making," Management Science, Vol. 30, No. 11, 1984.

Koksalan, M., Karwan, M. H., and Zionts, S., "Approaches for Discrete Alternative Multiple Criteria Problems for Different Types of Criteria," IIE Transactions, 18, September 1986, pp. 262-270.

Bahl, H. C., and Zionts, S., "Lot Sizing as a Fixed-Charge Problem," Production and Inventory Management (The Journal of the American Production and Inventory Control Society), Vol. 27, No. 1, pp. 1-11, 1986.

Korhonen, P., H. Moskowitz, J. Wallenius, and S. Zionts, "An Interactive Approach to Multiple Criteria Optimization with Multiple Decision Makers," Naval Research Logistics Quarterly, 33, pp. 589-602. 1986.

Ramesh, R., Karwan, M. H., and Zionts, S., "A Class of Practical Interactive Branch and Bound Algorithms for Multicriteria Integer Programming," European Journal of Operational Research, Vol. 26 (1986), pp. 161-172.

Zionts, S., (Book Review), White, D. J., "Optimality & Efficiency," *Interfaces*, Vol. 17, No. 1, Jan-Feb 1987, pp. 140-141.

Bahl, H. C. and Zionts, S., "Multi Item Scheduling by Benders Decomposition". Journal of the Operational Research Society, Vol. 38, No. 12, 1987, pp. 1141-1148.

Khairullah, Z. and Zionts, S., "An Approach for Preference Ranking of Alternatives," European Journal of Operational Research, Vol. 28, No. 3, 1987, pp. 329-342.

Zionts, S., (Book Review), Yu, P. L., "Multiple-Criteria Decision Making," Optima, No. 21, 1987, p. 10-11.

Ramesh, R., Karwan, M. H. and Zionts, S., "A Theory of Convex Cones in Multicriteria Decision Making," Annals of Operations Research, 16, 1988, pp. 131-147.

Koksalan, M. M., Karwan, M. H., and Zionts, S., "An Approach for Solving Discrete Alternative Multiple Criteria Problems Involving Ordinal Criteria," Naval Research Logistics, Vol. 35, No. 6, 1988, pp. 625-647.

Ramesh, R., Karwan, M. H., and Zionts, S., "Interactive Multicriteria Programming: An Extension of the Method of Zionts and Wallenius," Naval Research Logistics, Vol. 36, No. 3, 1989, pp. 321-335.

Ramesh, R., Karwan, M. H., and Zionts, S., "Preference Structure Representation Using Convex Cones in Multicriteria Integer Programming," Management Science, Vol. 35, No.9, 1989, pp. 1092-1105.

Zionts, S., and Lotfi, V., "Recent Developments in Multiple Criteria Decision Making, " ORION, (Published by the Operations Research Society of South Africa), Vol. 5, No. 1, 1989, pp. 1-23.

Palekar, U. S., Karwan, M. H., and Zionts, S., "A Branch and Bound Method for the Fixed Charge Transportation Problem", Management Science, Vol. 36, No. 9, 1990, pp. 1092-1105.

Ramesh, R., Karwan, M. H., and Zionts, S., "An Interactive Method for Bicriteria Integer Programming: Theory and Experimental Results". IEEE Transactions on Systems. Man and Cybernetics, Vol. 20, No. 2, March-April, 1990, pp. 395-403.

Diaby, M., Bahl, H., Karwan, M., and Zionts, S., "Large Scale Capacitated Lot Sizing by Lagrangean Relaxation", European Journal of Operational Research, Vol. 59, 1992, pp. 444-458.

Korhonen, P., Wallenius, J., and Zionts, S., "A Computer Graphics Based Decision Support System for Multiple Objective Linear Programming", European Journal of Operational Research, Vol. 60, No 3, August, 1992, 280-286.

Dyer, J. S., Fishburn, P. C., Steuer, R. E., Wallenius, J., and Zionts, S., "Multiple Criteria Decision Making, Multiattribute Utility Theory: The Next Ten Years", Management Science, Vol. 38, No. 5, May, 1992, pp. 645-654.

Reprinted in Derek Pugh and Samuel Eilon (eds.), The History of Management Thought, To be published by Dartmouth Publishing Company.

Lotfi, V., Stewart, T. J., and Zionts, S., "An Aspiration Level Interactive Model for Multiple Criteria Decision Making", Computers and Operations Research, Vol. 19, No. 7, 671-681, 1992.

Zionts, S., Invited Essay - "Some Thoughts on Research in Multiple Criteria Decision Making", Computers and Operations Research, Vol. 19, No. 7, 567-570, 1992.

Diaby, M., Bahl, H., Karwan, M. H., and Zionts, S., "A Lagrangean Relaxation Approach for Very Large Scale Capacitated Lot-Sizing", Management Science, Vol. 38, No. 9, 1329-1340, September, 1992.

Breslawski, S. T., and Zionts, S., "Some Results concerning the Quality of Vertex Solutions Found by a Method for Multiple Objective Linear Programming", Multi-Criteria Decision Analysis, Vol. 1, No. 3, 139-153, December, 1992.

Zionts, S., (Book Review) Bazerman, M. H. and Neale, M. A. 1992, Negotiating Rationally, Interfaces Vol. 23, No. 4, July-August, 1993, pp. 132-134.

Breslawski, S. T., and Zionts, S., "A Simulation Based Study of Modifications to the Zionts-Wallenius Algorithm for Multiple Objective Linear Programming", Computers and Operations Research, Vol. 21, No. 7, 757-768, 1994.

Teich, J. E., H. Wallenius, M. Kuula, and S. Zionts "A Decision Support Approach for Negotiation with an Application to Agricultural Income Policy Negotiations", European Journal of Operations Research, Vol. 81, 76-87, 1995.

Gopal, R. D., R. Ramesh, and S. Zionts, "Access Path Optimization in Relational Joins", ORSA Journal on Computing, Vol. 7, No. 3, 257-268, 1995.

Wang, C., Gopal, R. D., and S. Zionts, "Use of Data Envelopment Analysis in Assessing Information Technology Impact on Firm Performance", forthcoming in the Annals of Operations Research, 1995-1996.

Prasad, S. A., M. H. Karwan, and S. Zionts, "Use of Convex Cones in Interactive Multiple Objective Decision Making," Forthcoming in Management Science.

Lotfi, V., Y. S. Yoon, and S. Zionts, "Aspiration Based Search Algorithm (ABSALG) for Multiple Objective Linear Programming Problems: Theory and Comparative Tests," Forthcoming in Management Science.

Stanley Zionts is Alumni Professor of Decision Support Systems, Department of Management Science and Systems, School of Management, State University of New York at Buffalo (SUNYAB), Buffalo, New York 14260. He has been on the faculty there since 1967 and has served the school in various leadership roles. He served as faculty chairman of the Center for Entrepreneurial Leadership (CEL) at the school during the 1994-95 academic year. The CEL works with local entrepreneurs and helps them in various stages of business development. Dr. Zionts has an M.S. and Ph.D. in Industrial Administration as well as a B.S. in Electrical Engineering, all from Carnegie Mellon University. He has worked with the European Institute for Advanced Studies in Management in Brussels, Belgium, the Ford Foundation in India, and the U.S. Steel Corporation. In addition to his long-term positions in India (one-and-a-half years) and Belgium (two years), he has spent over a year cumulatively in China lecturing, doing research, and teaching in the SUNYAB School of Management's M. B. A. program. He spent the fall 1991 semester teaching in the SUNYAB program in Montpellier, France. He has also lectured in most western European countries, as well as several eastern European and Asian countries, and South Africa. He is fluent in French, and speaks Mandarin Chinese.

His areas of research include multiple criteria decision making, decision support systems, and mathematical programming. Prof. Zionts is the author/editor of several books and proceedings and over one hundred articles in journals, books, and proceedings. He has organized or co-organized and generated financial support for several international meetings. He currently serves as co-editor or associate editor of four academic journals. He has held leadership positions in national and international professional organizations. He teaches in the management science and management information systems areas, as well as courses in strategic management. He is an avid skier. A certified member of the Professional Ski Instructors of America, Inc., he is a part-time ski instructor at the Holiday Valley Ski Area in Ellicottville, New York.

MIX
Papier aus verantwortungsvollen Quellen
Paper from responsible sources
FSC® C105338

If you have any concerns about our products,
you can contact us on
ProductSafety@springernature.com

In case Publisher is established outside the EU,
the EU authorized representative is:
Springer Nature Customer Service Center GmbH
Europaplatz 3, 69115 Heidelberg, Germany

Printed by Libri Plureos GmbH
in Hamburg, Germany